110 Advances in Polymer Science

Responsive Gels:
Volume Transitions II

Editor: K. Dušek

With contributions by
J.H. Burban, E.L. Cussler, S.H. Gehrke,
O. Hirasa, S. Hirotsu, M. Irie, E. Kokufuta,
T. Okano, A. Suzuki, M. Suzuki, M. Tokita
P. Verdugo, K.L. Wang

With 144 Figures and 10 Tables

Springer-Verlag
Berlin Heidelberg GmbH

Volume Editor:

Prof. K. Dušek
Inst. of Macromolecular Chemistry
Czech Academy of Sciences
162 06 Prague 6, Czech Republic

ISBN 978-3-662-14937-9   ISBN 978-3-540-47836-2 (eBook)
DOI 10.1007/978-3-540-47836-2

© Springer-Verlag Berlin Heidelberg 1993
Originally published by Springer-Verlag Berlin Heidelberg New York in 1993
Softcover reprint of the hardcover 1st edition 1993
Library of Congress Catalog Card Number 61-642

Typesetting: Macmillan India Ltd., Bangalore-25

02/3020 5 4 3 2 1 0 Printed on acid-free paper

Preface

Gels are cross-linked networks of polymers swollen with a liquid. Softness, elasticity, and the capacity to store a fluid make gels unique materials. As our society becomes richer and more sophisticated, and as we increasingly recognize that natural resources are not unlimited, materials with better quality and higher functional performance become more wanted and necessary. Soft and gentle materials are beginning to replace some of the hard mechanical materials in various industries. Recent progress in biology and polymer sciences is unveiling the mystery of marvellous functions of biological molecules and promises new development in gel technologies. All these factors bring us to realize the importance and urgent need of establishing gel sciences and technologies.

Due to the cross-linking, various properties of individual polymers become visible on a macroscopic scale. The phase transition of gels is one of the most fascinating and important phenomena that allows us to explore the principles underlying the molecular interactions and recognition which exist in synthetic and biological polymers. The polymer network changes its volume in response to a change in environment; temperature, solvent composition, mechanical strain, electric field, exposure to light, etc. The prediction and finding of the phenomenon have opened the door to a wide variety of technological applications in chemical, medical, agricultural, electrical, and many other industrial fields.

The volume phase transition in gels has its history. It was theoretically predicted before it was discovered experimentally. However, the path from theory to experiment was not so straighforward because the conclusion of the theoretical analysis was that conditions for such a transition could hardly be met experimentally.

Among the participants of the IUPAC International Symposium on Macromolecular Chemistry in Prague in 1965, were the Editor of this volume (K.D.) and Donald Patterson (D.P.) of the CRM in Strasbourg and later McGill University in Montreal. D.P., well-known for his work in polymer solutions thermodynamics, presented a paper in this area, and K.D. presented a theoretical paper on phase separation in gels. This, however, concerned separation of a liquid from a swollen gel as a result of deterioration of polymer-solvent interaction or increasing crosslinking density during the crosslinking process where dilutions during crosslinking played an important role [1].

At the time of the conference, D.P. and K.D. discussed the possible peculiar shapes of the solvent chemical potential vs composition curves in swollen polymer networks prepared at different dilutions during network formation and values of the polymer solvent interaction parameter. Some of these curves exhibited a minimum followed by a maximum, a condition necessary for coexistence between two phases of different composition. Also at this symposium, a paper was given by Oleg Ptitsyn [2] on globule–coil transition in which he showed that a polyelectrolyte chain can undergo a collapse transition if the polymer-solvent interaction or degree of ionization were changed. All this inspired us in a deeper investigation of the phase equilibria in swollen polymer networks.

The result of analysis showed that a thermodynamic transition between two gels states differing in polymer concentration can be real and that the transition can be brought about not only by a change in the interaction parameter (temperature) but also by deformation. To exhibit this phase transition, the gel was to be prepared in the presence of a sufficient amount of diluent, its crosslinking density had to be sufficiently high, and the solvent in which it was swollen had to be rather poorer. The mechanistic explanation of the predicted phase transition was as follows: the network chains, after removal of the diluent after crosslinking, were rather supercoiled and had a tendency to assume more relaxed (expanded) conformation; this tendency was resisted by a strong tendency towards polymer segment association due to an unfavorable polymer – solvent interaction (poor solvent). The balance between these two strong and oppositely acting forces gave rise to the possibility of phase transition. However, it had turned out that preparation of such non-ionic gels at a high content of diluent and having high crosslinking density would be difficult due to a danger of gel-liquid phase separation during preparation. It was clear that a strong concentration dependence of the polymer solvent interaction parameter of the swelling liquid would greatly facilitate the occurrence of phase transition. Polyelectrolyte gels were not considered at all, although they could have been theoretically analyzed in view of the Ptitsyn's prediction of the globule-coil transitions.

The first report on the gel-gel transition was presented in September 1967 at the 1st Prague Microsymposium on Marcomolecules [3]. A paper was submitted to the *Journal of Polymer Science* and was published in 1968 [4]. One of the referees wrote that it was questionable whether a paper should be published on a phenomenon which could hardly be observed experimentally and recommended a reduction of the manuscript to about 50%. To meet, at least partly, his wishes, we reduced the manuscript to about 70% by removing all speculations about the possible concentration dependences of the interaction parameter.

These circumstances may explain why it took ten years for the phenomenon to be experimentally observed after the prediction. In 1973, prior to this finding, Lon Hocker, George Benedek, and Tanaka realized that a gel scattered light, and the light intensity fluctuated with time [5]. They established that the scattering is due to the thermal density fluctuations of the polymer network and derived a theory that explained the fluctuation. These fluctuations are similar to sound waves propagating in an elastic solid, which in this case is the polymer network. Since the network moves in water, however, the sound wave does not propagate, but decays exponentially with a relaxation time proportional to the square of the wavelength of the sound wave.

Time = Length^2/D

Here D is cooperative diffusion coefficient of the gel. Such a relationship applies to the random or diffusive motions of molecules in a fluid; for example, ink molecules in water. It is interesting that the same relation holds for a polymer network even though all the polymers are connected into a single network.

In 1977, while studying the light scattering from an acrylamide gel, Shin-ichi Ishiwata, Coe Ishimoto, and Tanaka found that the light intensity increased, and the relaxation time became longer as the temperature was gradually lowered [6]. They both diverged at a temperature of minus 17°C. Thus the critical phenomena were found in gels.

The finding raised a question of ice formation, although such a possibility was carefully checked and eliminated by the measurement of the refractive index of the gel. Such a question could be answered once and for all, if the temperature at which the scattering diverged was raised to much above the freezing temperature. So, many pieces of the gel were placed in acetone-water mixtures with concentrations ranging from 0% to 100%, hoping to find a proper solvent in which the gel would become opaque at room temperature. The next day, all the gel pieces were found to be transparent. But surprisingly, the gels in the lower acetone concentrations were swollen, and the gels in the higher acetone concentrations were collapsed. This meant that the gel volume changed discontinuously as a function of acetone concentration. The volume transition was found in gels 10 year after the first theoretical prediction [7].

The experiments were repeated but were not reproducible: Acrylamide gels were made anew with various recipes and their swelling curves were determined as a function of acetone concentration, but they were all continuous. It took a couple of months to recognize that the gels that showed the discontinuous transition were old ones, that is, gels prepared a month earlier and left within the tubes in which they were polymerized. Subsequent experiments were all carried out on "new" gels, and, therefore, underwent a continuous transition. At that time all the "old" gels were used up, and none were left in the laboratory.

Later the difference between the new and old gels was identified as ionization which induced an excess osmotic pressure within the gels leading to the discontinuous transition [8]. Hydrolysis was gradually taking place in the gel in a mildly high pH solution used at gelation. This explanation was experimentally proven by artificially hydrolyzing the gel and observing the increase in the discontinuity of the volume transition.

The theoretical formulation indicates that the gel transition should be universally observed in any gel. Many gels of synthetic and natural origin have been studied and the universality of the phase transition in gels seems to have been well established [9–11].

This volume contains the second part of short reviews with emphasis on the authors' work to show the present activity and state of knowledge in the field of volume transitions in gels. Part I was published in Volume 109. Unfortunately, a few of the leading groups were not able to prepare a review in time due to their overcommitments.

References

1. Dušek K (1967) J Polym Sci C 16: 1289
2. Ptitsyn OB, Kron AB, Eisner YE (1965) IUPAC International Symposium on Macromolecular Chemistry Prague, Preprint P747
3. Dušek K, Patterson D (1967) A transition in swollen polymer networks induced by intramolecular condensation, Microsymposium Polymer Gels and Concentrated Solutions, Inst. Macromol. Chem. Prague, Abstract F2
4. Dušek K, Patterson D (1968) J Polym Sci A-26: 1209
5. Tanaka T, Hocker LO, Benedek GB, (1973) J Chem Phys 59: 5151
6. Tanaka T, Ishiwata S, Ishimoto C, (1977) Phys Rev Lett 39: 474
7. Tanaka T (1978) Phys Rev Lett 40: 820
8. Tanaka T, Fillmore DJ, Sun S-T, Nishio I. Swislow G, Shah A (1980) Phys Rev Lett 45: 1636
9. Hrouz J, Ilavksý M, Ulbrich K, Kopeček J (1981) Eur Polym J 17: 361
10. Ilavský M, Hrouz J, Ulbrich K (1982) Polym Bull 7: 107
11. Amiya T. Tanaka T (1987) Macromolecules 20: 1162

Karel Dušek
Institute of Macromolecular Chemistry, Czechoslovak Academy of Sciences, 162 06 Prague 6, Czechoslovakia

Toyoichi Tanaka
Massachusetts Institute of Technology, Cambridge, MA, USA

Editors

Table of Contents

Coexistence of Phases and the Nature of First-Order Phase Transition in Poly-N-isopropylacrylamide Gels

S. Hirotsu

Department of Biological Sciences, Faculty of Bioscience and Biotechnology, Tokyo Institute of Technology, Ohokayama, Meguro-ku, Tokyo 152, Japan

The occurrence and the nature of first-order transition of gels are analyzed and discussed. First, the condition for the first-order transition and the crossover between the continuous and discontinuous transitions are reviewed within the Flory–Rehner and Erman–Flory phenomenological theories. Then, experimental investigations related to this subject are reviewed and compared with the predictions of the above theories. Our attention is focused on poly(N-isopropylacrylamide) (NIPA) gel and its ionized copolymers. It is shown that the continuous as well as slightly discontinuous transitions in neutral NIPA gels can be understood semiquantitatively on the basis of the above theories. On the other hand, it is pointed out that the strongly discontinuous transitions in ionized NIPA gels are less well understood, and that the real nature of these transitions has not been clarified at all. There is a summary of the results of new observations on the discontinuous transition process in NIPA gels, which have been undertaken to shed some light on this transition. From these results, an entirely new concept of the first-order transition of gels has emerged, in which the transition between the swollen and the shrunken phases proceeds via an intermediate state of coexistence of two phases. This state is considered to be not a transient, but a stable or metastable equilibrium one and has a number of unusual features. A close similarity is pointed out between this state and the coexisting-phases state observed in some metallic alloys around their order-disorder transition.

List of Symbols

d	diameter of cylindrical or spherical gels
k_B	Boltzmann constant
N_A	Avogadro's number
T	absolute temperature
V	volume of gel
V_0	volume of gel when it was formed
v_1	molar volume of solvent
N_c	number of chains
ΔF	free energy change induced when a solvent molecule is moved from the pure solvent phase into the pure polymer phase
Δh	enthalpy change induced by the same process as in ΔF
Δs	entropy change induced by the same process as in ΔF
α	linear swelling ratio
ϕ	volume fraction of polymer
ϕ_0	volume fraction of polymer when the gel was formed
π_m	osmotic pressure due to mixing
π_e	osmotic pressure due to network elasticity
π_i	osmotic pressure due to counter ions
π_s	osmotic pressure due to surface tension
π_T	total osmotic pressure
ρ	radius of curvature of the surface
γ	surface tension
χ	polymer-solvent interaction parameter

1 Introduction

The problem of discontinuous phase transition in polymer gels has a rather long history. About forty years ago, Hill [1, 2] studied this problem in relation to the electrostatic mechanism of muscle contraction, and concluded, on the basis of the statistical-thermodynamical theory of polymer networks, that polyelectrolyte gels under tension can exhibit a discontinuous transition in their stress-strain curve provided that an appropriate conditions are met. Later, Dusek and Patterson [3], stimulated by the theoretical work of Ptitsyn et al. [4] on a discontinuous coil-globule transition of a single polymer chain, reconsidered this problem and obtained a critical condition for a discontinuous transition. One of their conclusions was that a discontinuous transition will occur in a non-ionized gel under tension if the gel had been prepared by crosslinking polymer chains while swollen in a good solvent. Experimental verification was first made by Tanaka [5], who found that ionized polyacrylamide gels underwent a discontinuous volume change as functions of both solvent composition and temperature. This finding ensures that, in contrast to the previous theoretical predictions, a discontinuous phase transition can occur in free gels, which stimulated further research in this field. Since then, the study of volume phase transition has been developed rapidly both on experimental and theoretical sides.

The theoretical formulation of the collapse of a polymer chain [4, 6] and the volume phase transition of gels [1–3] has been developed by utilizing an analogy between the liquefaction of a real gas and the condensation of polymer segments. In fact, this analogy is quite helpful to understand the phenomenological aspect of the collapse of a polymer chain and of a polymer network. However, we do not know to what extent this analogy is valid in real cases. Because a gel is a solid, the elastic deformation of the network may play an important role in real phase transition processes.

In this connection, we admit that we know little of the real nature and the process of the discontinuous phase transition of gels. Although the phenomenological theory predicts that the whole sample transforms from one phase to the other at a specified temperature (the transition temperature), there has been some experimental evidence that the transition in real gels never occurs in such a manner. For example, a serious deformation of the sample [7] as well as the coexistence of phases [8] have been observed over a rather wide temperature range around the first-order transition point. A curious, and at the same time important point is that these states seem not to be transient but stable states of the gels [8].

The present article attempts to clarify the nature of the discontinuous transition of gels. First, in Sect. 2 we give an outline of the fundamental aspect of the volume phase transition on the basis of the Flory–Rehner theory of gels, with special attention to how the discontinuous transition comes about within the phenomenological treatment. Then, in Sect. 3 previous experimental results

on swelling and phase transitions are reviewed and compared with the calculation based on the phenomenological theory. The experimental results presented there are mostly confined to those obtained in the author's own
laboratory on neutral as well as on ionized N-isopropylacrylamide/water gels
(abbreviated as NIPA gels). It is pointed out, on the basis of this comparison,
that the present understanding of the discontinuous transition of gels is quite
inadequate. To shed some light on this situation, a new experiment was devised
to observe in detail the process of discontinuous transition of ionized NIPA gels.
The result is presented in Sect. 4 which includes several new findings on
anomalous features of the transition. An entirely new viewpoint for the first-
order transition of gels has emerged from this result, and it is outlined in Sect. 5.
In Sect. 6, a summary of the present article is given.

Before going on to the next section, it will be appropriate to comment on the
terminology. From the strict thermodynamical point of view, the term phase
transition should be used for gels only when the change of states occurs
discontinuously. In this article, however, we use this term more loosely by
regarding the continuous changeover between the swollen and shrunken states
of gels as a continuous phase transition so long as it occurs steeply enough as a
function of some intensive parameter.

2 Discontinuous Transition Within the Flory–Rehner
Phenomenological Theory

The Flory–Rehner phenomenological theory [9, 10] has been most widely used
to analyze the volume phase transition of gels. This theory, extended [11] to
take into account the concentration dependence of the polymer-solvent interaction parameter χ, can predict basic features of the phase transition.

Take a gel swollen in a solvent. We denote the number of chains contained in
a gel network as N_c, the number of solvent molecules contained as N_s, and the
concentration (volume fraction) of polymer as ϕ. To represent the degree of
swelling of the network, it is convenient to introduce the linear swelling ratio α
defined as

$$\alpha = (V/V_0)^{1/3} = (\phi_0/\phi)^{1/3} , \tag{1}$$

where V is the volume of the gel and the subscript 0 means the value of the
relaxed Gaussian network. If a gel network is prepared by a crosslinking
reaction in a solution, as in the case of our NIPA gels, ϕ_0 is equal to ϕ in the pre-
gel solution. If, on the other hand, the network is prepared by crosslinking
polymer chains without solvent, as was considered in Flory's original theory, we
must put $\phi_0 = 1$. Note that α can characterize the swelling state of a gel

provided that the whole gel is in a homogeneous and isotropic state. The Flory theory deals only with this idealized state of gels.

We can express the free energy of gel in terms of the above quantities, and calculate from it the osmotic pressure π acting on the network. For a gel to be in equilibrium with the outer solvent, π must be zero. Several different mechanisms are known to contribute to π, the mixing process, the rubber elasticity, and counter ions, etc. The osmotic pressure due to mixing is expressed as

$$\pi_m = -(N_A k_B T/v_1)\left[\phi + \ln(1-\phi) + \chi\phi^2\right], \qquad (2)$$

while that due to rubber elasticity is

$$\pi_e = (N_c k_B T/V_0)\left[(\phi/2\phi_0) - (\phi/\phi_0)^{1/3}\right]. \qquad (3)$$

In the above equations, χ is the polymer-solvent interaction parameter, k_B the Boltzmann constant, T the absolute temperature, N_A Avogadro's number, v_1 the molar volume of solvent. If the network is made up only of neutral monomers, π_m and π_e sum up to generate the total osmotic pressure π_T. On the other hand, when ionizable species are contained in the network, the osmotic pressure due to ions must be considered. The contribution from counter ions [5, 10, 11] is

$$\pi_i = (f N_c k_B T/V_0)(\phi/\phi_0), \qquad (4)$$

where f is the number of counter ions per chain. The electrostatic force due to charges fixed on the network is not taken into account in this theory. The equilibrium condition of gels is then written as

$$\pi_T = \pi_m + \pi_e + \pi_i = 0. \qquad (5)$$

In the present theory, the mechanism which drives the transition is entirely incorporated into χ. The meaning of χ [9, 10] is similar to the parameter expressing the contact energy change appearing in the Bragg–Williams theory of the order-disorder transition of alloys [12]. Writing the change in the free energy per solvent molecule (divided by $k_B T$) induced when a solvent-solvent contact is changed into a solvent-polymer contact as ΔF, we can formally write χ as

$$\chi = \Delta F/k_B T = (\Delta h - T\Delta s)/k_B T, \qquad (6)$$

where Δh and Δs are the corresponding changes in the enthalpy and the entropy, respectively. Usually, the same molecular interaction makes a dominant contribution to both Δh and Δs, and consequently the signs of both quantities are the same. Polymer-solvent systems possessing the upper-critical-solution temperature (UCST) is characterized by positive values of Δh and Δs, while the one possessing the lower-critical-solution-temperature (LCST) is characterized by negative values of these quantities.

In some polymer-solvent systems, χ has been determined experimentally [13, 14], according to which χ in general depends upon concentration in a nonlinear manner. Thus, expanding χ in ϕ, we obtain

$$\chi = \chi_1 + \chi_2\phi + \chi_3\phi^2 + \cdots. \tag{7}$$

It has been shown [11] that a particularly strong concentration dependence of χ is essential for a discontinuous transition to occur in neutral gels. Graphical presentation of the equilibrium condition (5) may be the most convenient way to understand the occurrence of the first-order transition. In Fig. 1, π_e and $-\pi_m$ are drawn as functions of ϕ. The intersection of these two curves is the root of Eq. (5), which corresponds to the equilibrium state for neutral gels. Numerical calculation with concentration independent χ always gives one root irrespective of temperature as shown in Fig. 1a, assuring that the change in state is continuous. On the other hand, when the concentration dependence of χ becomes strong enough, the convex shape of $-\pi_m$ in the intermediate concentration range becomes enhanced so that three roots appear as shown in Fig. 1b. The three roots correspond to the stable, unstable, and metastable states, respectively, thus predicting the occurrence of a discontinuous transition. The transition temperature can be determined by the Maxwell construction as in the case of a gas-liquid transition of the van der Waals fluid.

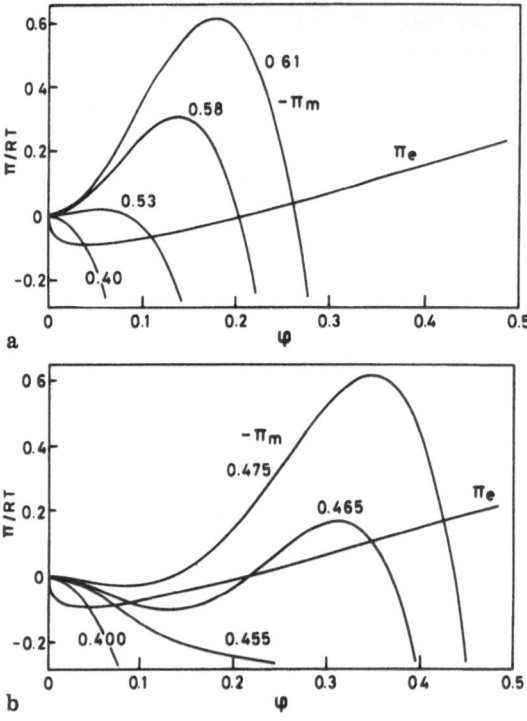

Fig. 1a, b. The elastic part (π_e) and the negative of the mixing part ($-\pi_m$) of the osmotic pressure as functions of polymer concentration ϕ. The intercepts of π_e and $-\pi_m$ correspond to the equilibrium state of neutral gels. Numbers beside each curve of $-\pi_m$ represent χ_1, which increases with temperature. (a) $\chi_2 = 0$. Only one root at all temperatures. (b) $\chi_2 = 0.56$. Three roots appear in the intermediate temperature range (around $\chi_1 = 0.465$), which correspond to stable, unstable, and metastable states, respectively. (Reproduced with permission from Ref. 20)

The critical value of χ_2 for the three roots to appear is shown [11] to be 1/3 for a gel with infinitely long chains. In actual gels, the critical value must be appreciably larger than 1/3. Thus, a discontinuous transition rarely occurs in neutral gels, because it requires an unusually large concentration dependence of χ. An NIPA gel with small crosslinking density is one such rare example.

In ionized gels, unlike neutral ones, the condition for a discontinuous transition can be realized relatively easily. Combining Eqs. (3) and (4), we obtain the effective osmotic pressure due to elasticity of ionized network as

$$\pi_e + \pi_i = (N_c k_B T / V_0) \left[(f + 1/2)(\phi/\phi_0) - (\phi/\phi_0)^{1/3} \right] . \tag{8}$$

Due to the presence of f in the first term on the right-hand side of the above equation, the gradient of $(\pi_e + \pi_i)$ vs the ϕ curve increases as compared with the case of neutral gels, so that three roots appear even when χ does not depend on ϕ. This is shown in Fig. 2. It is seen that the swelling ratio and the transition temperature become appreciably higher than those in neutral gels. The calculation of Fig. 2 does not represent a realistic case because in any real polymer-solvent systems, χ will depend more or less on the concentration. This calculation has been made only to emphasize the large effect of ions on swelling and phase transition in gels.

We see from the above argument that, within the Flory theory of gels, the concentration dependence of χ is the driving force for the transition in neutral gels. Hence, to understand the mechanism of the phase transition of gels on a molecular level, we must identify the microscopic interaction which makes χ depend on the concentration. For this purpose, we must specify not only the

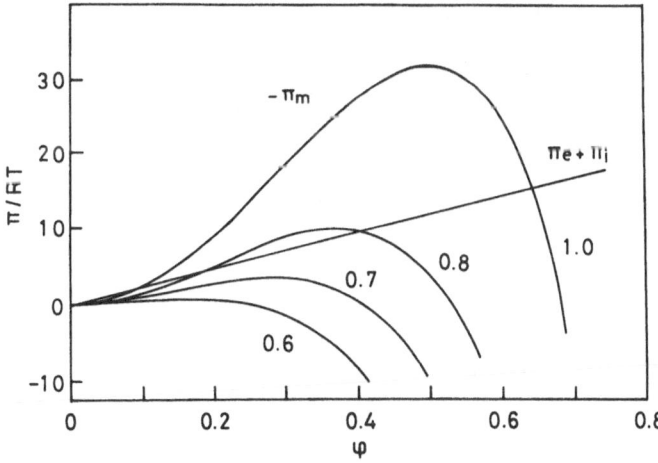

Fig. 2. The sum of the elastic and the ionic parts $(\pi_e + \pi_i)$ and the negative of the mixing part $(-\pi_m)$ of the osmotic pressure as functions of polymer concentration ϕ. The calculation was made for $f = 10$ and $\chi_2 = 0$. Note that the ordinate scale is fifty times larger than in Fig. 1. The intercepts of these curves correspond to the equilibrium states of ionized gels. Numbers besides each curve of $-\pi_m$ represent χ which increases with temperature

interaction between solvent molecules and polymer segments, but also solvent-solvent, and segment-segment interactions on a molecular level. Even for neutral gels (not considering ions), this is a very complicated problem and an attempt to develop really microscopic theory of volume transition of gels has hardly been made. It is certain, however, that the structure formation among water molecules in the vicinity of the polymer segments is the most important factor determining the swelling and phase transition properties of hydrogels. Moreover, some regular structures might be formed in the shrunken phase of gels among hydrophobic groups within chains. Thus, the investigation of the microscopic mechanism of phase transition in hydrogels is somewhat related to the problem of predicting the three-dimensional structure of proteins in an aqueous solution from its primary structure.

In some polymer-nonpolar solvent systems, χ has been calculated as a function of concentration on the basis of the statistical-thermodynamical theory called the "equation of state theory" [13, 14]. This semiempirical theory takes into account not only the interaction between solute and solvent, but also the characteristics of pure substances through the equations of state of each component. At present, however, we cannot apply this approach to such a complex case as the NIPA-water system. Thus, at the present stage, we must regard χ as an empirical parameter to be determined through a comparison between calculated and experimental results. The empirical estimation of χ for the NIPA-water system will be described in the next section.

We have outlined above the Flory–Rehner and Erman–Flory theories of the phase transition of gels. Other theoretical approaches to the same problem have also been developed and should be mentioned briefly. Prange et al. [15] applied the quasi-chemical method of mixtures developed by Guggenheim [16] to phase transition of gels. They extended the original theory by introducing an orientation-dependent interaction (hydrogen-bond formation) between polymer segments and solvent molecules. With several simplifying assumptions, they derived equations for chemical potentials of constituents, which contain three adjustable parameters, i.e. the exchange parameters for different types of contact pairs. By adjusting these parameters, they obtained a reasonably good agreement between the calculated and the experimental phase behaviors of aqueous polymer solutions and hydrogels. On the other hand, Marchetti et al. [17] developed an alternative approach, on the basis of the Flory–Rehner model, by introducing a vacant lattice site (hole) as the third component in addition to the polymer and solvent. The nonzero value of the mixing volume can be explained by the presence of holes. They derived a three-component (gel-solvent-hole) phase diagram and could explain the phase transition behavior of NIPA gels reasonably well.

These phenomenological theories are more complicated than the Flory-type theory, though they have certain advantages over the latter. In the quasi-chemical treatment, the molecular interaction responsible for the transition, which is hidden behind the parameter χ in the Flory theory, appears with clearer physical meaning. In the hole theory of gels, some properties of gels which are not easy to understand within the Flory theory can be explained. A hydrostatic pressure dependence of the transition temperature is one such example.

3 Comparison Between Theory and Experiment

We will now compare the prediction from the above phenomenological theory with the experimental results to see how well this theory can account for real volume transitions of gels. Our experiments were made on neutral and ionized NIPA gels. Neutral NIPA gel exhibits a nearly critical transition as a function of temperature in pure water [7, 8, 18 –20]. The order of the transition and the degree of discontinuity at the transition can be varied by changing the cross-linking density [20, 21] or adding ions [7] into the network. These simple and convenient properties of NIPA gels make them the most suitable systems for detailed studies of the crossover phenomenon between continuous and discontinuous transitions.

First, I will explain the experimental method to measure the swelling curve of gels. The preparation of NIPA gels has already been given [7, 8, 18–20], and will not be repeated here. Gels of desired shapes can be prepared by using appropriate molds, in which the gelation reaction takes place. Measuring the volume of gel with high accuracy under a strictly specified condition is not easy because we have to measure the spatial extension of the network without taking it out of solvent in a temperature-controlled bath. Moreover, it is important to realize that, as will be explained later, the observed phase transition behavior depends strongly on the method of measurement. The easiest way to measure the volume of gel is to use a gel of symmetrical shape (e.g. cylinder or sphere) and measure its characteristic length (e.g. diameter) from outside the temperature bath (in situ) with a microscope of long focal length. If the sample swells or shrinks homogeneously and isotropically, we can calculate the volume of gel from the measured length d as

$$V = V_0(d/d_0)^3 , \tag{9}$$

where d_0 is the length when the gel was formed, i.e. the inner size of the mold used to prepare the gel.

An alternative method to obtain a volume of gel is to measure its weight [22]. We can calculate the volume of gel from its weight provided that the weight of the dry network and the density of polymer are known. The advantage of this method over the first one is that the total volume of the sample can be obtained even if the sample becomes heterogeneous or distorts irregularly. On the other hand, the serious disadvantage is that we must take the gel out of water to measure its weight. Change in temperature, evaporation of water during measurement, and excess water attached on the surface of gel will introduce errors to the result, so that very precise measurements will be impossible by this method.

I adopted the first method using cylindrical gels with a few mm diameter and $10 \sim 30$ mm long, which were attached to the end of a thin glass capillary and set in a rectangular culture tube filled with pure water. The axis of the cylindrical gel was kept nearly vertical. The culture tube had a pair of flat glass windows, through which the diameter of gel was measured in situ with a microscope as a function of temperature.

I measured the diameter at one particular point (usually near the center) of the gel cylinder as a function of temperature. Thus, the result of this measurement does not represent the total volume of the sample but the local volume around a particular point in the sample. Of course, these two quantities are proportional to each other when the sample is homogeneous. However, if the sample becomes heterogeneous (splits into two phases, for example), the proportionality breaks down, so that we cannot obtain the total volume of sample from the local measurement. As will be mentioned in the next chapter, this point is of crucial importance in interpreting the experimental data in the transition region.

3.1 Neutral NIPA Gels

Swelling curves of two neutral NIPA gels with different initial polymer concentration ϕ_0 and the crosslinking density N_c/V_0 are shown in Fig. 3 [20]. One sample (call it gel A) had $\phi_0 = 0.075$ and $N_c/V_0 = 1.02 \times 10^{22}$, which undergoes a slightly discontinuous transition as shown in Fig. 3a, while the other (gel B) had $\phi_0 = 0.114$ and $N_c/V_0 = 2.40 \times 10^{22}$, which undergoes a continuous transition as shown in Fig. 3b. In the former sample, the coexistence of the swollen and the shrunken phases was observed within 0.05 °C of the transition temperature, which is concrete evidence that the transition is discontinuous. On the other hand, the latter sample was homogeneous, at least by visual inspection,

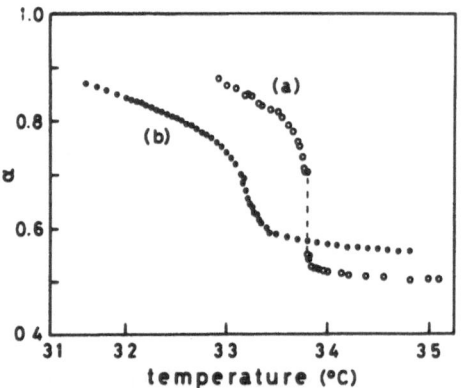

Fig. 3. Measured linear swelling ratio of neutral NIPA gels as functions of temperature. (a) gel A, (b) gel B. (Reproduced with permission from Ref. 20)

throughout the transition, which is the evidence that the transition is continuous. It has been confirmed [21] that increasing the crosslinking density leads to a crossover from the discontinuous to the continuous transition.

These results are in good accord with the prediction from the phenomenological theory. The calculation of the swelling curves was made in terms of Eqs. (2), (3) and (5)–(7). In Eq. (7), the terms higher than ϕ^2 were neglected. The structural parameters N_c/V_0 and ϕ_0 used in the calculation were determined from the amounts of monomers and a crosslinker used in preparing the gel. As noted in the last Section, the calculated transition behavior is determined essentially by the values of Δh, Δs, and χ_2. The values of Δh and Δs can be estimated [19] from the experimental swelling curves as follows. Substituting Eqs. (2) and (3) into Eq. (5) and putting $\pi_i = 0$, we obtain

$$\chi = \phi^{-2} \, [(N_c v_1/N_A V_0)\{(\phi/2\phi_0) - (\phi/\phi_0)^{1/3}\} - \phi - \ln(1 - \phi)] \,, \qquad (10)$$

which is an equation expressing χ as a function of ϕ when a neutral gel is in equilibrium with outer solvent. Substituting the experimental $\phi(T)$ in Eq. (10), we obtain χ at each temperature, which is shown in Fig. 4 as a function of $1/T$. Equation (6) shows that χ should be linear in $1/T$ as long as Δh and Δs are constant. In reality, the latter quantities depend on concentration, and thus we see that the large jump of χ at the transition is the result of jump of ϕ at this temperature. The low-temperature portion of Fig. 4, where ϕ is so small that we can neglect the ϕ dependence of χ, indeed shows a linear behavior. The slope of this portion gives Δh, and the extrapolation of this linear portion to $1/T = 0$ gives Δs. The values of these quantities thus determined are given in Table 1.

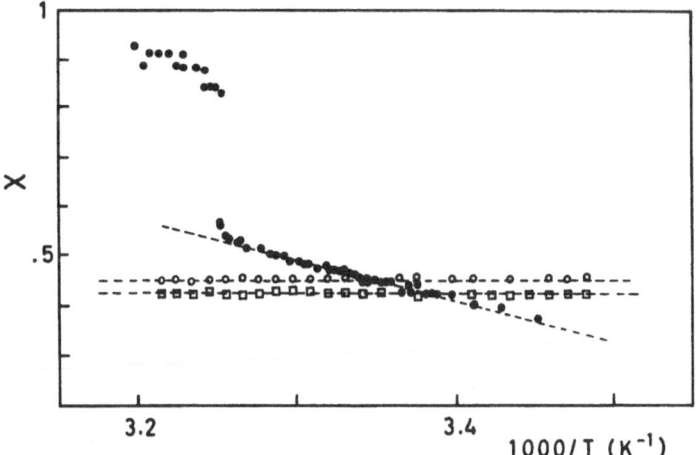

Fig 4. Polymer-solvent interaction parameter χ calculated from Eq. (10) using the experimental swelling ratio as a function of inverse absolute temperature. *Dashed lines* represent fits to straight lines which should be obeyed in a highly swollen state. ●: NIPA/water, ○: NIPA/ethanol, □: NIPA/n-propanol. (Reproduced with permission from Ref. 19)

Table 1. Changes in the enthalpy (Δh) and the entropy (Δs) parameters appearing in the polymer-solvent interaction parameter χ of NIPA-water system

Solvent	Δh (10^{-22} J)	Δs (10^{-23} J K^{-1})
Water	-130 ± 50	-4.5 ± 1.0
Methanol	-1.5 ± 1.0	-0.6 ± 1.0
Ethanol	-0.5 ± 1.0	-0.5 ± 1.0
n-Propanol	-3.5 ± 1.0	-0.7 ± 1.0

For comparison, χ vs $1/T$ curves for the same NIPA network with some alcohols as solvents [19] are also shown in Fig. 4, and the values of Δh and Δs determined are included in Table 1. It is seen that these quantities are all negative for the systems studied, and that the absolute values of these quantities in NIPA-water system are far larger than those in other systems. In fact, $|\Delta h|$ in NIPA-water system is much larger than $k_B T$, which explains the strong temperature dependence of the volume in this gel. Similarly, $T|\Delta s|$ is much larger than $k_B T$, which drives the transition to the phase with larger entropy as temperature rises.

The value of χ_2 was deduced from the jump at the transition of ϕ and that of calculated χ, because these two quantities should be proportional with each other. The value thus obtained was $\chi_2 = 0.6 \pm 0.1$. In the actual calculation, the values of Δh, Δs, and χ_2 were adjusted within the error limits so that the calculated swelling curve fits the measured one as closely as possible. The results of calculation [21] are shown in Fig. 5a and b, where they should be compared with Fig. 3a and b, respectively. The parameter values used are given in Table 2. Of course the values of Δh, Δs, and χ_2 include relatively large arbitrariness, although the fact that we can fit the observed swelling curves using reasonable values of the parameters shows that this theory captures an essential point of the phase transition of neutral gels.

It should be mentioned that the order of the transition depends not only on the interaction and the structural parameters appearing in the above theory but

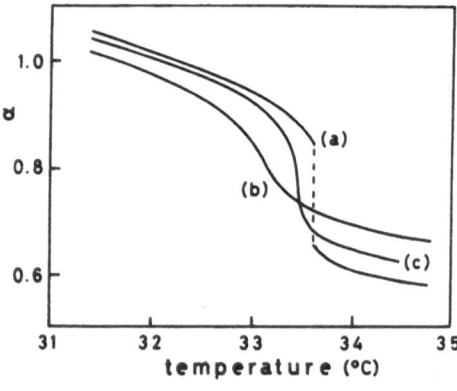

Fig. 5. Linear swelling ratio calculated for neutral NIPA gels on the basis of the Flory–Rehner theory. The structural parameters used in the calculation were determined from the preparation condition of (a) gel A, and (b) gel B. The curve (c) represents the critical transition. (Reproduced with permission from Ref. 20)

Table 2. Parameters used in the calculation of Figs. 5. A: gel A, B: gel B. The curves in Fig. 7 are all calculated using the same values as those of gel A

	A	B
ϕ_0	0.075	0.114
N_e/V_0	1.02×10^{22}	2.40×10^{22}

The other parameters take the common value listed below.

$\chi_1 = (\Delta h - T\Delta s)/k_B T$
$\chi_2 = 0.518$, $\Delta h = -12.462 \times 10^{-21}$, $\Delta s = -4.717 \times 10^{-23}$

also on more subtle conditions not expressible in the above theory. As a matter of fact, an apparent discrepancy as for the order of the transition exists among the previous results [7, 8, 18–20]. For example, in one of the earlier studies [7], the transition of neutral NIPA gel has been reported to be continuous, whereas in most of other studies [8, 18–20], it is reported to be slightly discontinuous. The initial composition of the samples are identical in all these studies, and the only difference is the accelerator used, i.e. in the former [7], sodium bisulfite (SB) was used, whereas in the latter [8, 18–20], N,N,N',N'-tetramethylethyl-enediamine (TEMED) was used as an accelerator.

In fact, the correlation deduced above between the accelerator used in the preparation of gels and the order of the transition they exhibit has been confirmed by the following experiment [23]. I have prepared NIPA gels from the same pregel solution using these two accelerators, one with SB and the other with TEMED (these we call SB-gel and TEMED-gel, respectively). The swelling curves of the gels obtained are shown in Fig. 6. It is clearly seen that the transition in SB-gel is continuous whereas that in TEMED-gel it is discontinuous verifying the above expectation. Moreover, the degree of swelling is appreciably larger in the former than in the latter.

Although it is impossible to specify the mechanism which causes this difference from these data alone, it is certain that some imperfections of the network play a decisive role. However, I do not insist from the above data alone on the explanation that defects of the network induce the crossover between continuous and discontinuous transition. We cannot deny the possibility that the seemingly continuous transition observed on SB-gels is due to the distribution of the discontinuous transition temperature. What I want to stress here is that the transition behaviour as well as the degree of swelling depend sensitively on the preparation condition of gels, and that the effect of imperfections on the swelling and the phase transition has scarcely been investigated.

Neutral NIPA gel is the most extensively studied among known gels from the standpoint of phase transition, and thus, various physical properties around the transition have been reported. These include the shear and bulk modulus [20, 24], the diffusion constant of the network [25], spinodal decomposition [26], specific heat [21], critical properties of gels in mixed solvents [8] and the effect of uniaxial [27] and hydrostatic [28] pressures on the transition, and so

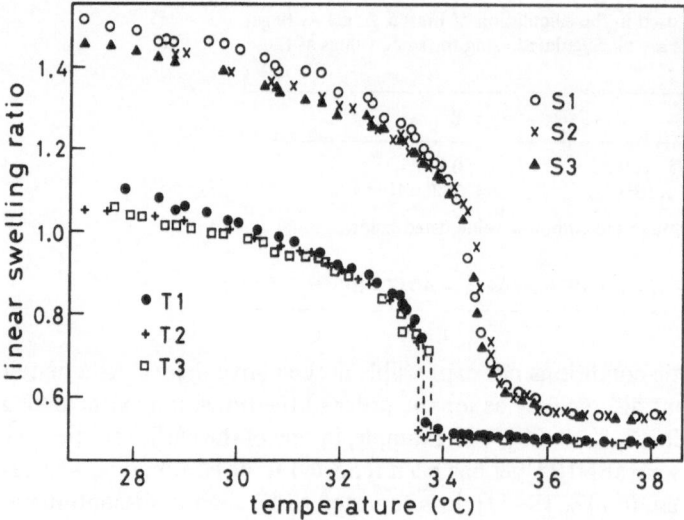

Fig. 6. Linear swelling ratio measured for neutral NIPA gels prepared from the same pregel solution with different accelerators. *S1–S3*: Three different samples prepared with SB as an accelerator. *T1–T3*: Three different samples prepared with TEMED as an accelerator

on. In connection with these studies, it should be mentioned that any measurement of gel properties with high precision is, in general, more difficult to make than in crystals or in liquids. Various technical problems arise from the necessity of doing measurements on gels kept in a solvent, and also from the fact that the gel changes its volume drastically and sometimes deforms irregularly as a function of temperature. These impose serious difficulties on making precise electrical, mechanical, thermal and optical measurements. The very long relaxation time is also unfavorable for measurements of equilibrium properties.

In addition to these technical problems, the complexity inherent to physical properties of gels is, as exemplified above, that they depend very sensitively on the preparation condition. This is because, in a formal language, a gel is a frozen system and we need two sets of statistical information, the "preparative ensemble" and the "final ensemble", to understand its equilibrium properties [29]. Hence, a gel is by nature more complex than the usual equilibrium systems. We should clarify the dependence of the properties of gels on preparation conditions, and also on structural defects of the network before going into precise investigations such as critical phenomena associated with the phase transition.

3.2 Ionized NIPA Gels

Swelling curves of NIPA-sodium acrylate (SA) copolymer gels have been measured [7] by the same method as applied to neutral gels using cylindrical samples. The concentration of NIPA (700–572 mM) and SA (0–128 mM) were varied, but their sum was fixed at 700 mM, which is equal to the concentration of monomers in the neutral NIPA gels studied in the last Section. Crosslinking was made by N,N'-methylenebisacrylamide (BIS, 8.6 mM). The results are shown in Fig. 7. It can be seen that α in the swollen phase, the transition temperature, and the magnitude of discontinuity at the transition all increase as the ionic content increases. The transition of neutral gels is seen to be continuous here, because SB was used as an accelerator in preparing these gels [7].

The qualitative features of the above experimental results can be reproduced by the phenomenological theory. However, the previous calculation [7] was unsatisfactory because we used χ which is independent of concentration, and as a result, we had to assume unreasonable values for structural parameters of gels to obtain a good fit between the theory and experiment.

This was because, at the time of the previous calculation, we had no idea how to improve the theory by taking into account the concentration dependence of χ. Now, however, we have determined the values of parameters χ_2, Δh, and Δs which yield a reasonably good fit to the swelling curves of neutral NIPA gels. Hence, to be consistent, we should use the same thermodynamical parameters also in the calculation for the ionized NIPA gels. Such a calculation has been made [30] and its result is shown in Fig. 8. In comparing the experimental swelling curves (Fig. 7) and the calculated ones (Fig. 8), it should be noted that $f = 1$ corresponds to 17.2 mM of ionic species. Then, if the dissociation of carboxyl group is complete, $f = 0.5, 2, 3$, and 4 in Fig. 8 correspond approximately to 8, 32, 50, and 70 mM in Fig. 7, respectively. As compared with the previous calculation, a reasonably good fit is obtained using structural parameters determined from the preparation condition of the samples.

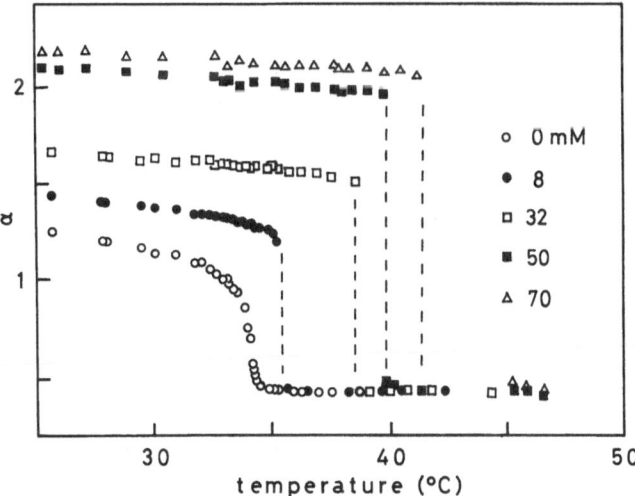

Fig. 7. Measured linear swelling ratio for ionized NIPA gels as functions of temperature. The curves have been replotted using the data reported in Ref. 7. The concentration of ionic species (SA) is shown in the figure

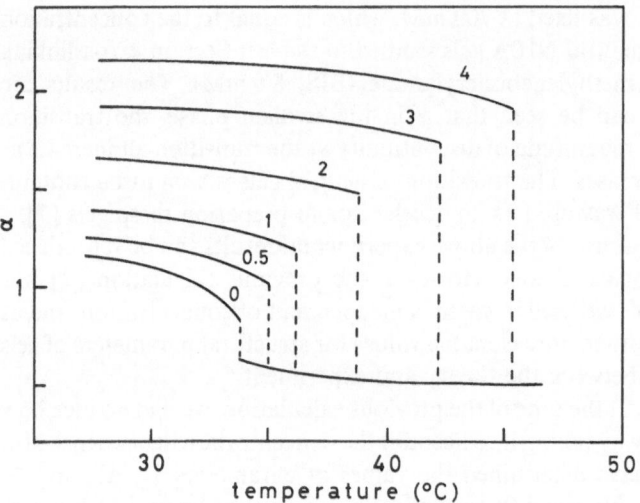

Fig. 8. Linear swelling ratio calculated for ionized NIPA gels. Numbers on curves represent the effective number of charges per polymer chain. Structural parameters were calculated from the preparation condition of the samples. Other parameters were identical with those used in the calculations for neutral NIPA gels

However, we should not overestimate this agreement between the experimental and the calculated results, because the underlying model leading to the expression of Eq. (5) for the ion osmotic pressure is too simple to be realistic. For example, the dissociation of ionic groups are assumed to be independent of both temperature and concentration. In reality, however, the dissociation of ionic groups depends on the concentration of ions in the solution. Considering that the dissociation constant of acrylic acid is fairly small [32], and that the solution within these ionized gels should be more or less acidic, the degree of dissociation of carboxyl groups must be smaller than 1, and it will depend on the degree of swelling of the gels. Hence, the use of constant f cannot be justified at least in the present analysis.

Complete neglect of the Coulomb interaction between charges fixed on the network is also a problem. It has been argued [33] that the contribution from fixed charges to the osmotic pressure plays an important role in determining the swelling equilibrium of gels. However, we will not discuss these theoretical problems further because the main concern of the present article is to clarify experimentally the nature of the first-order transition rather than to improve theoretical treatment.

3.3 Phase Coexistence Around the First-Order Transition

One very important phenomenon observed during the measurement of the discontinuous swelling curves should be mentioned here. It is the coexistence of different phases in a single sample around the transition. It has already been reported [7] that a gel with a large volume discontinuity distorts irregularly at the transition, and that this may be due to the coexistence of swollen and shrunken phases. Clear evidence of the coexistence of the swollen and the shrunken phases has been observed in NIPA cylinders [8]. The temperature width in which the phase coexistence persisted was less than 0.05 °C in a neutral gel [20], and of the order of 0.1 °C in an ionized gels with low ionic content, and thus could easily be overlooked in a rough measurement. On the other hand, in a gel with high ionic content, the width was more than 4 °C. In such cases, it was not possible to determine unambiguously the transition temperature of the sample. The transition temperatures shown by dashed lines in the experimental swelling curves of Fig. 7 roughly correspond to the center of the coexistence temperature width. However, there is no theoretical ground for a statement that the center of the coexistence-temperature region should correspond to the theoretical transition temperature determined by the Maxwell rule. Although the coexistence of phases around the transition has not been discussed seriously so far, it is considered to be an important feature of the first-order transition of gels. We will focus our attention on this phenomenon in Sect. 4.

3.4 Shape Dependence of the Swelling and Phase Transition

Another unique feature of ionized NIPA gel has been found recently; both of the equilibrium swelling ratio α and the first-order transition temperature T_0 depend strongly on the shape of samples [31]. The measurement of equilibrium α has been made on ionized NIPA gel rods of various diameters, and also on plates and cubes. The gel contained 680 mM NIPA, 20 mM acrylic acid (AA), and 8.6 mM BIS. All samples were prepared from the same pregel solution at the same time so as to guarantee that the composition and the structure of all samples were the same.

The measured α at room temperature of gel cylinders with various diameters are shown in Fig. 9. It is seen that α increases with the diameter of cylinders, and also α measured on plates and cubes are larger than that of cylinders. This result is surprising because it shows quite unambiguously that the equilibrium degree of swelling depends on the shape of samples. The transition temperature also depends strongly on the shape of samples, as shown in Fig. 10. The difference between the T_0 of the thinnest (0.2 mm) and of the thickest (5 mm) cylinders measured in this study differed by more than 8 °C.

In neutral gels, in contrast, neither α nor T_0 depend on the shape of samples at all, as shown in Fig. 11. Thus, it is certain that the unusual shape-dependent properties shown in Figs. 9 and 10 are due to ions contained in gels. Although the mechanism underlying the shape dependence is not clear at present, one possible mechanism has been proposed [31] on the basis of the surface tension of ionized gels. Denoting the surface tension as γ and the radius of curvature of the gel surface as ρ, an additional contribution to the osmotic pressure due to

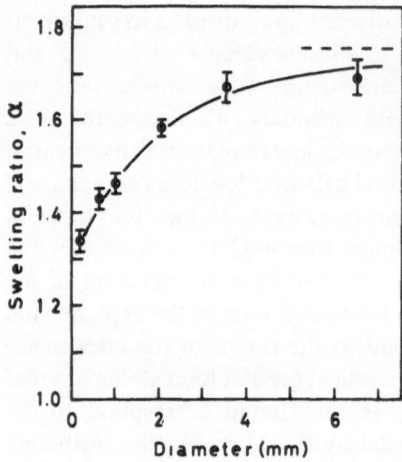

Fig. 9. Linear swelling ratio of NIPA-AA co-polymer gel cylinders at 20 °C as a function of their diameter. The *solid line* is a guide for the eye. The *dashed horizontal line* denotes the value obtained on cubes and plates. (Reproduced with permission from Ref. 31)

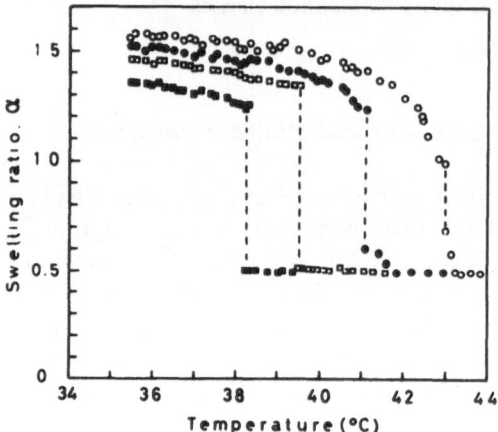

Fig. 10. Linear swelling ratio of NIPA-AA copolymer gel cylinders as a function of temperature measured on heating. Diameters are 0.6, (■), 1.0 (□), 2.0 (●) and 3.5 mm (○). (Reproduced with permission from Ref. 31)

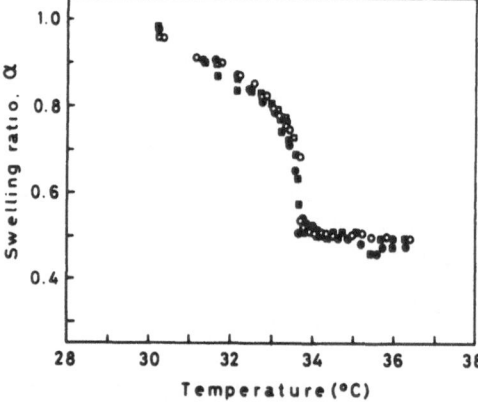

Fig. 11. Linear swelling ratio of neutral NIPA gel cylinders as a function of temperature on heating. Diameters are 0.2 (■), 0.6 (□), 1.5 (●), and 3.5 mm (○). (Reproduced with permission from Ref. 31)

the surface tension is

$$\pi_s = 2\gamma/\rho \ . \tag{11}$$

Numerical calculation [31] including this osmotic pressure component into the equilibrium condition Eq. (5) shows that the observed shape dependence of α and T_0 can be explained provided that γ is large enough. It is desirable to investigate the surface state of ionized gels in detail and to measure the surface tension whether it gives such a large influence on the swelling equilibrium of ionized gels.

4 Process of Discontinuous Transition: Triphasic Equilibrium

The phase coexistence observed around the first-order transition in NIPA gels cannot be interpreted by the Flory–Rehner theory because this theory tacitly assumes that the equilibrium state of a gel is always a homogeneous one. Heterogeneous structures such as two-phase coexistence are ruled out from the outset in this theory. Of course, if the observed phase coexistence is a transient phenomenon, it is beyond the thermodynamical theory. However, as will be described below, the result of the detailed experiment strongly indicates that the coexistence of phases is not a transient but rather a stable or metastable equilibrium phenomenon. At any rate, we will focus our attention in this article only on static equilibrium phenomena.

I have investigated experimentally [34] the detailed process of the first-order transition of ionized NIPA gels by observing the evolution of the phase-separated structure in the transition region. For this purpose, it is crucial that the temperature must be changed extremely slowly and must be regulated precisely. In our experiment, the rate of the temperature change was less than 0.1 °C/day. The long-term temperature stability was within $\pm$ 0.015 °C/day and the short-term temperature stability was within $\pm$ 0.007 °C/h. In some cases, the gel was kept at constant temperature in the transition region for over a month to see if any change occurs during this period. The samples were NIPA-acrylic acid (AA) copolymer gels containing 680 mM NIPA, 20 mM AA, and 8.6 mM BIS. Shapes of the samples were cylinder, plate, and cube of various sizes.

Below, I will first describe the observation on thin cylinders because the phase coexistence can most clearly be observed in these samples and, moreover, samples of this shape are most frequently used in various experiments. The results of the observation are depicted schematically in Fig. 12. As the temperature was increased from the swollen phase, the sample gradually shrunk following the swelling curve (Fig. 7) and the onset of the transition region was manifested by the appearance of a nucleus of the high-temperature (shrunken) phase at the end of the cylinder. We denote this temperature as T_1. As long as

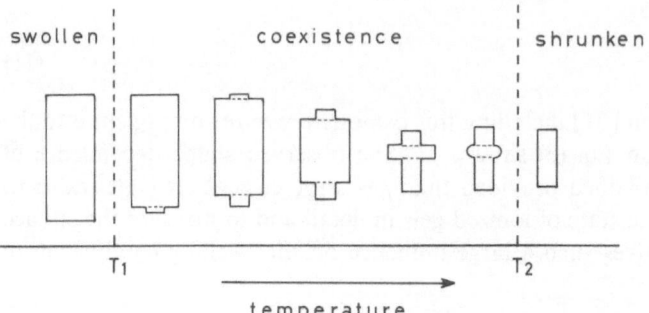

Fig. 12. Process of the first-order transition observed on an ionized NIPA gel rod on heating. The phase-coexistence starts at temperature T_1 and ends at T_2. Between these temperatures, the volume fraction of the swollen (shrunken) phase decreases (increases) with temperature, and the gel never becomes homogeneous. On cooling from the shrunken to the swollen phase, the process similar to the above occurs in the reverse direction. This time, the nuclei of the swollen phase appear at both ends of the shrunken rod and grow towards the center

the temperature was kept constant at T_1, the nucleus remained still, i.e. it neither grew nor vanished for a long time, by which I mean a time very long for a usual laboratory experiments, say, weeks to months. As the temperature was raised slightly again, the shrunken phase grew quickly (within an hour) to some extent and the phase boundary stood still again at the new position. Moreover, if the temperature was reduced to the previous value, the phase boundary went back to the previous position. i.e. the volume of the shrunken phase was diminished. A slight temperature hysteresis accompanied this process. The essential points of these observations are such that the coexistence of the two phases is stable, and that the volume fraction of each phase within the coexistence temperature range is not arbitrary but has a definite value as a function of temperature.

An example of this phase-coexistence is shown in Fig. 13. We will call this macroscopic phase boundary a "bottleneck" due to its shape. In fact, this coexistence includes three phases, i.e. swollen gel, shrunken gel and pure solvent phases surrounding the gel, and has been called "triphasic equilibrium" in the

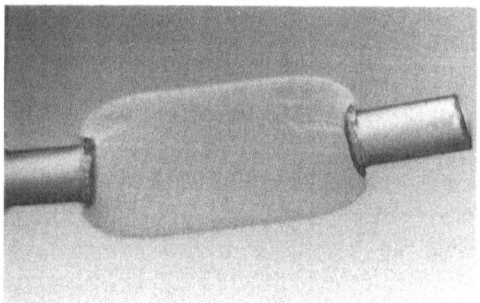

Fig. 13. A typical domain structure observed in an ionized NIPA gel rod in the first-order transition region. The sample is immersed in water and the both phases of the gel are optically translucent and homogeneous. The diameter of the rod in the swollen part is ~ 3 mm

literature [11]. As the temperature was increased further, the high-temperature phase grew at the expense of the low-temperature one, and eventually the transition completed over the whole sample at a temperature, say, T_2. When the sample was cooled from the shrunken to the swollen phase, the same process as above occurred in the reverse direction with a temperature hysteresis of a few degrees.

These observations were repeated many times with various NIPA gels having different ionic content. The qualitative behavior was the same for all the samples as that depicted in Fig. 12. The temperature width in which the tripasic-phasic equilibrium was observed, i.e. $\Delta T = T_2 - T_1$, depended on the degree of discontinuity at the transition. The larger was the discontinuity, so was ΔT. Let us express the degree of discontinuity by the ratio R of α at the discontinuity as

$$R = (\alpha \text{ in the swollen phase})/(\alpha \text{ in the shrunken phase}), \qquad (12)$$

where α in both phases should be measured at the same position of the sample before and after the transition. In neutral NIPA gel, which exhibited only very slight discontinuity at the transition, R was about 1.8 and the triphasic equilibrium was seen only within $0.05\,°C$. On the other hand, in NIPA-AA copolymer gels, which underwent a distinctive first-order transition, the observed values were $\Delta T = 1.0\,°C$ for a gel with R = 2.4, and $\Delta T > 4\,°C$ for a gel with R = 12.

The observation of the transition process was also made on samples with the shape of plate and cube. However, these samples distorted irregularly in the transition region, and no clear phase boundary such as the bottleneck was formed. On close inspection, it turned out that the nuclei of the new (shrunken) phase were created on the surface (usually at some surface defects such as scars), and grew anisotropically along the surface but did not go into the bulk. Consequently, the shrunken phase formed a surface layer with the shape of a stripe, and elongated irregularly along the surface. Thus, the coexistence of phases was realized in the gel in which some portions of the surface belonged to the shrunken phase, while the remaining part of the surface as well as the bulk belonged to the swollen phase. As a result, the sample with the shape of plate, cube, and thick rod always deformed irregularly in the transition region.

In this way, the first-order transition in plates, cubes, and thick rods proceeds in a manner quite different from that in thin rods. Because of the irregular deformation of the sample, it is clearly impossible to measure the swelling curve on these samples with the first method (measuring the characteristic length of the sample with a microscope) as explained in Sect. 3. Even though no clear phase boundary is formed, the coexistence of the swollen and shrunken phases is also realized in these samples.

The phase coexistence of gels at the first-order transition is accompanied by a number of unusual features, of which a few will be mentioned below. First, the fact that the triphasic equilibrium persists over a wide temperature range is an apparent contradiction to the Gibbs phase rule. This rule predicts that the number of thermodynamic degrees of freedom for a two component-three phase system, to which a gel in triphasic equilibrium belongs, is only one. Accordingly, the coexistence of three phases should be observable only at one specific temperature (at the transition temperature determined from the Maxwell rule) under a constant external pressure. The violation of the Gibbs phase rule is a characteristic feature of the first-order transition of gels and, in the author's view, is a key to understanding the real nature of it. This will be discussed in the next Section.

The second anomalous feature is that, as can be clearly seen in Fig. 12, the volume fraction of the respective phases in the coexistence-temperature range is a monotonic function of temperature. This indicates that the state of gel in this temperature range is a distinct thermodynamic state. This state is characterized by mixing of the two phases, and is not identifiable to any of the homogeneous phases. The mixing ratio is not arbitrary but takes a definite value determined as a function of temperature.

The third anomalous characteristic is that the shape of the phase boundary depends on the macroscopic shape of the sample, viz. a clear phase boundary like the bottleneck can be formed only in thin cylinders. This is equivalent to saying that the distribution of domains depends on the shape of the sample. This may be related to the fact that the two coexisting phases have concentrations differing markedly from each other, and consequently, very large strains are induced near the interface between the two phases. The strain field will extend over a long distance and overlap with that due to other domains. Thus, the free energy of the whole sample depends on the distribution of domains. The domains rearrange themselves so as to make the total free energy minimum. The distribution of domains will depend on the shape of sample through its coupling to the macroscopic deformation of the whole sample. Then, it is possible that the transition temperature and the equilibrium swelling ratio depends on the sample shape. The shape-dependent properties described in Sect. 3.4 might be related to this situation.

5 The Nature of the First-Order Transition of Gels

5.1 Can We Call These Phase Transitions First-Order?

We have seen above that the transition between the swollen and the shrunken phases of NIPA gels does not occur in a single step at one temperature, but proceeds over a considerably wide temperature interval in which the gel is split into two phases. This fact has so far been overlooked, but is of crucial importance in characterizing the transition. None of the existing theories of gels can account for this phase-coexisting state, which belongs to none of the homogeneous phases, and should be regarded as a new state of gels.

From an experimental point of view, it is important to realize that within the region of phase coexistence, different methods of measuring the swelling curve give completely different results. Suppose, for example, we are to measure the swelling curve of NIPA-AA copolymer gel with the two different methods mentioned in Sect. 3. In the first method we measure the diameter of a cylindrical gel and calculate its volume by Eq. (9), while in the second we measure the weight of the same sample and determine its volume therefrom. The first method will give the swelling curve like shown in Fig. 14a, which is similar to those in Fig. 7. It should be noted once more that the discontinuity in these curves represents not that the total volume of the sample changes discontinuously at that temperature, but that the phase boundary (bottleneck) traverses the point at which the measurement was made. On the contrary, the second method will give the result shown schematically in Fig. 14b, which is completely different from the first one in the transition region. The method of measuring the weight of gels has actually been applied [22] to measure the swelling curves of a series of positively ionized NIPA gels (NIPA-MAPTAC copolymer gels). The samples used were disc-shaped gels. The swelling curves obtained are all continuous even for gels which should undergo highly discontinuous transition, thus showing clear discrepancy with the theoretical prediction. Although the authors [22] did not comment on this apparent contradiction between theory and experiment, I believe that the above explanation resolves the puzzling results they obtained.

The continuous swelling curve such as depicted in Fig. 14b will be obtained even with the first method if we measure not only the diameters at one particular

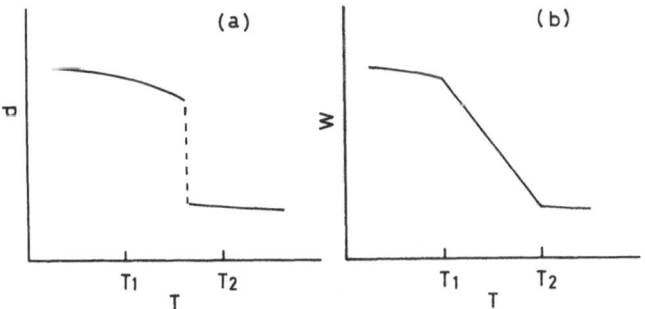

Fig. 14a, b. Two different methods of measuring the swelling curve of gel can give completely different phase transition behavior on the same sample. The meaning of T_1 and T_2 are the same as in Fig. 12. (**a**) Measuring the diameter d of a cylindrical sample at a particular point. When the bottleneck domain boundary traverses the measuring point, the discontinuity of the diameter at that point is observed. The ordinate is the diameter which is proportional to the total volume below T_1 and above T_2, but not between these temperatures. (**b**) Measuring the weight W of the sample. The ordinate is the weight which is almost proportional to its volume. Although the linear variation of W between T_1 and T_2 is assumed here, it is only tentative. The essential point is that no discontinuity of the total volume is observed

point, but also the lengths and the diameters of the swollen and shrunken portions on the bottleneck-shaped sample, and calculate from these the total volume of the sample. The essential point is that, at the so-called "first-order transition" on NIPA gels, the total volume of the sample does not exhibit discontinuity.

Thus we note that the "first-order" transition of gels (at least that of NIPA gels) has a curious characteristic that the volume changes discontinuously on a local scale, but the total volume of the sample changes in a continuous manner. Thus, it is even questionable to call this transition "first-order", though we still call it so in this article following the previous convention. However, it is not the first-order transition in the usual sense as has been predicted on the basis of the existing theories of gels. To understand the nature of the first-order transition of gels, it is crucial to investigate the structure and stability of the phase-coexistence state of gels.

5.2 Violation of the Gibbs Phase Rule

We have already noted in the last Section that the bottleneck structure (or the triphasic equilibrium) persists over a wide temperature range, and that this is a violation of the Gibbs phase rule. In this connection, it is instructive to compare the present result with the theoretical prediction on the phase equilibrium in coherent solids, which has been developed over the past decade mainly in relation to the order-disorder transition in metallic alloys [35–39]. It has been shown that the phase diagram of coherent solids are quite different from that of fluids or of incoherent solids, and that the Gibbs phase rule does not hold for the former [39]. Moreover, at the first-order transition of these solids, Cu_3Au for example, the stable state of two-phase coexistence exists over a wide temperature range in between the ordered and disordered phases. The volume fraction of the two phases in this mixed-phase state changes nearly linearly with temperature, just as has been observed in ionized NIPA gels.

We note here that gel is a coherent solid because its structure is characterized by a polymer network, and hence, the above theoretical considerations on crystalline alloys should be applicable to gels without essential alteration. It is expected that the curious features of the first-order transition of NIPA gels will be explained within the concept of the coherent phase equilibrium if the proper calculation of the coherent energy and the elastic energy of the gel network is made. This may be one of the most interesting unsolved problems related to the phase transitions of gels.

In the present article, we have dealt only with the phase transition of NIPA gels. Hence, it is not clear at present whether the anomalous phenomena observed on NIPA gels are universal features of the discontinuous transitions of gels or not. According to the theory of coherent-phase equilibria, it depends on the values of the elastic constants whether the phase-coexistence structure is preferred over the homogeneous structure or not. Gels with elastic constant values quite different from those of NIPA gels might exhibit different transition processes. Then, it might be possible to find a gel in which the total volume of sample changes discontinuously at one temperature, just like the first-order transition visualized in the conventional theory of gels. More complex cases might also be found in which the permissible values of the volume fraction of each phase in the coexistence region are limited, so that the volume change occurs in three steps, i.e. discontinuous-continuous-discontinuous changes [37]. Thus, in the first-order phase transition of gels, even in its equilibrium properties, there still remain unsolved problems which are basically important from the standpoint of statistical-thermodynamics.

6 Summary

The Flory–Rehner theory, and the Erman–Flory theory which is an extended version of the former to take into account the concentration dependence of the interaction parameter χ, can predict the main features of the continuous and slightly discontinuous phase transitions in NIPA gels. On the other hand, phase transitions with a large volume discontinuity as observed in ionized NIPA gels are basically quite different from those predicted from the phenomenological theory. They are characterized by an unusual coexistence of the swollen and shrunken phases, which is believed to be a stable or metastable equilibrium state. This state has a number of unusual features, e.g. the volume fraction of each of phase in each sample is a function of temperature, the existence of this state itself violates the Gibbs phase rule, the distribution of domains depends on the macroscopic shape of sample, etc. It is suggested that these features of the first-order transition of NIPA gels may be understood in terms of the coherent nature of the network structure and the so-called coherent-phase equilibrium.

In addition, the unusual shape-dependent swelling and phase transition properties observed recently in ionized NIPA gels have been reviewed. Although the mechanism responsible for these phenomena is still open to discussion, these results show that an important role is played by the surface state of ionized gels in determining the swelling equilibrium.

Although the present work is confined to NIPA gels, it is suggested that unique and unusual properties presented in this article are more of less common to first-order phase transitions of other ionized gels as well. It is highly desirable in this connection to make detailed investigations of the actual process of phase transitions in various other gels.

Acknowledgements. I am indebted to Professor K. Sekimoto for informing me of the problem related to the phase rule in coherent solids and for discussions on this topic. I also thank Professor A. Onuki for valuable discussions on the problems covered in this article.

7 References

1. Hill TL (1952) J Chem Phys 20: 1259
2. Hill TL (1960) An introduction to statistical thermodynamics, chap 21. Addison-Wesley, Reading
3. Dusek K, Patterson D (1968) J Polym Sci A-2 6: 1209
4. Ptitsyn OB, Kron AK, Eizner YE (1965) J Polym Sci C16: 359
5. Tanaka T, Fillmore D, Sun ST, Nishio I, Swislow G, Shah A (1980) Phys Rev Lett 45: 1636
6. De Gennes PG (1975) J Physique Lett 36: L55
7. Hirotsu S, Hirokawa Y, Tanaka T (1987) J Chem Phys 87: 1392
8. Hirotsu S (1988) J Chem Phys 88: 427
9. Flory PJ, Rehner J (1943) J Chem Phys 11: 512, ibid 521
10. Flory PJ (1953) Principles of polymer chemistry, chaps 12, 13. Cornell Univ Press, Ithaca
11. Erman B, Flory PJ (1986) Macromolecules 19: 2342
12. Guggenheim EA (1952) Mixtures. Clarendon Press, Oxford
13. Hocker H, Shih H, Flory PJ (1971) Trans Faraday Soc 67: 2275
14. Eichinger BE, Flory PJ (1968) Trans Faraday Soc 64: 2053
15. Prange MM, Hooper HH, Prausnitz JM (1989) AIChE J 35: 803
16. Guggenheim EA (1944) Proc Roy Soc A 183: 203, ibid 213
17. Machetti M, Prager S, Cussler EL (1990) Macromolecules 23: 1760, ibid 3445
18. Hirokawa Y, Tanaka T (1984) J Chem Phys 81: 6379
19. Hirotsu S (1987) J Phys Soc Jpn 56: 233
20. Hirotsu S (1991) J Chem Phys 94: 3949
21. Li Y, Tanaka T (1989) J Chem Phys 90: 5161
22. Beltran S, Hooper HH, Blanch HW, Prausnitz JM (1990) J Chem Phys 92: 2061
23. Hirotsu S (1989) (unpublished)
24. Hirotsu S (1990) Macromolecules 23: 903
25. Tanaka T, Sato E, Hirokawa Y, Hirotsu S, Peetermans J (1985) Phys Rev Lett 55: 2455
26. Hirotsu S, Kaneki A (1987) In: Komura S, Furukawa H (eds) Dynamics of ordering processes in condensed matter. Plenum, New York, p 481
27. Hirotsu S, Onuki A (1989) J Phys Soc Jpn 58: 1508
28. Lee KK, Cussler EL, Marchetti M, McHugh MA (1990) Chem Engin Sci 45: 766
29. De Gennes PG (1979) Scaling concept in polymer physics. Cornell Univ Press, Ithaca
30. Hirotsu S (1992) (to be published)
31. Hirotsu S (1992) Macromolecules 25: 4445
32. Ricka J, Tanaka T (1984) Macromolecules 17: 2916
33. Ilavsky M (1981) Polymer 22: 1687
34. Hirotsu S (1992) (to be published)
35. Cahn JW (1962) Acta Metall 10: 907
36. Williams RO (1980) Metall Trans 11A: 247
37. Williams RO (1984) CALPHAD 8: 1
38. Cahn JW, Larche F (1984) Acta Metall 32: 1915
39. Johnson WC (1987) Metall Trans 18A: 1093

Friction Between Polymer Networks of Gels and Solvent

Masayuki Tokita*
Department of Chemistry for Materials, Faculty of Engineering,
Mie University, Tsu, 514, Japan

Establishment of the physics of the volume phase transition of gels requires independent information about the collective diffusion constant of gel, the elastic property of gel, and the friction between the polymer network of gel and the fluid. Such information has been obtained using the dynamic light scattering and the rheological techniques so far. The behavior of frictional properties of polymer networks of gels are revealed recently, due to the development of a newly designed apparatus for the measurement of the friction coefficient of gel. The present article reviews recent studies on the frictional properties of gels. The difficulties of the friction measurement and the way they are solved will also be described in detail.

* On leave from the Department of Polymer Science, Faculty of Science, Hokkaido University, Sapporo, 060, Japan

1 Introduction

Gels consist of a three-dimensional polymer network and solvent. Although the fraction of the polymer network is small, the gel shows solid-like mechanical properties. Because of such unique physical properties, gels play important roles in a wide variety of biological and chemical systems. Viscoelastic properties of gels are uniquely represented by three parameters, the osmotic bulk modulus K, the shear modulus μ, and the friction coefficient f between the polymer network and the solvent. It has been established that the viscoelastic properties of gels can be determined by dynamic light scattering from the collective mode of the density fluctuations of the polymer network of the gel [1]. The decay time of the scattered light by the collective diffusion mode of the polymer network is given as follows

$$\tau = \frac{\lambda^2}{4\pi D} \tag{1}$$

where λ is the wave length of the scattered light in the gel and D is the collective diffusion constant of the polymer network,

$$D = \frac{E}{f} = \frac{K + \frac{4}{3}\mu}{f} \tag{2}$$

Here, E represents the longitudinal modulus only of the polymer network. The amplitude of the scattered light intensity from the gel is expressed by

$$I = Const\frac{kT}{E} \tag{3}$$

The equations above suggest that it is possible to determine both the longitudinal modulus and the collective diffusion constant of gel from the intensity and the decay time of fluctuations of light, and this has been proven experimentally by many researchers using different gels [1–5]. With the combined experiment on the shear modulus, all three parameters, which describe the physical properties of the gel, i.e., K, μ, and f, can be uniquely determined using dynamic light scattering spectroscopy.

The elastic properties of gels as well as their dynamic light scattering of gels have been investigated by many researchers with various gels [2, 3]. These results are in good agreement with the prediction of the scaling theory of the semi-dilute polymer solution [6]. In contrast to these studies, there have not been many systematic experimental and theoretical studies on the frictional property of gels. The concept of the friction is simple but there are many problems which must be solved for the actual and accurate determination of the friction between the polymer network of the gel and the solvent.

The principle of the mechanical measurement of the friction coefficient of a gel is schematically shown in Fig. 1. A thin slab of a gel of thickness d is held in

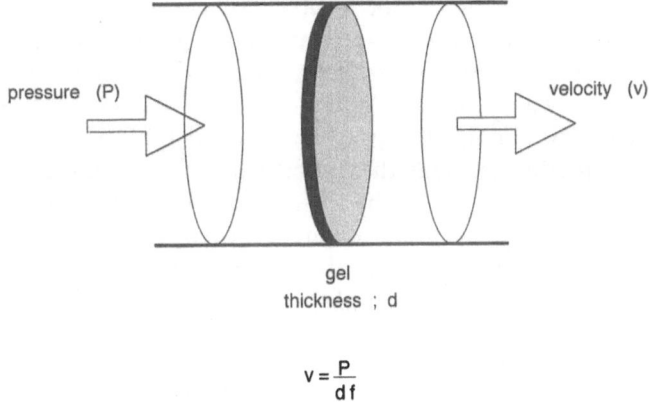

$$v = \frac{P}{d\,f}$$

Fig. 1. Principle of the mechanical measurement of the friction f between a polymer network and solvent. The gel is assumed to be chemically fixed at the rim

a tube with the rim chemically fixed on the tube so that fluid cannot leak past the side of gel. The fluid is put under a small pressure P. The velocity of the permeation flow of the fluid through the gel v_s is determined by measuring the rate at which the fluid flows out of the gel in a steady state. The friction coefficient of the gel f is defined as

$$f = \frac{P}{dv_s} \tag{4}$$

This equation indicates that the steady state velocity of the fluid is a linear function of the pressure.

Although there have been several pioneering studies on the frictional properties of gels, these studies, compared with the present state of knowledge, were not carried out under ideal experimental conditions [1, 7–9]. In these previous studies on the frictional property of gels, relatively large thicknesses of the gels were used to determine the friction coefficient. If this dimension of the gel becomes larger, the characteristic time for the mechanical deformation by the applied pressure becomes longer. A characteristic time for such a gel with a typical thickness of 1 cm is of the order of 10 days, indicating that it takes more than 10 days to determine the friction coefficient of the gel in a stationary state. However, such long term measurements have not been made so far. Furthermore, the apparent velocity of the fluid is slow for this thickness and a relatively high pressure is necessary to secure the accurate determination of the velocity. When the applied pressure is high, however, other effects become pronounced, namely, the destruction of the gel structure, the nonlinear relationship between the velocity of the fluid and the applied pressure, and the shrinkage of the gel due to the frictional force. These problems seem to have been solved recently for a reliable and accurate determination of the friction coefficient of gels. Accurate measurements of friction coefficient were made on poly(acrylamide) and poly(N-isopropylacrylamide) gels.

In this paper, we will review these experimental results on the frictional properties of gels. The details of a newly designed apparatus which is critical for the accurate determination of the friction coefficient of the gel will be given in Sect. 2. Then in Sect. 3, we will discuss the frictional properties of non-critical poly(acrylamide) gel and that of poly(N-isopropylacrylamide) gel [10, 11]. The relationship between the friction constant, the collective diffusion constant, and the elastic constant are discussed on the basis of the scaling theory of the semi-dilute polymer solution using the results obtained for the poly(acrylamide) gel. The effects of the static fluctuations will be also discussed. In the case of the poly(N-isopropylacrylamide) gel, an unusual phenomenon is found. Although the density of the gel is constant, the friction of the gel undergoing the volume phase transition, shows a reversible decrease by three orders of magnitude and appears to diminish as the gel approaches a certain temperature.

2 Measurement of Gel-Solvent Friction

We shall first describe the details of a newly designed apparatus that is critical for the accurate determination of the friction coefficient of gels. As shortly stated in the previous section, there are some difficulties to be solved for an accurate determination of the friction coefficient of gels. In the following, we will describe such difficulties in detail and the way they are solved.

Since the gels we study are fragile, the pressure applied upon the gel for the friction measurement should be small enough not to brake the gel. In addition, the small applied pressure has other advantages because the velocity of the fluid inside the gel should be small enough not to impose a substantial frictional resistance that induces a shrinkage of the gel. The osmotic compressibility of only the polymer network is expected to be small, typically 10^3–10^4 Pa, the applied pressure should be comparable or less than these values to assure no shrinking of the gel. Furthermore, the applied pressure should be stable because the friction measurement at such a low applied pressure takes several hours. Taking into account the points described above, the hydrostatic pressure of a water column, which is also a reservoir for the water, is used as a pressure source.

A simple calculation using the previously available results of the dynamic light scattering and the mechanical measurements suggests that the friction coefficient of typical poly(acrylamide) gel is expected to be of order of 10^{10}–10^{11} Pa s cm^{-2} because $D \sim 10^{-7}$ cm^2 s^{-1} E $\sim 10^3$ – 10^4 Pa, and $f = E/D$. The velocity of the flow of fluid inside a gel having a thickness of 1 cm becomes, therefore, of the order of 10^{-7}–10^{-8} cm s^{-1} at a pressure of 10^3 Pa. In order to determine the velocity of fluid, a capillary with a homogeneous and well calibrated diameter should be used to amplify the velocity of fluid.

As described above, the flow of the fluid is slow and it is important to avoid any leakage of fluid around the gel sample. This is achieved by using a plastic film with a special surface treatment to which the gel is chemically attached. The chemical bonding of the gel to the film ensures that there is no leakage of solvent.

The apparent velocity of a fluid in a capillary tube is, as shown below, fast at first, slows down and reaches a constant value in a steady state. This apparently fast initial velocity seems to be due to the bending deformation of the gel and the gel folder, and the shrinking of the gel due to the frictional resistance. Although the deformation of the gel due to the frictional pressure is small, it contributes to the apparent flow of fluid out of the capillary. The velocity of the fluid is, therefore, time dependent. The characteristic time of the bending deformation of the gel holder is estimated to be about 30 min by control experiments, in which the gel is cast between two gel bond films with no opening, so that no fluid flow is allowed. In addition to that, it is also important to use a stainless steel gel holder rather than plexiglass to reduce the amplitude of bending deformation of the gel holder. The former has ten times less deformation than the later. On the other hand, the characteristic time of the shrinking deformation is much longer than that of the gel holder which practically determines the overall time interval to achieve the steady state of the fluid flow. The order of the characteristic time at which the flow of fluid reaches a steady state due to the shrinking of the gel can be estimated using the kinetic theory of the swelling of a gel [12]. It has been shown that the characteristic time of swelling and shrinking of a gel is proportional to the square of a typical linear size of the gel. In the present case, the typical size of the gel is the thickness d. The characteristic time of the gel is then estimated using time $= d^2/\pi^2 D$. For instance, the characteristic time of the swelling and shrinking of a gel having a thickness of $d = 0.1$ cm and the typical collective diffusion constant of $D = 10^{-7}$ cm^{2}s^{-1} is the order of 10^4 s $= 3$ h which is a reasonable time interval for the actual friction experiment. The smaller the thickness of the gel, the shorter is the relaxation time. The thinner gel has a further advantage, namely, the velocity of the fluid at a constant pressure is inversely proportional to the thickness of the gel making it possible to have more accurate determination of the velocity of the fluid.

An apparatus has been constructed taking into account all the aspects described above. The design of the apparatus is schematically illustrated in Fig. 2. The connectors, pipes, and valves are made of stainless steel to avoid the expansion due to the applied hydrostatic pressure. The chromatography column of 50 cm length is used as a reservoir for the water to generate the hydrostatic pressure. The range of the height of the water column can be changed from 20 to 60 cm, which corresponds to the pressure from about $2-6 \times 10^3$ Pa. The temperature of the cell is controlled to within an accuracy of better than 0.1 °C using a water circulating system. The apparatus is set up on a vibration free optical table to avoid any external mechanical disturbances.

The structure of the gel holder and the cell are shown in Fig. 3(a) and (b). The gel holder is made of stainless steel or plexiglass. The gel bond films, which

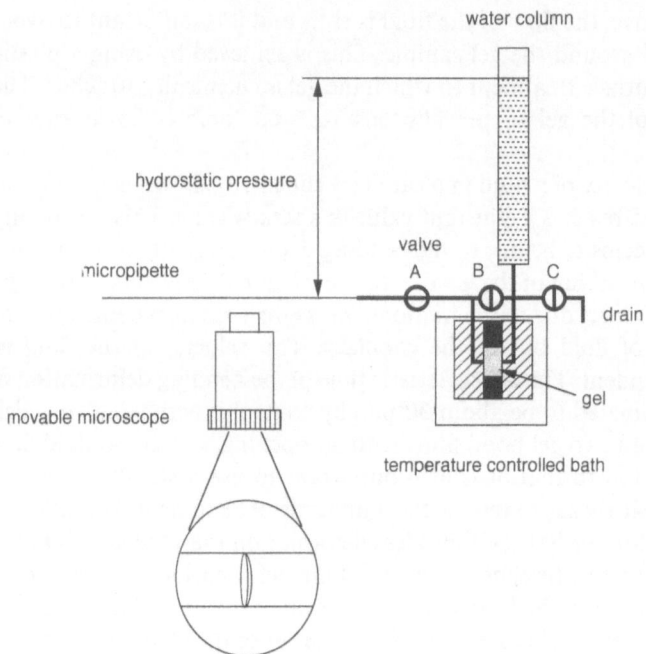

Fig. 2. Schematic illustration of a newly designed friction apparatus. The valve A is opened immediately after the valve B is closed when the measurement begins. The velocity of water flow in the micropipette is measured by a microscope which is set on a mechanical taranslational stage with micrometers with an accuracy of 0.001 mm. After the measurement, the value A is closed and the valve B is opened to equalize the pressures of the either sides of the sample gel. The valve C is used to change the hydrostatic pressure

has chemically active site on the surface, are stuck on both sides of the holder using silicone glue. The final thickness of the gel holder including gel bond film is about 1 or 2 mm. The reaction between the active site of the gel bonding film and the monomer and/or the cross-linker in the pre-gel solution occurs during the gelation reaction. The leakage of fluid around the gel is completely avoided since the gel is chemically bound to the gel bonding films.

After the gel has been cast, the gel holder is taken out of the reaction bath and glass plates are removed carefully under the water so as not to damage the surface of the sample gels. Then, the gel holder which contains the sample gel is tightly held between two plexi plates of the cells (thickness is about 1 cm) shown in Fig. 3(b). The O-rings are almost completely embedded in the plexi plate cells to minimize the bending of the gel holder. It is worth noting that, because of the structure of the cell and the gel holder, the friction coefficient of gel can be measured at a constant volume. Usually, the gels studied here tend to swell when they are left freely under water. The surface of the sample gel is, however, pressed by the cell and the rigid paper filters in the present apparatus. Therefore, the swelling of the sample gel in the cell is prevented. A calibrated micropipette is set

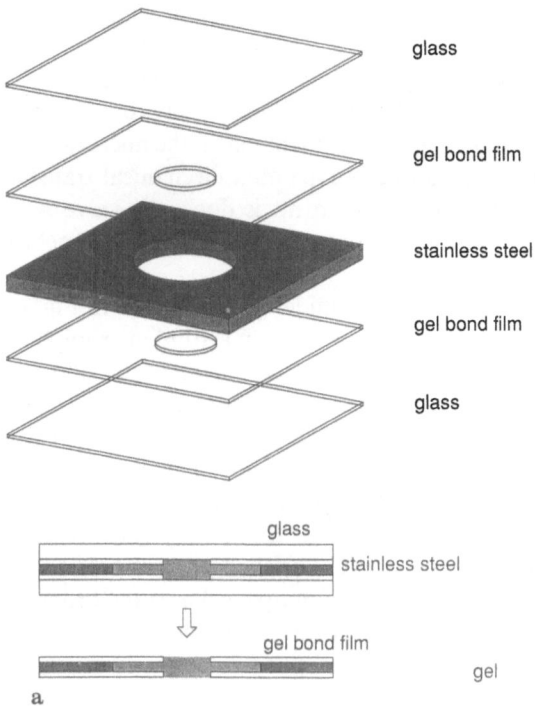

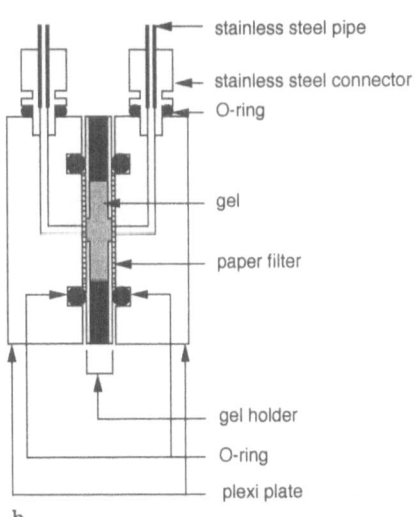

Fig. 3a. The structure of the gel holder. Two gel bond films, which have a circular opening at the center, are stuck on both sides of a stainless steel gel mold by silicone glue. This gel holder is lightly held between two glass plates and immersed into the pre-gel solution. After degassing the pre-gel solution, the gelation reaction is initiated. The gel holder is taken out of the gel after the completion of the gelation reaction and the glass plates are removed carefully under water so as not to damage the surface of the gel. **b.** Schematic representation of the cross section of the cell. The gel holder, which contains the sample gel, is tightly held between two plexi plates. The larger O-ring is almost completely embedded in the plexi plates to avoid substantial bending of the gel holder. The rigid paper filters are packed between the plexi plate and the gel holder not to damage the surface of the sample. It also prevents the gel from swelling

at the end of the stainless steel pipe of the outlet side to measure the flow rate of the fluid. The inner diameters of the micropipettes are 0.34 and 0.50 mm which are commercially available and indicated as 5 and 10 µl micropipettes, respectively. Different micropipettes are used depending on the applied pressure and the friction coefficient of gel.

The gel is left overnight without pressure so that equilibrium in the solvent at a fixed temperature is reached. The valve A in Fig. 2 is closed and the valve B is left open so that no pressure is exerted upon the gel. At the time zero of the measurement, the valve B is closed and A is opened immediately to apply the hydrostatic pressure to the gel. The position of the meniscus in the micropipette is measured as a function of time using a microscope on a mechanical transla- tional stage with micrometers. After the measurement is done, the valve A is closed and the valve B is opened to release the pressure, which allows the gel to cover its initial state.

Typical experimental results for a poly(acrylamide) gel are shown in Fig. 4. The time course of the position of the meniscus of water in a 10 µl micropipette is given in this figure. Different curves denote the measurement at different pressures. The position of the meniscus rapidly moves as soon as the pressure is applied on the gel. The velocity of meniscus, which is given by the slope of the curves shown in Fig. 4, decreases with time and approaches to an asymptote in the time region of 1×10^4 s. The velocity of the fluid in the micropipette at the stationary state, v_{sc}, is obtained from the slope in this steady state region. Although the measurement of the water flow at the applied pressure takes 5 h, the total volume of the water that flows out of the sample gel is of the order of 10 µl. The height change of the water column due to this water flow is negligible. Therefore, we can reasonably assume that the applied pressure is constant during a friction experiment.

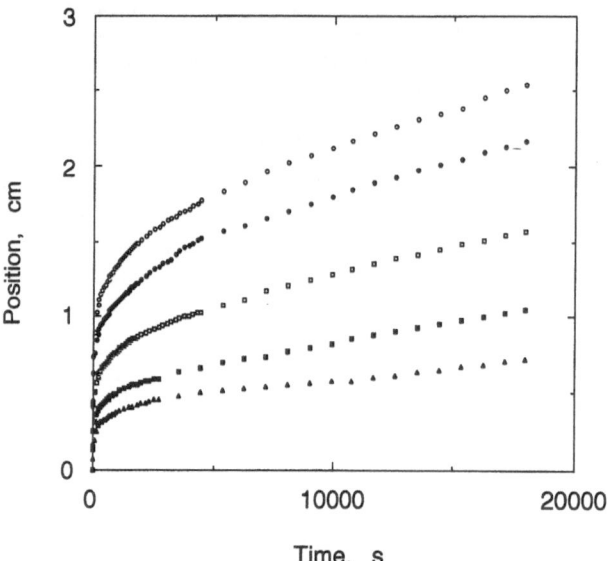

Fig. 4. The time course of the position of the meniscus in the micropipette. The applied pressure is 5.8_8(○), 4.9_0(●), 3.9_2(□), 2.9_4(■), and 1.9_6(△)(10^3 Pa) which correspond to 60, 50, 40, 30 and 20 cm of the height of the water column. The volume of the micropipette used in these measurements are 10 µl. The thickness of the gel is 1.95 mm

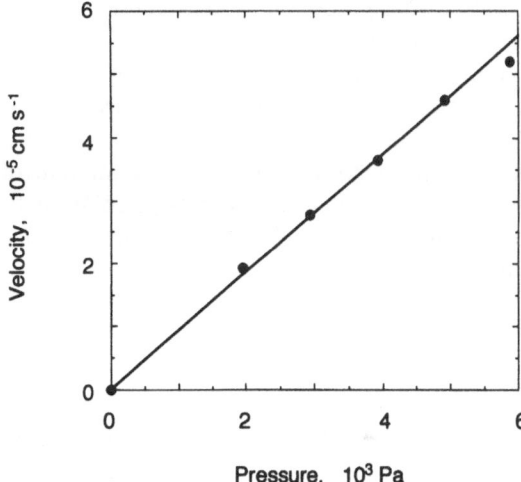

Fig. 5. The pressure dependence of the steady state velocity of water flow in the micropipette

The steady state velocity of water flow thus obtained is plotted as a function of the applied pressure in Fig. 5. The relationship between the applied pressure and the velocity is linear as expected. The friction coefficient, $f = 1.0 \times 10^{11}$ Pa s cm^{-2}, is obtained from the slope of the straight line in Fig. 5 using the following equation

$$ f = \left(\frac{dv_{sc}}{dP}\right)^{-1} \frac{1}{d} \left(\frac{R}{r}\right)^2 \tag{5} $$

where $dv_{sc}/dP = 8.95 \times 10^{-9}$ cm Pa^{-1} s^{-1} is the slope of the straight line in Fig. 5. The thickness of the gel d is d = 0.19 cm. The factor $(R/r)^2$ is the ratio of the area of the micropipette of radius r = 2.5 $\times 10^{-2}$ cm and the area of the hole on the gel bonding film with radius R = 3.3 $\times 10^{-1}$ cm, which is necessary to convert the velocity of the flow of water in the micropipette, v_{sc}, to the velocity of the flow of water in the gel, v_s. The value of the friction coefficient obtained here is in a reasonable agreement with the one expected from the dynamic light scattering studies. These results indicate that it is possible to determine accurately the friction coefficient of a gel using the present apparatus.

3 Frictional Properties of Gels

3.1 Friction of Non-Critical Gels

In this section, the major results of the frictional studies on the non-critical gels are discussed. The sample used in this study was the poly(acrylamide) gel. The samples were prepared using a standard free radical polymerization. All the reagents used here were of the electrophoresis grade (Bio-Rad) and used without further purification. The predetermined amounts of acrylamide (main constituent) and N,N'-methylenebisacrylamide (cross-linker) were dissolved in 100 ml of distilled water. Then 240 μl of tetramethylenediamine, which is an accelerator, was added and stirred for about 20 min. The gel holder was immersed into this pre-gel solution and degassed in vacuum for 30 to 40 min. A 1-ml amount of 4 wt.% ammonium persulfate solution was added to the pre-gel solution to initiate the gelation reaction. The solution was kept at room temperature undisturbed for one day to ensure the complete reaction.

3.1.1 Temperature Dependence

The temperature dependence of the friction coefficient of poly(acrylamide) gel was studied at two different concentrations of acrylamide (AAm) and N,N'-methylenebisacrylamide (Bis); AAm : Bis = 1.24 M : 22.4 mM and AAm : Bis = 693 mM : 7 mM. The weight concentrations of these sample gels was about 8.8 and 5 wt.%, respectively. The composition of the latter gel is almost the same as the composition typically recommended for acrylamide gel electrophoresis. The friction coefficient of these gels was measured at a fixed pressure of 5.88×10^3 Pa, which corresponds to 60 cm of the height of the water column. The temperature was varied from about 0 to 60 °C.

The friction coefficient of these gels is plotted as a function of the temperature in Fig. 6. The open symbols in the figure indicate the results obtained

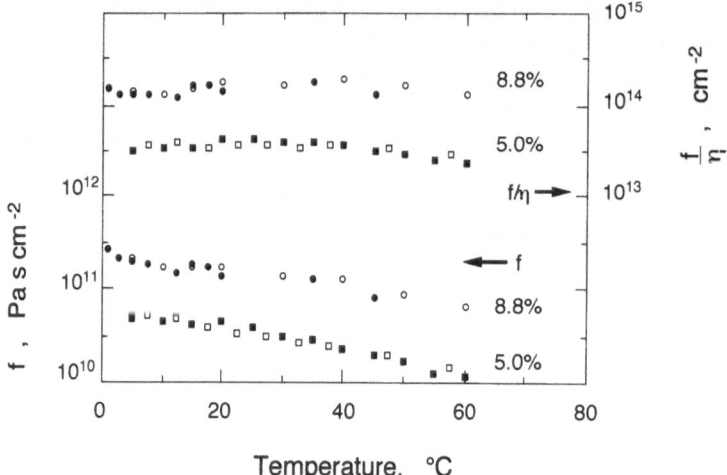

Fig. 6. The temperature dependence of the friction coefficient of the poly(acrylamide) gels. The total concentrations of acrylamide are 1.24 M (8.8% mass fraction), and 693 mM (5% mass fraction), respectively. Open symbols are used for the results obtained in the cooling process and closed symbols represent the results taken upon increasing the temperature. In the upper part of this figure, the temperature dependence of the ratio $f(T)/\eta(T)$ is shown. The values of the viscosity are taken from a table. Symbols are the same as those used in the raw value of the friction coefficient

upon lowering the temperature and the solid symbols represent the ones obtained while the system is heated step by step. The hysteresis is not found between the heating and the cooling processes. The results indicate that the friction coefficient of gels decreases slightly but monotonically with the temperature.

The application of the hydrostatic pressure to the solvent of gel causes the permeation flow of the solvent. In this process, the gel behaves as a molecular sieve and imposes a frictional resistance on the flowing water. The permeation flow of water through a gel is a process analogous to the capillary flow of a viscous fluid. The frictional resistance of a single capillary is well described by the Hagen-Poiseuille's law by which the relationship between the dimension of the capillary, the applied pressure, and the flow rate is given as follows

$$ v = \frac{Q}{\pi r^2} = \frac{P}{L(8\eta r^{-2})} = \frac{P}{Lf} \tag{6} $$

where Q, r, P, L, η, and f denote the flow rate, the radius of the capillary, the applied pressure, the length of the capillary, the viscosity of the fluid, and the friction coefficient of the capillary, respectively. The above equation indicates that the friction coefficient of the capillary is proportional to the viscosity of the fluid and is inversely proportional to the square of the radius of the capillary, i.e. the area of the cross section of the capillary. If we view a gel as consisting of N pores of average radius r, the relationship between the friction coefficient of a gel, the viscosity of water, and the pore size of the gel is expected to be given by the same equation. The average pore size of a gel is reasonably assumed to be proportional to the correlation length ξ of the gel, which, in a good solvent, represents the average distance between the neighboring contact points of the polymer chains. Therefore, we finally obtain the following relationship for the friction coefficient of a gel by substituting the radius of capillary, r, to the correlation length of the gel, ξ, [13, 14]

$$ f = \text{Constant} \frac{\eta}{\xi^2} \tag{7} $$

Here, both the viscosity of water and the correlation length of the gel are a function of the temperature. Therefore, in order to discuss the temperature dependence of the correlation length of the gel, it is convenient to use the ratio $f(T)/\eta(T)$ rather than the raw value of the friction coefficient $f(T)$ since the ratio directly represents the effective size of the pores and their distribution.

The temperature dependence of the friction coefficient of poly(acrylamide) gels are analyzed according to the above equation. In our analysis, the values of the viscosity of water is taken from the table. The results thus obtained are also shown in Fig. 6. It can be seen from this figure that the friction of the poly(acrylamide) gel normalized with the viscosity of water is independent of the temperature. It indicates that the pore size of the poly(acrylamide) gel is stable in the temperature range studied.

The dynamic light scattering studies showed that the density fluctuations of the dilute poly(acrylamide) gel increased with decreasing temperature below room temperature and appeared to diverge as they approached the spinodal line [4]. On the spinodal line, the osmotic modulus and the friction coefficient of the gel are expected to vanish. The difference between the results obtained from the dynamic light scattering study and the present mechanical study is most likely due to the difference of the concentration of the gels studied in the two experiments. In the present mechanical study, more concentrated gels (5 and 8.8 wt.%) are used rather than the dilute gels (2.5 wt.%) which were used in the dynamic light scattering studies. The critical fluctuations become dominant in the immediate vicinity of the phase boundary and they are more pronounced in the dilute gels. The phase boundary of a 5 wt.% poly(acrylamide) gel is expected to lie much below the 2.5 wt.% poly(acrylamide) gel. It is, therefore, necessary to lower the temperature below $0\,^{\circ}C$ in order to observe the critical behavior of the friction coefficient. However, the measurements could not be made at such a lower temperature because of ice formation in the cell.

3.1.2 Effects of Total Concentration

The dependence of the friction coefficient of the poly(acrylamide) gel on the total concentration of polymers of the gel, including both the main constituent and cross-linker, has been studied under the experimental conditions of a constant molar concentration of the cross-linker at 1 mol%. The total concentration was changed from 400 mM to 2.8 M which corresponds to a concentration from about 3 to 20 wt.%. The temperature was fixed at $20 \pm 0.1\,^{\circ}C$.

The results are shown in Fig. 7 in which the friction coefficient f is plotted as a function of the weight polymer concentration ϕ in the double logarithmic

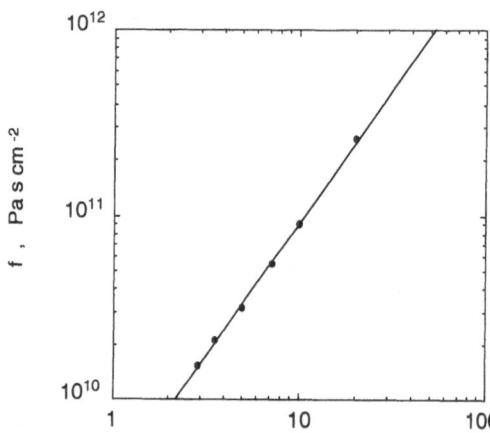

Fig. 7. The double logarithmic plot of the friction coefficient of poly(acrylamide) gel as a function of the total concentration of the gel. The cross-linker concentration is constant at a mole fraction of 1%. The straight line in this figure indicates the result of the least squares analysis. The slope of the straight line is 1.5

Total concentration ϕ, % mass fraction

scale. The data are well explained by a straight line with a slope of 1.5 which is determined by the least square fitting. The results indicate that the concentration dependence of the friction coefficient is expressed by a power law relationship,

$$f = C\phi^{1.5} \tag{8}$$

Here ϕ is the total weight concentration of gel and C is a numerical constant.

Since the cross linking density of the gel is low enough, there are many contact points between polymer chains. The structure of the gel, therefore, is analogous to that of the semi-dilute polymer solution suggesting that the concentration dependence of the correlation length is explained by the following power law relationship [6]

$$\xi = C_\xi \phi^{-3/4} \tag{9}$$

Here C_ξ is a numerical constant. Substitution of Eq. (9) into Eq. (7) gives the following results for the concentration dependence of the friction coefficient of the gel

$$f = C_f \eta \phi^{3/2} \tag{10}$$

The experimentally obtained value of the exponent 1.5 is, therefore, in a good agreement with the scaling theory which predicts the exponent of 3/2.

The results obtained by the present mechanical measurements are also consistent with the previous experimental results of the dynamic light scattering studies of the collective diffusion coefficient of gels and the rheological studies of the shear modulus of gels. The studies published by different researchers indicate that the concentration dependence of the collective diffusion constant of the polymer networks of gel and that of the elastic modulus are well represented by the following power law relationships: [2, 3, 5]

$$D = C_D \phi^{3/4} \tag{11}$$

$$E = C_E \phi^{9/4} \tag{12}$$

The above experimental results as expressed in Eqs. (11) and (12), together with our experimental results Eq. (8), confirm the relationship between the collective diffusion coefficient, the elastic modulus, and the friction coefficient which is given in the Eq. (2).

3.1.3 Static Fluctuation and Friction

The density of the polymer network of the gel fluctuates with time and space in microscopic scale. Such fluctuations are frozen in the gel under certain gelation conditions. As a result of the frozen density fluctuations, opacity develops in the gel. Although quantitative studies have not been made so far, the qualitative studies of the opacity development in poly(acrylamide) gel have been reported by many researchers [15, 16, 17]. In this section, we will discuss the relationship between the static fluctuation and the friction coefficient of the gel.

The density fluctuations of the polymer network are easily frozen in the poly(acrylamide) gel by increasing the cross-linker concentration to a mole fraction of more than 3% in the pre-gel solution. The pre-gel solution, which is clear immediately after the beginning of the gelation reaction, becomes opaque when gelation reaction proceeds. Since the fluctuations are chemically fixed, the opacity of the poly(acrylamide) gel thus obtained is stable against the external stimuli such as the temperature change. It suggests that the density fluctuations that have developed in the poly(acrylamide) gel are static in time and space. To study the relationship between the effects of the static fluctuations on the frictional properties of the poly(acrylamide) gel, the friction coefficient of poly(acrylamide) gels prepared at various cross-linker concentrations are measured at a constant temperature of 20°C. The cross-linker concentration is changed from 1.4 to 70 mM under a constant total concentration of the acrylamide and N,N'-methylenebisacrylamide at 700 mM: this corresponds to a mole fraction of 0.2 to 10% of the cross-linker in the entire network constituent.

The results of the friction measurements are shown in Fig. 8. The friction coefficient of the poly(acrylamide) gel decreases with an increase of the cross-linker concentration in the major part of the concentration range studied, except that it sharply decreases at a mole fraction of 0.2% of the cross-linker concentration. The measurements cannot be performed below this concentration because the gel becomes too soft to be used in the present experimental setup. It suggests that the system is close to the gelation threshold. The sharp decrease of the friction observed in the gel with a mole fraction of 0.2% may be due to the effect of the sol-gel transition. It is also found from visual inspection that the gels becomes opaque above a cross-linker concentration of 3 mole-% fraction.

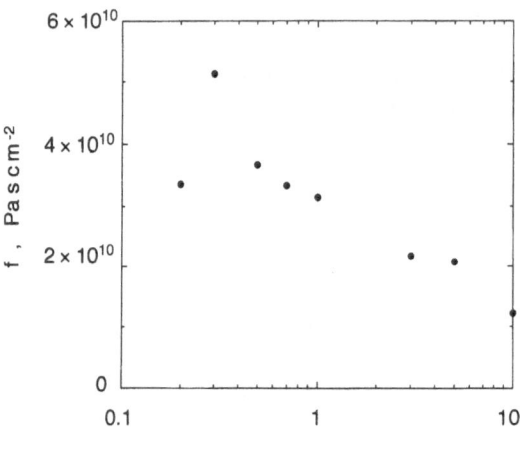

Fig. 8. The cross-linker concentration dependence of the friction coefficient of the poly(acrylamide) gel. The total concentration of the gel is constant at 700 mM

Bis concentration, % mole fraction

It may be natural to expect from Eq. (7) that the friction coefficient of the gel should increase with an increase of the cross-linker concentration because the correlation length decreases with the cross-linker density, and thus the pore size of the polymer network of gel is expected to be smaller. The results observed here, however, are entirely different from our expectation, i.e., the friction coefficient decreases upon increasing the cross-linker concentration. The appearance of the opacity of the gel suggests that the structure of the polymer network become increasing inhomogeneous with the concentration of cross-linker. The cross-linking agent is not very soluble in water when it is polymerized. The pre-critical density fluctuations that appear in the polymer and crosslinking density are frozen to form an inhomogeneous polymer network upon gelation. In such an inhomogeneous gel, some regions are more dilute or swollen, and the other regions are denser or shrunken. These density fluctuations give rise to a non-uniform refractive index within the gel and scatter light. This may be the reason for the opacity of the gel prepared with a larger concentration of the cross-linker.

The friction is expected to be small for such an inhomogeneous gel, since water flows choose mainly the dilute regions that serve as the more open and free paths, while avoiding the denser regions that block the flow. Namely, the opaque gel behaves as a porous sponge for water flow, and the more opaque, the larger the pore size. The opaque gel, therefore, has a much smaller friction coefficient than a transparent gel.

It is worth noting that the effect of inhomogeneities on the friction coefficient appears at much lower cross-linker concentrations than the appearance of the visual opacification in the gel. Indeed, the friction coefficient of the gel decreases with the increasing cross-linker concentration even in the lower concentration region of the cross-linker where the gel looks completely transparent. This result suggests that the seeds of the concentration fluctuations are already formed in the lower cross-linker concentration region, which have not been detected by optical measurements. Thus the friction coefficient of a gel reveals in an extremely sensitive way the spatial inhomogeneity of the gel.

It is important to note that in the inhomogeneous gel, the average crosslinking density is not a relevant parameter for determining the frictional pore size of the gel. It is the spatial correlation length of the density fluctuations that determines the bulk frictional behavior of water in the gel.

3.2 Friction of Gel and Volume Phase Transition

It has been established that the volume phase transition of gels is an universal phenomenon [17]. Dynamic light scattering studies indicate that the dynamic fluctuations of the density correlation diverges in the vicinity of the volume phase transition point of the gel. It has also been shown that the time scale of the density fluctuations become slow in the vicinity of the volume phase transition point. The gel becomes opaque as a result of this critical behavior of the density fluctuation. The increase of the opacity of the gel indicates that both the longitudinal modulus of the polymer network and the cooperative diffusion constant of the gel become smaller at the volume phase transition point Eqs. (2) and (3). Although the friction coefficient is expected to disappear at the volume phase transition point, the actual measurements have yet to be done. The information about the frictional property of a gel in the vicinity of the volume phase transition point, therefore, promotes a better understanding of the volume phase transition. In this section, we will show the frictional property of a gel undergoing volume phase transition where a drastic change of the frictional property is observed.

The sample used to study the relationship between the volume phase transition and the frictional property is poly(N-isopropylacrylamide) gel which shows a small discontinuous volume phase transition at 33.6 °C. The sample gel is prepared by free radical polymerization: 7.8 g of re-crystallized N-iso-propylacrylamide (main constituent, Kodak), 0.133 g N,N'-methylenebis-acrylamide (cross-linker, Bio-Rad), 240 ml tetramethylenediamine (accelerator, Bio-Rad), and 40 mg ammonium persulfate (initiator, Mallinckrodt) are dissolved in distilled water (100 ml) at 0 °C. The gel mold is immersed in the pre-gel solution and then degassed for 40 min at 0 °C. The temperature is raised to 20.0 °C after this treatment to initiate the gelation reaction. The sample gel thus obtained is homogeneous and transparent, at least by visual inspection.

The swelling behavior of poly(N-isopropylacrylamide) has been studied extensively [18, 19]. It has been shown that this gel has a lower critical point due to the hydrophobic interaction. Such a swelling curve is schematically illustrated in Fig. 9. The gel is swollen at a lower temperature and collapses at a higher temperature if the sample gel is allowed to swell freely in water. The volume of the gel changes discontinuously at 33.6 °C. The swelling curves obtained in this way correspond to the isobar at zero osmotic pressure. On the other hand, the friction coefficient is measured along the isochore, which is given in Fig. 9,

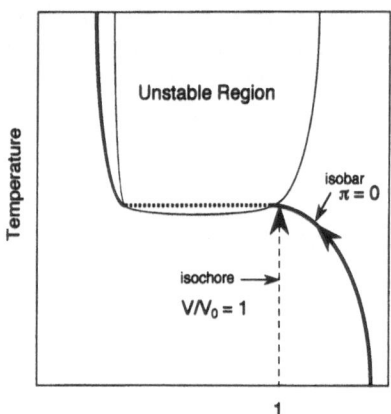

Fig. 9. The swelling curve of the poly(N-isopropylacrylamide) gel is schematically shown. The isobar curve (*thick line*) corresponds to the zero osmotic pressure. The *dotted line* indicates the experimental path at which the volume is fixed at the initial volume V_0 (the volume at which the gel is prepared)

because the gel tends to swell but the volume of the gel is fixed at a constant volume in the present apparatus.

When the temperature is raised along the isochore, the gel is presumably brought into the negative osmotic pressure region and then into the coexistence regime. The temperature dependence of the swelling and shrinking behavior of the gel is checked on a sample which is prepared in the same mold. Without outer cell and rigid paper filters, the gel swells slightly in the open portion at lower temperatures. No substantial shrinking of the gel is observed up to the highest temperature of the friction experiment. The attachment of the gel bond film remain intact in the entire temperature range studied.

During the friction measurement, a test tube containing a small piece of gel is placed in the same water bath and the appearance of the gel is continuously monitored. The decrease of the diameter of the free gel from the isochore diameter is less than 10% at the highest temperature of the present friction experiment. This observation confirms that the gel swelling or shrinking as a whole should not have a significant effect on the friction measurement. A slight opacity of the gel develops in the gel in the vicinity of the temperature at which the friction of the gel diminishes and is presumably due to the dynamic density fluctuations.

The temperature dependence of the friction coefficient normalized by the viscosity of the water, f/η, is given in Fig. 10. The solid symbols are used in the increasing of the temperature and the open symbols are used in the lowering of the temperature. The values of the viscosity of the water, $\eta(T)$, are taken from the literature. For the chemically cross-linked gels, such as the poly(acrylamide) gel, the friction, f/η, is independent of the temperature which has already been shown in previous section. It is, however, found from this figure that the friction

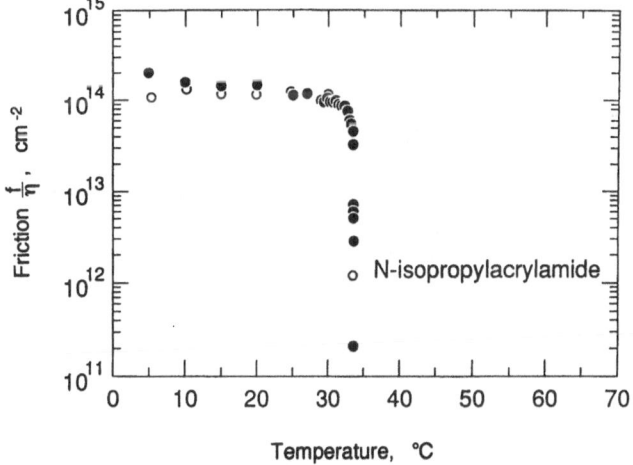

Fig. 10. The temperature dependence of the friction coefficient f of poly(N-isopropylacrylamide) gel normalized by the viscosity of water η. *Solid circles* are used in the increasing of the temperature and *open circles* are used for the lowering of the temperature

of the poly(N-isopropylacrylamide) gel reversibly decreases by three orders of magnitude as the temperature approaches 33.6 °C. This reversible decrease of the friction occurs in spite of the fact that the network density is unchanged. It should be noted that, if the friction is measured along the swelling curve, i.e. along the isobar, the friction is expected to increase with the temperature since the density of the gel increase due to the collapse of the gel.

This phenomenon of the reversible diminishing of friction may be intuitively explained as follows. The friction between the polymer network of gel and the solvent is primarily determined by the average pore size of the polymer network and the viscosity of the solvent as given in Eq. (7). The viscosity of the solvent should not show any singular behavior in the temperature range studied. When the polymer network is homogeneous, the pore size should be given by the average distance between the nearest polymer-polymer contact points. The polymer network, however, undergoes substantial fluctuations in space and time under certain conditions which makes the gel inhomogeneous. Although the density of some portions of the gel are dilute while others are dense, the total gel volume remains constant. The effective pore size of such an inhomogeneous gel is then given by the distance over which the network density fluctuations are correlated. The water passes through the swollen open space avoiding the shrunken regions. The overall friction between the polymer network of gel and the solvent thus decreases due to the density fluctuations of the polymer network.

The pore size practically diverges as the gel shifts the coexistence curve into the two-phase region. The divergence of the pore size of the gel is probably described by the one of the following reasons. First, the gel may remain in the metastable single phase as a superheated gel. Such metastable state is observed in the hysteresis of the swelling curves of various gels including poly(N-isopropylacrylamide) gels. The temperature gap at the hysteresis can be several degrees [19]. In this case the density fluctuations of the polymer network are dynamic and should diverge on the spinodal line. Second, the gel may undergo phase separation which creates domains of swollen and shrunken phases. The density fluctuations are static in this case and would diverge at or in the vicinity of the coexistence curve. In both cases, the effective pore size of the gel diverges, making the friction diminish. The pore structure is not a permanent one, but is reversibly changed or reduced with temperature.

The temperature dependence of the dynamic fluctuations which contribute to the effective pore size may be estimated by means of mode coupling theory, which views a gel as consisting of N pores of diameter ξ over which the density fluctuations are correlated [4, 20, 21].

$$f = 6\pi\eta\xi N \tag{13}$$

The theory further suggests

$$N = N_0 \left(\frac{\xi_0}{\xi}\right)^2 \quad \text{and} \quad \xi = \xi_0 \left(\frac{T_{f=0} - T}{T}\right)^{-\nu} \tag{14}$$

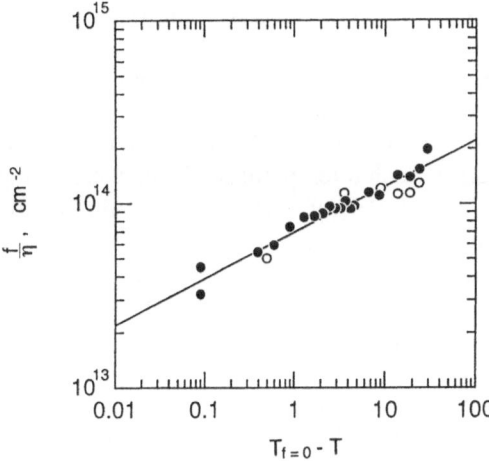

Fig. 11. The double logarithmic plot of the normalized friction coefficient of the poly(N-isopropylacrylamide) gel and the temperature difference from $T_{f=0}$. The *straight line* given in this figure is the result of the least square analysis. The slope of the straight line is 1/4, which is much smaller than the value 5/8 predicted for dynamic spinodal fluctuations

where $T_{f=0}$ and ν denote, respectively, the spinodal temperature, at which the density fluctuations diverge, and the critical exponent of the correlation length. Thus, by combining the two equations above we obtain the temperature dependence of the friction as follows

$$\frac{f}{\eta} = 6\pi N_0 \xi_0 \left(\frac{T_{f=0} - T}{T}\right)^{\nu} \tag{15}$$

The least-squares analysis of the friction according to the above equation yields $\nu = 1/4$ and $T_{f=0} = 33.59\,°C$. The results are shown in Fig. 11 in which the logarithm of the friction, f/η, is plotted as a function of the logarithm of the distance from the spinodal temperature. The slope of the straight line gives the critical exponent ν.

The determination of various critical exponents indicated that the phase transition of the poly(N-isopropylacrylamide) gels belongs to the same universality class with the three-dimensional Ising models [22]. The exponent $\nu = 1/4$ obtained here is much smaller than the theoretical value 5/8 expected for the three-dimensional Ising model [23, 24]. This discrepancy between the exponents may be due to the following reasons: Under the present experimental conditions, the isochore is not critical. Alternatively, the pore size may diverge upon phase separation in the metastable region before the gel reaches the spinodal line. In this case, $T_{f=0}$ should be considered as the temperature at which the domain size grows to infinity rather than the spinodal temperature.

4 Concluding Remarks

The frictional interaction between the polymer networks of gels and the solvent discussed above is one of the relevant parameters to describe the physical properties of gels. The detailed studies of the frictional properties of gels, however, have not been performed so far because of the difficulties in the accurate determination of the frictional properties of gels. The advance of the studies on the frictional properties of gels, which we have reviewed here, is mainly due to the improvement of the apparatus.

In poly(acrylamide) gels, the effective pore size of the polymer network is independent of the temperature change. Such a temperature dependence of the mesh size of the polymer network of the poly(acrylamide) gel is one of the advantages of using this gel as a matrix of the electrophoresis since the pore structure is stable. The total concentration dependence of the friction coefficient of poly(acrylamide) gel is well described by the scaling theory of the semi-dilute polymer solution which, together with the dynamic light scattering and the rheological studies, confirms the relationship between D, E, and f given by Eq. (2). Although this equation was proposed in 1973, the experimental verification has not been made so far because of the lack of the frictional studies of the polymer networks of gels. The relationship between the collective diffusion constant, the elastic modulus, and the friction coefficient is established in a reasonable way by the present experimental study on the frictional property of the gel. The poly(acrylamide) gel becomes increasingly inhomogeneous due to the density fluctuations frozen in the gel. This phenomenon occurs when the cross-linker concentrations is increased. According to this, static density fluctuations of the polymer network cause an opacity to develop in the gel. In such an inhomogeneous gel, the frictional property is mainly governed by the spatial correlation length of the density fluctuations rather than by the average cross-linking density of the gel; hence, the friction decreases upon increasing the cross-linker concentration. Furthermore, it is found that the friction of the gel decreases even with lower concentrations of the cross-linker where the gel looks completely transparent. The results obtained here suggest an important role played by the density fluctuations of the polymer network within the gel in the transport properties of the gel. The detailed study on the frictional properties of the opaque gel will promote a better understanding of the delicate relationship between the structure and the properties of gels.

In spite of the constant density of the gel, the friction of the poly(N-isopropylacrylamide) gel reversibly decreases by three orders of magnitude and appears to diminish as the gel approaches a certain temperature. This phenomenon should be universal and may be observed in any gel under optimal experimental conditions of the solvent composition and the temperature because the unique parameter describing the friction is the correlation length which tends to diverge in the vicinity of the volume phase transition point of gels. The exponent ν for the correlation length obtained from the frictional experiment is far from the theoretical value. It will, therefore, be important to study a poly(N-isopropylacrylamide) gel prepared at the critical isochore where the frictional property of gel may be governed by the critical density fluctuations of the gel.

The pore size of the poly(N-isopropylacrylamide) gel can be estimated by a simple calculation using Eqs. (13), (14), and (15). The pore size of the poly(N-

isopropylacrylamide) gel prepared under the same conditions is expected to be of the order of 90 Å at room temperature while it becomes of the order of 9000 Å at the highest temperature of our experiment at which the friction coefficient approaches the minimum value. This drastic and reversible change in friction may have applications in separation technology and it may be also significant in some biological transport phenomena. For instance, detailed studies of the transport in the gel will bring a general insight into the transport of substances such as the proteins through membranes and tissues which is vital for various biological activities. It will also open a variety of experiments of diffusion and transport that require more careful theoretical considerations.

Acknowledgement. This work was carried out at the Department of Physics, Massachusetts Institute of Technology in collaboration with Professor Toyoichi Tanaka. The author is extremely grateful to Professor Toyoichi Tanaka for his continuous encouragement and discussions. The author also thanks the Kao Corporation for financial support.

References

1. Tanaka T, Hocker LO, Benedek GB (1973) J Chem Phys 59: 5151
2. Munch JP, Candau S, Herz J, Hild G (1977) J Phys (Paris) 38: 971
3. Munch JP, Lemarechal P, Candau S (1977) J Phys (Paris) 38: 1499
4. Tanaka T, Ishiwata S, Ishimoto C (1977) Phys Rev Lett 38: 771
5. Takebe T, Nawa K, Suehiro S, Hashimoto T (1989) J Chem Phys 91: 4360
6. de Gennes PG (1979) Scaling concepts in polymer physics. Cornell University Press, Ithaca, New York
7. Weiss N, Silberberg A (1975) Polym Prepr Am Chm Soc Div Polym Chem 16: 289
8. Weiss N, van Vliet T, Silberberg A (1979) (J Polym Sci Polym Phys Eds) 17: 2229
9. Chiarelli P, de Rossi D (1988) Progre Colloid Polym Sci 78: 4
10. Tokita M, Tanaka T (1991) J Chem Phys 95: 4613
11. Tokita M, Tanaka T (1991) Science 253: 1121
12. Tanaka T, Fillmore JD (1979) J Chem Phys 70: 12114
13. Brochard F, de Gennes PG (1977) Macromolecules 10: 1157
14. Allain C, Amiel C (1986) Phys Rev Lett 56: 1501
15. Richards EG, Temple CJ (1971) Nature (Phys Sci), 220: 92
16. Bansil R, and Gupta MK (1980) Ferroelectrics 30: 63
17. See for instance, Tanaka T, Nicoloni C (eds) (1987) Structure and dynamics of biopolymers. (NATO ASI series), Martinus Nijhoff, Dordrecht and references therein
18. Hirokawa Y, Tanaka T (1984) J Chem Phys 81: 6379
19. Hirotsu S, Hirokawa Y, Tanaka T (1987) J Chem Phys 87: 1392
20. Tanaka T (1981) Sci Amer 244: 124
21. Tanaka T (1978) Phys Rev A17: 763
22. Li Y, Tanaka T (1989) J Chem Phys 90: 5161
23. Li Y (1989) Structure and critical behavior of polymer gels. Thesis, Massachusetts Institute of Technology, Cambridge MA
24. Stanley HE (1971) Introduction to phase transitions and critical phenomena. Clarendon, Oxford and references therein

Received 12, August 1992

Stimuli-Responsive Poly(N-isopropylacrylamide). Photo- and Chemical-Induced Phase Transitions

Masahiro Irie
Institute of Advanced Material Study, Kyushu University, Kasuga-Koen 6-1,
Kasuga, Fukuoka, Japan

Several attempts have been made to construct stimuli-responsive polymer solution and gel systems which undergo isothermal phase transitions by external stimulation, such as photons or chemicals. Aqueous solutions of poly(N-isopropylacrylamide) having photoisomerizable chromophores or host molecules in the pendant groups showed reversible phase separations by photoirradiation or by the addition of specific metal or ammonium ions. The gels made of the polymers also underwent photostimulated or chemical-induced volume phase transitions.

Advances in Polymer Science, Vol. 110
© Springer-Verlag Berlin Heidelberg 1993

1 Introduction

The conformation of polymers governs their various physicochemical properties. When the conformation is reversibly controlled by external stimulation, such as photons, or chemicals, the conformation change should produce a concomitant change in polymer properties in solution as well as in the gel. Polymers which change their properties reversibly by such external stimulation may be referred to as stimuli-responsive polymers.

So far, various kinds of polymers which change their conformation reversibly by photoirradiation have been reported [1-6]. The polymers contain pendant or backbone photoisomerizable chromophores, and the molecular property changes, such as geometrical structure or dipole moment changes, control the conformation. The polymers change their conformation in proportion to the number of photoisomerized chromophores. Thus, when the polymers contain more photoisomerizable chromophores and absorb more photons, the conformation changes more. Physical and chemical properties associated with the conformation changes also vary with the number of absorbed photons.

The efficiency of energy conversion in the polymers, i.e. how much the change in shape of the chain is induced by a photon, is, however, rather low. They need many photons and a high content of pendant photochromic chromophores to induce a large conformation change. To make a sensitive photoresponsive polymer, i.e. one which responds more efficiently to fewer photons, it is necessary to introduce an amplification mechanism to the system. One possible way of achieving this is to utilize the phase transition of polymers [7, 8].

The phase transitions, such as a phase separation of polymer solutions, a sol-gel transition, or a volume phase transition of gels, are always accompanied by conformation changes of polymers. Therefore, when the phase transitions are induced isothermally by external stimulation, the transitions cause efficient conformation changes. This contribution describes how such efficient stimuli-responsive polymer systems can be constructed.

2 A Principle of Molecular and System Design

Figure 1a shows a general schematic illustration of the stimuli-responsive phase transition of a polymer system from the state X to the state Y. In the absence of external stimulation, the polymer system changes the state at a temperature T_a. We assume that the phase transition temperature will rise to T_b in the presence of external stimulation. Then, if the external stimulation is applied to the system at T ($T_a < T < T_b$), the state will change from Y to X isothermally at a certain value of the external stimulation, C_c, as shown in Fig. 1b. This principle is useful for constructing efficient stimuli-responsive polymers.

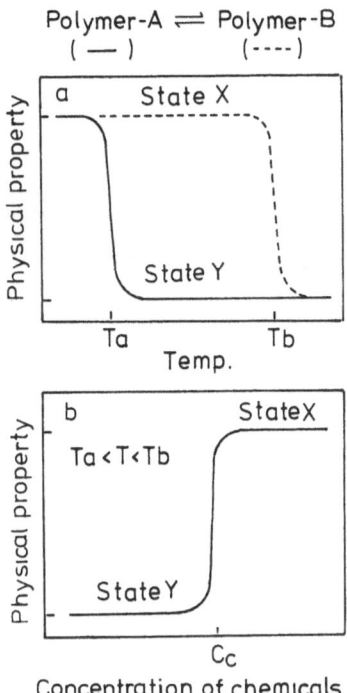

Fig. 1a, b. Schematic illustrations of chemical-induced phase transition of a polymer system. Thermal phase transition from state X to Y in the presence and absence of chemicals (**a**), and isothermal phase transition by chemicals (**b**)

To realize the above system, it is required to design a polymer which reversibly changes the molecular properties, such as hydrophilicity, by the external stimulation. Many molecules are known to be reversibly transformed to other isomers by external stimulation, such as photons, electrons or chemicals. Table 1 lists a few examples. Azobenzene shows the property change by photoirradiation. It isomerizes from the *trans* to the *cis* form by ultraviolet irradiation, and the dipole moment increases from 0.5 to 3.1 debye. The polar *cis* form returns to the less polar *trans* form by visible irradiation. Electrochemical oxidation of ferrocene changes the hydrophilicity. When it is oxidized from Fe(II) to Fe(III), the hydrophilicity increases. The Fe(III) state returns to the Fe(II) state by either electrochemical or chemical reduction. Host molecules also change the properties in the presence of suitable guest ions. Benzo[18]crown-6, for example, captures potassium ions in the cavity, and increases the hydrophilicity.

These molecules are useful for detecting external stimulation, and referred to as receptor molecules. When they are incorporated into the pendant groups of a polymer-A, the polymer changes the properties and converts to polymer-B by photons, electrons or chemicals.

$$\text{Polymer-A} \underset{}{\overset{\substack{\text{Photons}\\\text{Electrons}\\\text{Chemicals}}}{\rightleftharpoons}} \text{Polymer-B}$$

Table 1. Receptor molecules

Consequently, the stimulation alters the phase transition temperature from T_a of polymer-A to T_b of polymer-B.

Several attempts will hereinafter be described to construct such stimuli-responsive polymer solution and gel systems, which undergo isothermal phase transitions by photoirradiation or chemicals. We used poly(N-isopropyl-acrylamide), PNIPAM, as the model polymer and incorporated the receptor molecules into the pendant groups. An aqueous solution of PNIPAM is known to undergo the phase separation upon heating above 31 °C [9–12]. The solution is homogeneous and the polymer chain has a random conformation below 31 °C, while the well solvated random coil shrinks to a state of tightly packed globular particle above 31 °C [12]. This is then followed by the onset of aggregation of individual chain molecules and the solution turns inhomogeneous. A large conformation change occurs in the phase separation process. The phase separation temperature T_c depends on the subtle balance between the ability of the polymer to form hydrogen bonds with water and the inter- and intramolecular hydrophobic forces. The hydrophobic interactions are expected to be controlled by the property changes of the pendant receptor groups. Thus, the temperature at which this system undergoes phase separation is expected to be altered by external stimulation and an efficient conformation change is induced by the stimulation.

3 Photostimulated Phase Separation of Aqueous PNIPAM Solutions [13]

With the aim of controlling the phase separation temperature T_c of the PNIPAM aqueous solution azobenzene chromophores were incorporated into the pendant groups. The azobenzene chromophore is known to isomerize from the *trans* to the *cis* form upon UV irradiation, and the hydrophilicity increases as described before. The *cis* form again returns to the *trans* form by visible irradiation. Such property change is expected to affect the phase separation behavior. The most convenient method of detecting the phase separation of an aqueous polymer solution is to measure the transmittance change. Figure 2 shows the transmittance changes at 750 nm, which were observed when a 1 mass % aqueous solution of the copolymer with pendant azobenzene groups (2.7 mole %) was heated. In the dark before photoirradiation, the solution turned lactescent at 18.5 °C, and the transmittance decreased to a half the initial value at 19.4 °C. The solution became completely opaque above 22 °C. The solution again became transparent when cooled below 18 °C.

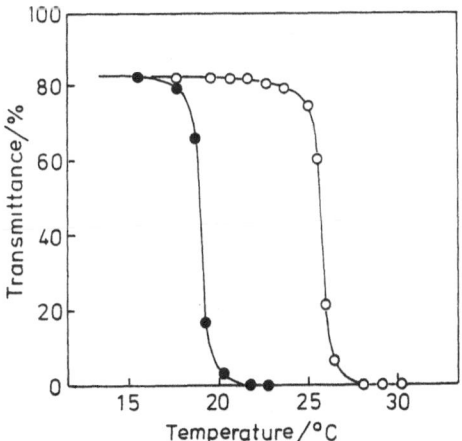

Fig. 2. Transmittance changes at 750 nm of a 1 mass % aqueous solution of PNIPAM with pendant azobenzene groups (2.7 mole %) when heated at a rate of 2 °C/min (●) before photoirradiation and (○) in the photostationary state by irradiation with UV (350 < λ < 410 nm) light

The solution was irradiated with UV light ($350 < \lambda < 410$ nm) for 10 min at 10 °C. The preirradiated solution showed the phase separation at 26.0 °C. On irradiation with visible light ($\lambda > 470$ nm), T_c again returned to 19.4 °C. T_c increased as much as 6.6 K by UV irradiation.

The phase separation temperature shift by photo-isomerization implies that in the temperature range between 19.4 and 26.0 °C, ultraviolet irradiation solubilizes the polymer or expands the polymer chain, while visible irradiation decreases the solubility, or shrinks the chain. Figure 3 shows the photo-stimulated phase separation at 19.5 °C accompanied by the conformation change of the polymer chain. Upon exposure to UV light ($350 < \lambda < 410$ nm), the opaque solution became transparent, while visible irradiation ($\lambda > 470$ nm) again decreased the transmittance of the solution. The polymer chain conformation changed concomitantly with the phase separation.

Here, it is worthwhile to note the difference in the response times of the dissolution and phase separation processes. This is important from the view point of energy conversion efficiency. At 19.5 °C, which is very close to T_c of the polymer solution with all *trans* azobenzene chromophores, the isomerization of a small number of chromophores, in other words, a small number of photons, was enough to raise T_c above 19.5 °C. Therefore, the transmittance increase took place immediately by irradiation for a very short time. The polymer chain was efficiently expanded by a small number of photons.

In the phase separation process, however, it needed some induction period for the polymer to start the phase separation. Almost complete isomerization of the azobenzene pendant groups from the *cis* to the *trans* form is required to decrease the phase separation temperature below 19.5 °C. The phase separation process exhibited a non-linear response to the irradiation time or the number of photons. When the number of absorbed photons reached a critical value, the system underwent the phase separation and the polymer chain was shrunk. The photo-stimulated phase separation/dissolution cycle was not observed below 19.4 and above 26.0 °C.

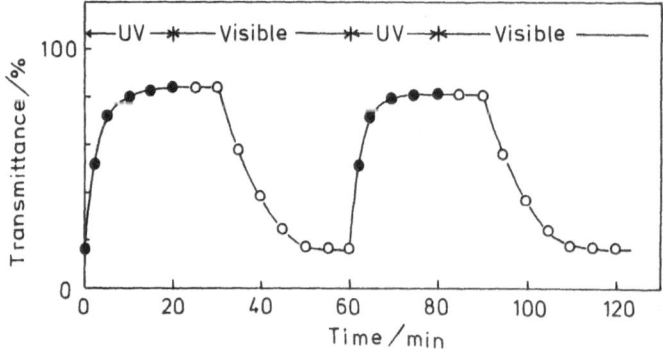

Fig. 3. Photostimulated isothermal phase separation of the 1 mass % aqueous solution of PNIPAM with pendant azobenzene groups (2.7 mole %) by alternate irradiation with UV ($350 < \lambda < 410$ nm) and visible ($\lambda > 470$ nm) light at 19.5 °C

With the intention of finding a large photo-effect on T_c, the dependence of azobenzene content on the increase in T_c by UV irradiation was measured. The maximum difference in T_c was observed at a very small azobenzene content of 2.7 mole %. Below and above this content the phase separation was not affected by photoirradiation.

T_c depends on the balance between the ability of the polymer to form hydrogen bonds with water and on the inter- and intramolecular hydrophobic forces. The former raises T_c, while the latter lowers T_c. By incorporating 2.7 mole % azobenzene chromophores into the polymer, the hydrophobic interaction was enforced, and T_c decreased from 31 °C to 19.4 °C. Conversion of the *trans* form to the hydrophilic *cis* form decreased the hydrophobicity, and raised T_c as much as 6.6 K. At a content higher than 3.4 mole %, the hydrophobic property decrease was not enough to weaken the existing inter- and intramolecular interaction, and T_c remained constant.

PNIPAM containing pendant triphenylmethane leuconitrile groups also showed photostimulated phase transition from the phase separated state to the homogeneous state [13]. The triphenylmethane leuconitrile group is known to change the polarity more pronouncedly than the azobenzene chromophore by the ionic photodissociation of C–CN bond [14].

Figure 4 shows the phase separation temperature before and after UV irradiation of the aqueous solution of PNIPAM polymer containing leuconitrile groups. Before irradiation in the dark, the T_c decreased with increasing leuconitrile content in the pendant groups. This decrease is due to hydrophobic property of the leuconitrile groups. Upon exposure to UV light, T_c increased with increasing leuconitrile content up to 0.8 mole %. Photogenerated triphenylmethyl cations in the pendant groups decreased the hydrophobicity of the pendant groups and raised T_c. T_c increased as much as 5.8 K upon UV irradiation.

When the leuconitrile content was 1.0 mole %, the 1 mass % aqueous polymer solution showed T_c at 29.2 °C. Upon UV irradiation, the polymer did not show any more clear phase separation. The transmittance decreased gradually with increasing temperature above 36 °C. The absence of the phase separation behavior indicates that the hydrophobic interaction of the main chain is not strong enough to overcome the hydrophilicity of the photogenerated triphenylmethyl cations and the polymer chain cannot shrink any further even at a higher temperature.

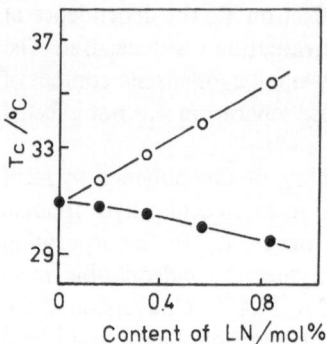

Fig. 4. Phase separation temperature (●) before and (○) after UV irradiation of the aqueous solutions of PNIPAM polymers containing leuconitrile (LN) groups

4 Chemical-Induced Reversible Phase Separation of Aqueous PNIPAM Solutions [15]

The concept of photostimulated phase separation can be applied to construct chemical-induced phase transition systems, which change the conformation reversibly in response to special chemicals. For the systems, host molecules are used as the receptor groups instead of photoisomerizable chromophores. Host molecules, such as crown ethers or cyclodextrins, are known to change the property by capturing guest chemicals in their cavity [16]. We employed benzo[18]crown-6 as the receptor molecule and incorporated it into the pendant groups of PNIPAM.

When the pendant crown ether groups bind metal ions, the phase separation temperature is expected to rise because the hydrophilicity of the polymer increases. The concentration change of special metal ions in solution can be detected as the phase transition of the polymer solution, or the conformation change of the polymer chain.

The transmittance change of a 1 mass% aqueous solution of PNIPAM containing pendant crown ether groups (11.6 mole%) was measured in the presence and absence of potassium chloride. Upon being heated with the rate of 0.3 °C/min, the solution turned lactescent at 30 °C in the absence of the metal ion and the transmittance at 500 nm decreased to a half the initial value at 31.5 °C. In the presence of potassium chloride (1.05×10^{-1} M), on the other hand, the aqueous solution remained transparent even when the solution temperature was raised to 31.5 °C. The solution became opaque above 37 °C. T_c was observed at 38.9 °C. T_c increased as much as 7.4 K by the addition of potassium chloride.

The increase in T_c depended on metal ions, as shown in Fig. 5. A large temperature increase was observed for potassium chloride, while the increase was only 1.5 °C for sodium chloride. The temperature increase was not observed by the addition of lithium and cesium chlorides. The binding affinity of a crown ether with metal ions depends on the cavity size [16]. When the ion diameter fits in the cavity size, the ion is captured by the crown ether. The cavity size of benzo[18]crown-6 is known to accommodate the diameter of K^+ [17]. The relative T_c increase correlates well with the binding affinity of benzo[18]crown-6 to metal ions. K^+ which efficiently binds to pendant benzo[18]crown-6 most pronouncedly increased T_c, while Li^+ and Cs^+ which are hardly captured by the crown ether groups could not increase T_c.

If T_c increases from T_a to T_b by the ion binding, the phase transition is induced isothermally at T ($T_a < T < T_b$) by the addition of metal ions, as illustrated in Fig. 1. The polymer chain conformation is also expected to expand in the process. As shown in Fig. 5, T_c was raised from 31.5 °C to 38.9 °C by the addition of potassium chloride, and from 31.5 °C to 33.0 °C by sodium chloride. These results imply that in the temperature range between 31.5 °C and 33.0 °C,

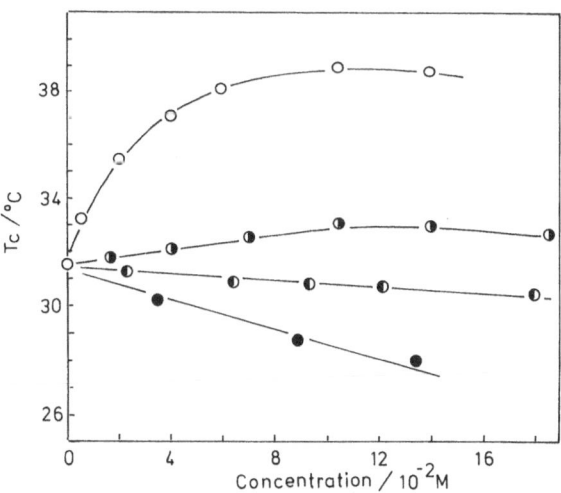

Fig. 5. Phase separation temperature changes of the aqueous solutions of PNIPAM containing pendant 11.6 mol % crown ether groups by the addition of potassium chloride (○), sodium chloride (◑), lithium chloride (◐), and cesium chloride (●). Polymer concentration was 1 mass %

the phase transition is induced by the addition of both potassium and sodium chlorides, while in the temperature range between 33.0 °C and 38.9 °C only potassium chloride can induce the isothermal phase transition.

Figure 6 shows the phase transitions at 32 °C and 37 °C by the addition of potassium and sodium chlorides. At 32 °C, both K^+ and Na^+ induced the phase transition from the phase separated to homogeneous state. The polymer chain expanded by the addition of K^+ and Na^+. The ion concentration necessary to induce the phase transition, however, depended on the ions. A very small amount was enough for K^+ to induce the transition, while 5.0×10^{-2} was necessary for Na^+. The system was very sensitive to K^+.

When the measuring temperature was raised to 37 °C, the phase transition was not observed by the addition of sodium chloride. Only K^+ could induce the transition. As observed at 32 °C for Na^+, a critical ion concentration which can induce the phase transition existed. K^+ with a concentration higher than 4.0×10^{-2} could induce the transition, while the ion at a concentration lower than that given above did not cause any effect.

Figure 7 shows the reversible phase separation by alternative addition of potassium chloride and low molecular weight [18]crown-6 at 37 °C. The addition of K^+ caused the increase of transmittance at 500 nm, while

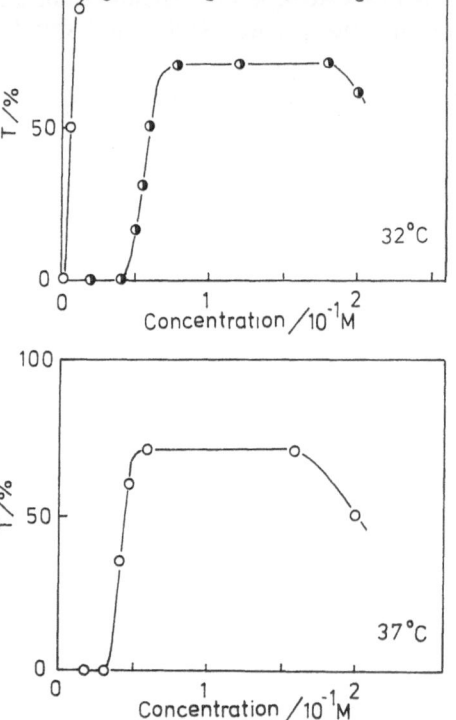

Fig. 6. Isothermal phase transitions from the phase separated to homogeneous state of the aqueous solution of the copolymer containing 11.6 mol % pendant crown ether groups by the addition of (O) potassium and (◑) sodium chlorides at 32 °C and 37 °C. Polymer concentration was 1 mass %

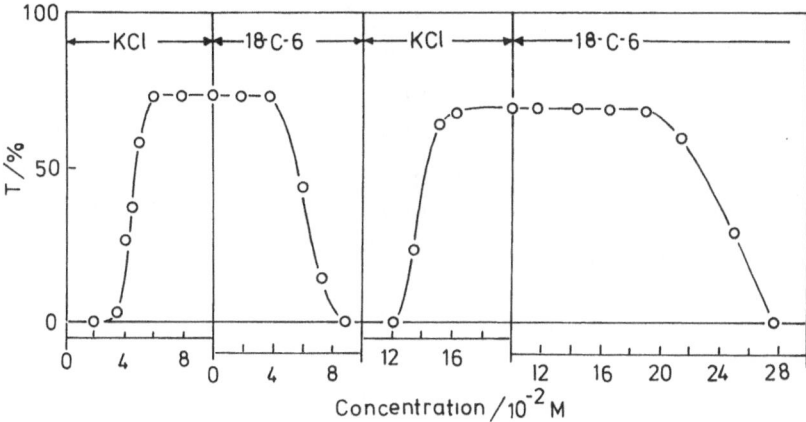

Fig. 7. Transmittance changes by alternate addition of potassium chloride and [18] crown-6 to the aqueous solution of the copolymer containing 11.6 mol % pendant crown ether groups at 37 °C. Polymer concentration was 1 mass %

[18]crown-6 decreased the transmittance. The phase transition from the inhomogeneous to the homogeneous state is due to the increase of hydrophilicity of pendant groups by the ion binding. Subsequent addition of low molecular weight [18]crown-6 extracted the metal ion from the pendant groups, and this resulted in the decrease of hydrophilicity and the shrinkage of the polymer chain.

5 Photostimulated Volume Phase Transition of PNIPAM Gels

By applying the concept of photostimulated phase separation of aqueous polymer solutions to a gel system, it becomes possible to construct sensitive photodeformable gels. PNIPAM gels containing azobenzene or triphenylmethane leuconitrile groups were synthesized with the aim of making gels which show the photostimulated volume phase transition.

Figure 8 shows the volume change of a polymer gel containing pendant azobenzene groups (10.8 mole %) before and after UV irradiation. The aqueous solution of PNIPAM having pendant azobenzene groups showed sharp phase separation upon heating, as shown in Fig. 2. In the gel phase, however, the volume change of the gel upon heating was continuous. Introduction of the azobenzene groups removed the discontinuous deswelling property of the gel in the dark as well as under UV irradiation. The UV irradiated gel showed a higher swelling degree compared with that swollen in the dark below 30 °C. This indicates the possibility of controlling both swelling and deswelling processes by UV and visible irradiation. In fact, the volume of the gel was reversibly

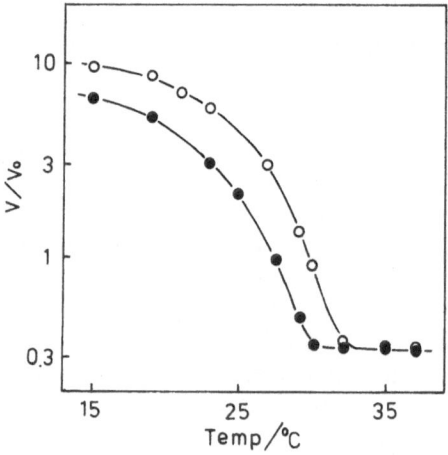

expanded as much as 25% at 15 °C by alternate irradiation with UV (350 nm < λ < 410 nm) and visible (λ > 450 nm) light.

A characteristic feature of a PNIPAM gel with ionizable groups is that it undergoes a phase transition as the temperature is varied. The gels having sodium acrylate ranging from 0 to 0.3 mole % undergo continuous transitions, whereas the gels with more than 0.6 mole % show discontinuous volume phase transition [18]. This suggests the possibility of photostimulated volume phase transition by introducing photoionizable chromophores, such as triphenylmethane leuconitrile, into PNIPAM gels.

Fig. 8. Volume changes of a PNIPAM gel containing pendant azobenzene groups (10.8 mole %) as a function of temperature (●) before and (○) after UV irradiation (350 < λ < 410 nm)

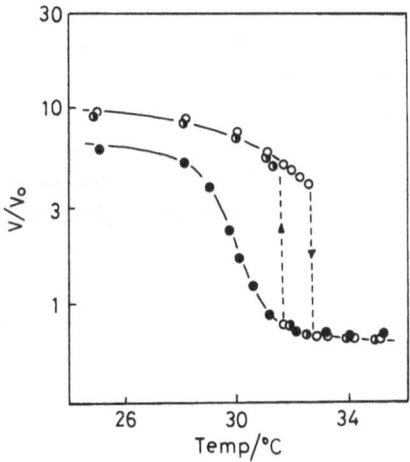

Fig. 9. Volume changes of a PNIPAM gel containing pendant triphenylmethane leuconitrile groups as a function of temperature. *Filled circles* show the swelling of the gel without UV irradiation. *Open circles* denote the swelling curve on UV irradiation when raising the temperature. *Half filled circles* are for the curve on UV irradiation when lowering the temperature

Figure 9 shows equilibrium volumes of the gel containing pendant triphenylmethane leuconitrile groups measured as a function of temperature in the dark and under UV irradiation [19]. The leuconitrile groups are known to dissociate into ions by UV irradiation. In the dark before photoirradiation, it underwent a continuous volume change at 30 °C. Upon irradiation with UV light, the gel showed a discontinuous volume transition. At 32.6 °C the volume of the gel suddenly decreased approximately 10 times. Above this temperature, the gel volume did not change markedly. When the temperature was lowered, starting from 35.0 °C the gel swelled discontinuously 10 times its original size at approximately 31.5 °C. The presence of the hysteresis confirms that this is the phase transition. When the temperature was fixed at 32.0 °C, the gel underwent a discontinuous swelling–shrinking switching upon exposure to UV light and removal of the light.

Another approach to induce the volume phase transition by photoirradiation utilizes the direct heating effect of the gel by light [20]. Gels containing

NIPAM and the light-absorbing chromophore, trisodium salt of copper chlorophyllin shown above, were prepared. In the absence of light, the gel underwent a
sharp but continuous volume change at around 35 °C. At a light intensity of
60 mW, the volume change became more pronounced and the transition
temperature was lowered to 33 °C. At an intensity of 120 mW the gel showed
discontinuous volume transition at 31.5 °C. The result indicates that at an
appropriate temperature the gel should undergo a discontinuous transition
when the light intensity is varied. Figure 10 shows the diameter of the gel as a
function of light intensity at 31.5 °C. A discontinuous volume change was
observed at 85 mW.

The interesting features of the photo-heating phase transition are that
irradiation causes the originally continuous transition to become discontinuous
and the transition temperature to be lowered. These can be explained using the
Flory–Huggins equation of state:

$$
\frac{1}{T} = \frac{\Delta S}{\Delta H} + \frac{k}{\Delta H}\left[\frac{v_1 v_e}{N\phi^2}\left\{(2f+1)\left(\frac{\phi}{\phi_0}\right) - 2\left(\frac{\phi}{\phi_0}\right)^{1/3}\right\}\right.
$$
$$
\left. - \frac{2}{\phi} - \frac{2\ln(1-\phi)}{\phi^2}\right]
$$

The equation can be expressed as follows:

$$T = T_{gel}(\phi)$$

where T is the absolute temperature and ϕ is the polymer network density. This
function increases monotonically with network density ϕ. When the gel is
irradiated, the temperature of the irradiated part of the gel is locally raised. The
temperature increase should be proportional to the light intensity and the
chromophore concentration and, therefore, to ϕ. Thus,

$$T = T_0 + \alpha I \phi$$

where T_0 is the ambient temperature, I is the light intensity, and α is a constant.

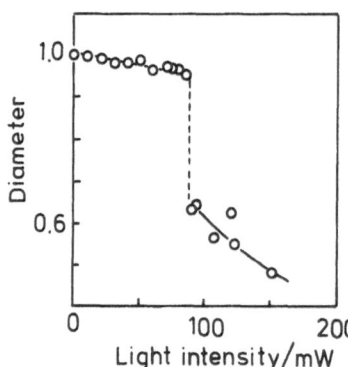

Fig. 10. Diameter of a PNIPAM gel containing copper
chlorophyllin as a function of light intensity at 31.5 °C

Combining the above two equations

$$T_0 = T_{gel}(\phi) - \alpha I \phi$$

This equation indicates that the second term owing to photo-heating effect lowers the transition temperature and finally brings the gel state to an unstable region, resulting in the discontinuous transition.

6 Chemical-Induced Volume Phase Transition of PNIPAM Gels

A gel system that swells and shrinks in response to specific ions or molecules has many potential applications, such as sensors, drug delivery devices and actuators. There are two possible approaches to attain this end. One approach is to utilize host molecules as the receptor part, the other utilize enzymes.

The PNIPAM containing benzo[18]crown-6, described in Sect. 4, which underwent the chemical-induced phase transition in aqueous solution can be applied for constructing chemicals-responsive polymer gels. When the PNIPAM is crosslinked to form a gel network, the conformation change accompanied by the phase transition is expected to result in the volume phase transition.

PNIPAM microsphere gels with diameter of 100–200 μm were prepared by emulsion polymerization [21]. The gel containing 12 mole % benzo[18]crown-6 was immersed in water and the diameter change of the gel was measured during heating at a rate of 0.3 °C/min. The gel was swollen below 25 °C. In the absence of metal ions, it started to shrink at 26 °C and showed a sharp volume change at 28.4 °C. Finally, the volume decreased by as much as 10 times the original volume.

The deformation process was remarkably affected by the addition of metal ions, as expected from the results of phase separation of the aqueous solution of the linear polymer. Figure 11 shows the temperature at which the gel underwent

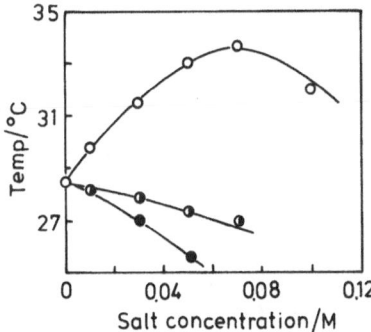

Fig. 11. Concentration dependence of the transition temperature at which a PNIPAM gel containing benzo[18]crown-6 (12 mole %) undergoes a sharp volume change in the presence of (○) potassium, (◑) sodium and (●) lithium sulfates

a sharp volume change in the presence of various metal ions. The addition of potassium sulfate increased the temperature, while other metal sulfates, such as sodium and lithium sulfates, decreased the temperature.

The swelling and shrinking changes of the gel are presumably determined by a balance between the hydrogen bonding ability of the polymer and the attractive hydrophobic interaction among NIPAM residues. The former acts to raise the transition temperature, while the latter lowers the temperature. When potassium ions are bound to the pendant crown ether groups, the hydrophobic interaction among pendant groups is diminished because of the hydrophilic nature of the ions. Consequently, the ions cause a raise in the transition temperature. Na^+ and Li^+ only act to break the hydrogen bonding networks, resulting in the decrease of the temperature.

The increase in the transition temperature suggests that in the temperature range between 28.4 and 33.8 °C the addition of potassium ion expands the gel. This was confirmed at 31 °C. The PNIPAM gel expanded by as much as 10 times on the addition of 3.0×10^{-2} M potassium sulfate.

The gel volume is dependent on the solvent composition. When the solvent composition in the gel is changed by enzymatic reaction, the gel is expected to undergo the volume phase transition. This is another approach of chemicals-induced volume phase transition [22]. Rabbit liver esterase was immobilized in the gel by an entrapped method. The esterase catalyses the following reaction.

$$BuCOOEt + H_2O \rightarrow BuCOOH + EtOH$$

The volume phase transition temperatures T_c for collapsing and T_s for swelling changed depending on the activity of the entrapped enzyme. When the enzyme was active in the presence of 47.3 mM of ethyl butyrate, $T_c = 29.8$ °C and $T_s = 29.1$ °C, while $T_c = 28.4$ °C and $T_s = 27.6$ °C with inactivated enzyme. The change in the phase transition temperature was presumably caused by the change in the substrate and product composition induced by enzymatic reaction within the gel phase.

When the gel with entrapped enzyme was placed in water containing 47.3 mM ethyl butylate at 28.9 °C, a swollen state was reached from the beginning of the measurement, while it took more than 80 min for the enzyme free gel to swell in enzyme/substrate solution. The rapid swelling is due to the change of solvent composition within or at the very vicinity of the gel phase by the enzymatic reaction shown above.

At 28.9 °C the gel containing inactivated enzyme was in the collapsed state. Therefore, it is safe to say that the expansion of the gel is induced by the hydrolysis of ethyl butylate.

The idea of changing the volume of gels in response to specific type of molecules has also been examined by using gels containing glucose oxidase [23]. The enzyme catalyzes the oxidation of glucose to gluconic acid. The reaction changed the ionization of the gel and subsequently the degree of swelling. This was applied to release insulin in response to glucose concentration. The change of the gel volume, however, was gradual as a function of glucose and was not a discontinuous transition.

The gel containing lectin, concanavalin A, was also found to undergo distinct swelling behavior in response to different saccharides (ionic and non-ionic saccharides) [24].

The concept of photostimulated and chemical-induced phase transitions described in this chapter can be applied to other systems, such as proton-induced or electrochemical phase transitions. All of these systems have high energy conversion efficiency in principle and the stimuli-responsive polymer systems are sensitive. A serious disadvantage of these systems is that the working temperature is limited to a rather narrow range near the phase transition of the parent polymer. This point should be kept in mind when used in practice.

7 References

1. Irie M, Menju A, Hayashi K (1979) Macromolecules 12: 1176
2. Menju A, Hayashi K, Irie M (1981) Macromolecules 14: 755
3. Irie M, Hirano S, Hashimoto K, Hayashi K (1981) Macromolecules 14: 262
4. Kumar GS, Neckers DC (1989) Chem Rev 89: 1915
5. Irie M (1990) Adv Polym Sci 94: 27
6. Irie M (1990) Pure Appl Chem 62: 1495
7. Kungwatchakun D, Irie M (1988) Makromol Chem Rapid Commun 9: 243
8. Irie M, Iga R (1986) Macromolecules 19: 2480
9. Heskins M, Guillet JE (1968) J Macromol Chem A2: 1441
10. Chiantore O, Guaita M, Trossarelli L (1979) Makromol Chem 180: 969
11. Fujishige S (1987) Polym J 19: 297
12. Fujishige S, Kubota K, Ando I (1989) J Phys Chem 93: 3311
13. Irie M, Kungwatchakun D (1992) Proc. Jpn Acad 68 Ser B: 127
14. Irie M, Kungwatchakun D (1986) Macromolecules 19: 2476
15. Irie M, Misumi Y, Tanaka T (1993) Polymer (in press)
16. Weber E, Vögtle F (1981) In: Vögtle F (ed) Host guest chemistry. Springer, Berlin Heidelberg New York, p 18
17. Kopolow S, Hogen Esch TE, Smid J (1973) Macromolecules 6: 133
18. Hirotsu S, Hirokawa Y, Tanaka T (1987) J Chem Phys 87: 1392
19. Mamada A, Tanaka T, Kungwatchakun D, Irie M (1990) Macromolecules 23: 1517
20. Suzuki A, Tanaka T (1990) Nature 346: 345
21. Irie M, Tanaka T (1992) Polym. preprint, Jpn: 2550
22. Kokufuta E, Tanaka T (1991) Macromolecules 24: 1605
23. Ishihara K, Kobayashi M, Shinohara I (1983) Makromol Chem Rapid Commun 4: 27
24. Kokufuta E, Zhang YQ, Tanaka T (1991) Nature 351: 302

Received July 29, 1992

Hydrogels as Separation Agents

K. L. Wang, J. H. Burban and E. L. Cussler
University of Minnesota, Department of Chemical Engineering and Materials
Science, 421 Washington Avenue SE, Minneapolis, Minnesota 55455, USA

We have summarized our experimental and theoretical work in the study of pH, temperature and pressure dependent phase transitions in hydrogels. Our focus, on the mechanisms of swelling and collapse, suggests using the gels for size-selective separations. An application of poly(N-isopropylacrylamide) to produce soy protein isolate is described in detail.

Advances in Polymer Science, Vol. 110
© Springer-Verlag Berlin Heidelberg 1993

1 Separation Scheme Using Hydrogels

A gel is a polymer network which swells when immersed in solvent, but which is prevented from dissolving by the presence of crosslinks which hold the structure intact. Gels which swell in aqueous solvents are called hydrogels. Tanaka [1] discovered that a partially hydrolyzed polyacrylamide gel would suddenly collapse with small changes in pH, solvent composition, or temperature. Since his discovery, many research groups have studied aspects of this swelling behavior. Our group has studied the behavior of such gels because we have been intrigued with the possibility of using such gels as a renewable concentrating agent [2, 3]. We have studied gels whose volume changes with pH [4, 5], temperature [6–8] and pressure [9]. Our work with the temperature sensitive gels includes a thermodynamic model of the swelling [10–12] and a commercial evaluation of the gel for soy protein concentration [13–14].

A general depiction of our gel separation process is shown in Fig. 1. In our process, collapsed gel is introduced into the aqueous solution which is to be concentrated. Typical solutions to be concentrated might be cheese whey containing nutritionally valuable proteins or a fermentation broth containing an antibiotic. The gel swells in the solution, absorbing small species but excluding large ones. When the gel reaches its equilibrium swelling, it is physically separated from the unabsorbed raffinate and subjected to a second environment which induces the collapse of the gel. The solution released by the gel is separated from the collapsed gel. The gel is then ready to be introduced into fresh aqueous solution or into the raffinate solution if further concentration is desired.

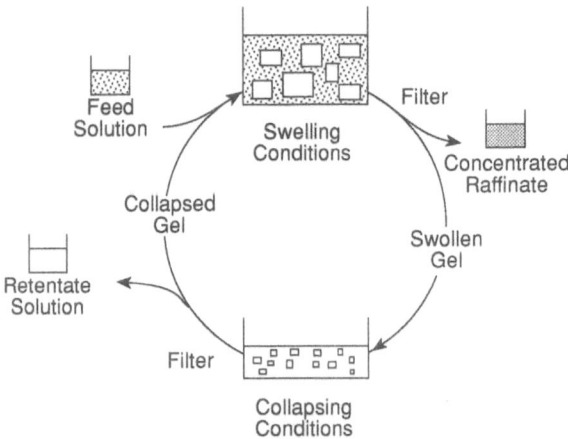

Fig. 1. Gel separation process. The gel is alternately swollen and collapsed to produce a concentrated raffinate and dilute retentate from a feed solution

Such a gel process can require less energy than evaporation and can be operated under mild conditions which will not damage the solutes. The key feature is the reuse of the gel. Previous workers [15] suggested concentrating solutions using a swellable gel which had to be discarded after each use, an economically and environmentally unattractive requirement.

The effectiveness of the separation can be quantified by an efficiency, η, which is defined as the actual increase in raffinate concentration divided by the increase expected from the altered solution volume. For example, if half of the feed solution is absorbed by the gel and the raffinate has twice the feed concentration, then the efficiency would be 100%. The efficiency is primarily a measure of how well the solute is excluded from the polymer network of the gel. The efficiency can be reduced, however, by physical entrainment of raffinate solution around the gel particles. The entrained solute can be recovered by washing the swollen gel and adding the wash liquor to new feed.

To determine the feasibility of this gel process, we studied various gels to elucidate their swelling behavior. We were concerned about the swelling kinetics, the conditions for swelling or collapse, the effect of gel composition on swelling, and the size of molecules which are excluded from the gel. These investigations are the focus of the following sections.

2 pH Sensitive Swelling

One class of hydrogels undergoes dramatic changes in swelling due to changes in pH. We have studied acrylamide-based gels, the synthesis of which is detailed elsewhere [5]. These gels contain a small amount of sodium methacrylate which undergoes an ion exchange that sparks the swelling or collapsing behaviour.

2.1 Mechanism of Swelling

At a pH below the pK_a of the methacrylate group (about 4.7), the methacrylate is in the nonionic, protonated form. Upon immersion of the gel in aqueous sodium hydroxide, the carboxyl group is ionized:

$$[RCOOH]_{gel} + [OH^-]_{aq} \rightarrow [RCOO^-]_{gel} + H_2O \qquad (1)$$

The electrostatic repulsion between the anionic groups within the gel causes it to swell. If the swollen gel is then placed in an acidic environment, it will collapse because of the protonation of the ionic groups:

$$[RCOO^-]_{gel} + [H^+]_{aq} \rightarrow [RCOOH]_{gel} \qquad (2)$$

Hence, the volume changes of the gel are most often due to changes in the ionic charge within the gel.

The volume of a charged gel can also change with ionic strength. This change is demonstrated by a set of experiments in which a sample of swollen gel is immersed in aqueous solutions of alkali metal chlorides. Initially, the ion concentration outside of the gel is higher than that inside. In order to achieve osmotic equilibrium, ions must diffuse into the gel, or water must diffuse out of the gel. Since the former is prevented by Donnan equilibrium, the gel volume decreases as shown in Fig. 2 [16]. Because the decrease in gel swelling comes from an osmotic effect, it is independent of the specific alkali metal ion and depends only on the osmotic force. Because of these ionic interactions, these gels can be used to concentrate small negatively charged solutes, as well as large uncharged solutes. The separation efficiencies of the small negatively charged solutes can be accurately predicted from Donnan equilibria [16] and are found to be highest for dilute, multivalent solutes present in a feed solution without added salt.

2.2 Kinetics of Swelling

The swelling kinetics of pH sensitive gels depend on the rates of diffusion of two species in the gel structure: ions to trigger the change in equilibrium swollen volume and water to accomplish the volume change. The mechanisms for swelling and collapse are different, since collapse occurs more quickly than swelling. Hydrochloric acid diffuses into the gel much faster than the rate at which the gel collapses, indicating that water diffusion, rather than ion exchange, limits the collapse. By contrast, during swelling, water and sodium hydroxide uptake occur at the same rate suggesting that ion exchange is the rate limiting step during swelling. The swelling rate can be increased by stirring the solution around the gel particles, which points to boundary layer resistance to mass transfer around the gel particles [5].

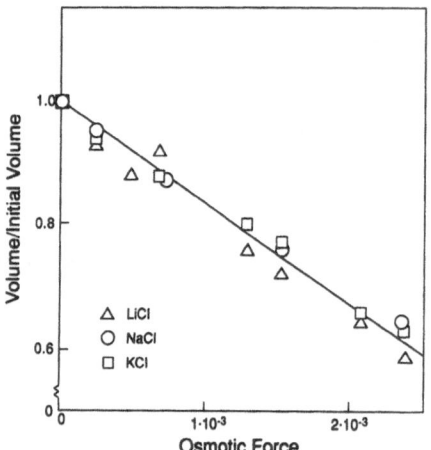

Fig. 2. Hydrolyzed polyacrylamide gel swelling vs ionic strength. The salt increases the osmotic force in the water outside the gel. The gel's swelling is reduced because Donnan exclusion prevents small ions from entering the gel. (Reprinted with permission from [16], 1986 Pergamon Press)

Consider the gel collapse in more detail. As hydrogen ions diffuse into the gel, they will rapidly react to form neutral carboxylic acid groups. Thus, a nonionic shell of collapsing gel will develop around a still-swollen ionized core. The diffusion of ions occurs freely in this nonionic shell, so that the collapse is limited by the Fickian diffusion of water out of the gel. We have confirmed this Fickian behavior by measuring the collapse of cylindrical gel samples of differing radii and ionic compositions [5]. The data for the fractional approach to equilibrium fall on a single curve against $[Dt/R^2]^{1/2}$, where D is the diffusion coefficient and R is the initial radius of the gel cylinder.

Gel swelling kinetics are more complex than the gel collapse kinetics, since several factors can be limiting [5]. At low concentrations of NaOH (below pH 11-12), a Donnan potential builds in the swelling gel, preventing the hydroxide ion from diffusing into the nonionic core. The ion exchange will be limited by diffusion of dissociated hydrogen ions from the gel's carboxylic groups to the surface, where they react with the hydroxide. At higher concentrations, the Donnan potential will be diminished; the nonionic core will shrink progressively as NaOH diffuses more easily into the gel. This process will be controlled by the diffusion of hydroxide through the ionized shell. In other cases, external boundary layer resistances limit the swelling. This occurs when the acidic hydrogen concentration in the gel is much larger than the hydroxide concentration in the solution, or when the gel particle size is comparable to the boundary layer thickness. Our results show that diffusional resistances both in and around the gel can affect the swelling rate.

2.3 Degree of Swelling

In studying the gel behavior, we are also concerned with the amount of swelling, as this determines the amount of gel required to achieve a given separation. As expected intuitively, we find that the gel swelling is inversely proportional to the degree of crosslinking, as shown by the inset of Fig. 3 [16]. The main part of Fig. 3 shows that the gel volume to the (− 2/3) power is proportional to the degree of crosslinking. The same proportionality between the gel volume to the (− 2/3) power and the amount of added acid was also observed. These relationships are consistent with theoretical predictions which describe gel swelling as a function of gel elasticity (related to degree of crosslinking) and gel ionization (related to amount of added acid).

Decreasing the degree of crosslinking will increase the water uptake for a mass of dry gel, though compromises in the efficiency will result. The effect of crosslinks on the separation of vitamin B-12, a nonionic solute of molecular weight 1355, is shown in Fig. 4 [16]. As the crosslink density decreases, the polymer chain length between crosslinks increases, yielding a looser structure which vitamin B-12 can more easily penetrate. The behavior fits well with the prediction from Flory excluded volume theory [16]:

$$\eta = 1 - e^{-cv} \tag{3}$$

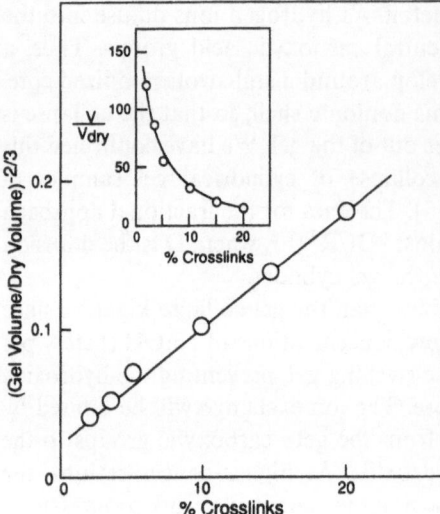

Fig. 3. Hydrolyzed polyacrylamide gel swelling vs degree of crosslinking. The increased crosslinks reduce the gel's elasticity, thereby reducing its swollen volume. (Reprinted with permission from [16], 1986 Pergamon Press)

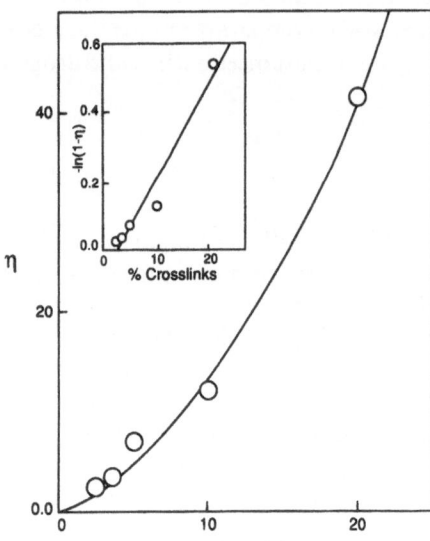

Fig. 4. Separation efficiency vs degree of crosslinking. The solute is vitamin B-12 and the gel is hydrolyzed polyacrylamide. The *solid line* in the inset is the prediction from Flory excluded volume theory (Eq. 3)

where v is the solute volume and c is the crosslink density. Using this relationship, we can synthesize gels of specific crosslink densities to achieve a target efficiency, keeping in mind that the solute volume is a function of both the molecular weight and configuration.

3 Temperature Sensitive Swelling

A second class of gels shows sudden volume changes with small changes in temperature. These gels are affected relatively little by changes in pH, ionic strength, and divalent ion concentration. As a result, they are much more suitable for chemical separations, and have been the major focus of our research.

3.1 Experimental Studies

These temperature sensitive gels, first found by M. Ilavsky [17], can be synthesized from the monomers N,N-diethylacrylamide or N-isopropylacrylamide. Details are given elsewhere [8]. At low temperatures, the gel is in its swollen state. As the temperature is increased, the gel undergoes a rapid decrease in volume, shown in Fig. 5 [12] for poly(N-isopropylacrylamide).

The volume change in these gels is not due to ionic effects, but rather to a thermodynamic phenomenon: a lower critical solution temperature (LCST). The uncrosslinked polymer which makes up the gel is completely miscible with water below the LCST; above the LCST, water-rich and polymer-rich phases are formed. Similarly, the gel swells to the limit of its crosslinks below the LCST, and collapses above the LCST to form a dense polymer-rich phase. Hence, the kinetics of swelling and collapse are determined mostly by the rate of water diffusion in the gel, but also by the heat transfer rate to the gel.

As with the pH sensitive gels, the degree of swelling will decrease as the amount of crosslinking increases since the polymer network becomes less flexible. Furthermore, the efficiency of separating solutes decreases with solute size and increases with amount of crosslinking. The introduction of charged groups into the gel, for example through an acrylamide or sodium methacrylate

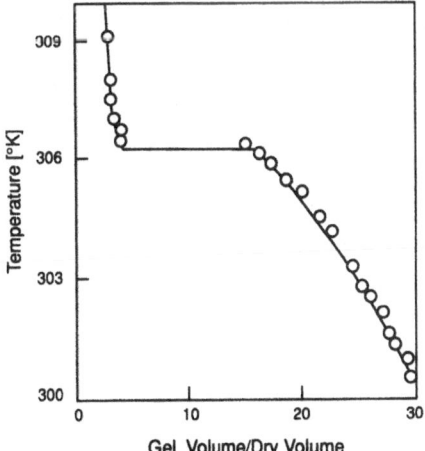

Fig. 5. Poly(N-isopropylacrylamide) gel swelling vs temperature. *Circles* are data points and the *solid line* is the lattice model prediction with one fitted parameter. (Reprinted with permission from [12], 1990 American Chemical Society)

copolymer, significantly increases the separation efficiency of charged solutes through the Donnan exclusion effects described for pH sensitive gels.

3.2 Thermodynamic Model

In designing applications for these gels, we must accurately predict the swelling behavior. To accomplish this goal, we have developed a model for predicting the volume changes in these gels [10, 11]. This model combines the Sanchez and Lacombe theory of mixing and the Flory-Rehner theory of rubber elasticity. The gel is approximated as a lattice, the sites of which can be occupied by polymer segments or by water molecules, or left vacant to allow for compressibility. The parameters required in the model include the cohesive energy densities for the water and gel, the close packed volume of the water, the volume of an empty lattice site, the crosslink density in the gel and an interaction energy. Except for the interaction energy, values for the other parameters are known or can be experimentally determined.

Such a model seems to us superior to those based on physics or on multiple chemical equilibria. While our model cannot be solved analytically, it is easily solved numerically. The solution begins by recognizing that the free energy change is a sum of that due to mixing and that due to elasticity. The mixing contribution includes the usual Flory contribution from the mole fraction and the usual lattice correction. It also includes an interaction energy per lattice site, which is in turn a function of the polymer-polymer interaction, the polymer-solvent interaction, and the interactions of the polymer and of the solvent with vacant sites. The elastic free energy is taken to be the normal form due to Flory and Rehner. The minimum free energy at a given concentration can then be easily calculated, yielding in turn the phase diagram of the system.

This simple lattice model yields a rich harvest of phase behavior. It easily gives both the binodals and the spinodals, which meet at the critical temperatures. Interestingly, there may be up to four critical temperatures. The first and coldest, an upper critical solution temperature, is that predicted by the Flory-Huggins theory. The next highest, a lower critical solution temperature, is that used in this work for gel based separations. The third, another upper critical solution temperature, represents renewed mixing at still higher temperatures. The fourth, a lower critical solution temperature, occurs first below the solvent's critical point. It represents polymer precipitating from a solvent which is increasingly gas-like, and has been carefully studied by D. Patterson [18], an early associate of Professor Dušek, the editor of this volume.

An example of the model prediction is shown by the solid line in Fig. 5 for poly(N-isopropylacrylamide) gel at 1 atm. In this case, three parameters were adjusted to fit the experimental data: the cohesive energy density of the gel, the crosslink density and the interaction energy. However, the first two are within 2% and 6%, respectively, of independently measured experimental values. Hence, the model prediction is based on a single fitted parameter, the interaction energy. Given one set of experimental data for a gel, the model is a useful tool for predicting the effect of changing variables such as crosslink density or temperature.

4 Pressure Sensitive Swelling

The lattice model for temperature sensitive gels predicts that the swelling behavior will also be a function of pressure as shown in Fig. 6 [12]. As the pressure increases, the transition temperature will increase. Thus at a constant temperature, the gel is expected to swell more as pressure increases. Although this effect may seem counterintuitive, it is the result of minimizing the free energy of the gel-water system by increasing the gel-water interactions. This prediction has been experimentally verified using poly(N-isopropylacrylamide) gel as shown in Fig. 7 [9]. The solid line in Fig. 7 gives the model prediction based on the same parameters as in Fig. 5 and is qualitatively consistent with the experimental data. Such a pressure effect could allow cheaper separations than are possible with other gels.

5 An Application: Soy Protein Isolate Production

The swelling behavior of these gels has spurred a search for applications. Albin et al. [19] have immobilized glucose oxidase enzyme within pH sensitive gels. As

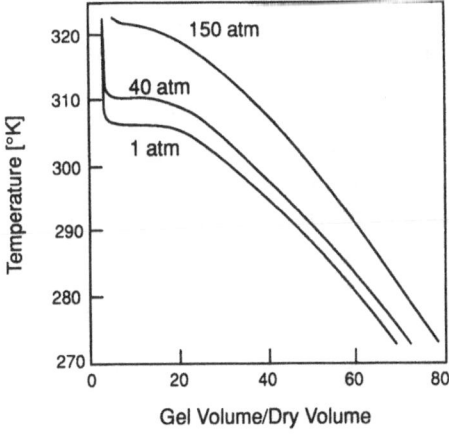

Fig. 6. Poly(N-isopropylacrylamide) gel swelling at various pressures. The model predicts more gradual swelling at higher transition temperatures as the pressure increases. Parameter values are the same as Fig. 5. (Reprinted with permission from [12], 1990 American Chemical Society)

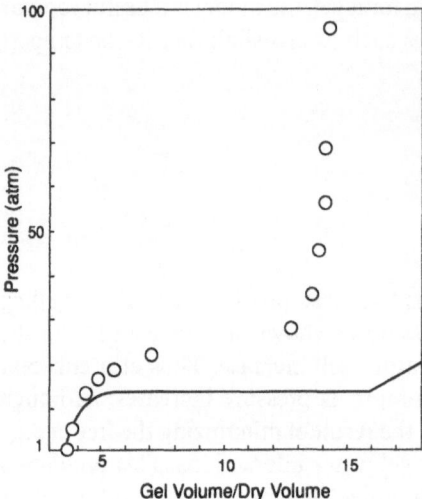

Fig. 7. Poly(N-isopropylacrylamide) gel swelling vs pressure. (T = 307.5 K) *Circles* are data points; *solid line* is the lattice model prediction with the same parameters as Fig. 5. (Reprinted with permission from [9], 1990 Pergamon Press)

the enzyme acts on glucose to lower the pH in the gel, the gel swells and delivers insulin into the bloodstream. Hoffman [20] has also investigated biomedical applications, but with temperature sensitive gels. Marchetti and Cussler [21] have surveyed the possibility of using hydrogels for ultrafiltration. Vasheghani-Farahani et al. [22] have concentrated solutions of proteins and enzymes with acrylamide-based gels without loss of enzyme activity. Badiger et al. [23] have used a column of lightly crosslinked ion exchange resin beads to concentrate biological macromolecules.

We have applied our gel technology most extensively to the production of soy protein isolate, an edible protein product that contains at least 90% protein on a dry basis. Our process is based on the concepts outlined in Fig. 1, and has been the subject of several patents [24–26]. We have primarily used poly(N-isopropylacrylamide) gel particles, although in principle, sheets of gel could also be used [27]. Currently, soy protein isolates are commonly made by the acid precipitation process shown in Fig. 8 [14] using a feed of defatted soybean flakes. Soybean flakes are made by cleaning, dehulling, and extracting the oil from raw soybeans. In the acid precipitation process, one part of soybean flakes is slurried with 6 to 20 parts water at a pH of 8.5 to disperse the globular and albuminous proteins and other water-soluble solutes. These solutes are mainly carbohydrates, but include low molecular weight phytins and salts. The slurry is centrifuged to separate the protein solution, referred to as the mother liquor, from the spent flakes. The spent flakes are dried and used as animal feed and more recently as an increasingly important source of human dietary fiber.

The mother liquor is then mixed with acid to precipitate the globular proteins as a curd. The albumins, which are the most valuable solutes in the feed and which account for approximately 10% of the total soluble proteins, do not precipitate and remain in the liquid whey accompanying the curd. The phytins,

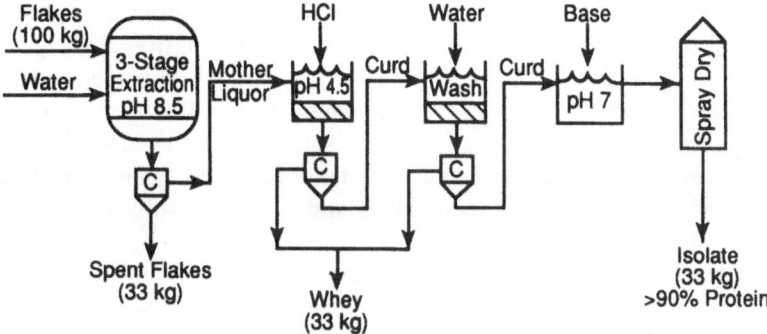

Fig. 8. A typical commercial soy protein isolate process. The soy proteins are extracted with alkali, precipitated with acid, washed with water, redispersed and dried. (Reprinted with permission from [14], 1989 Institute of Food Technologists)

which are toxic in large amounts, do precipitate. This protein curd is separated from the whey, washed with water, and redispered at a pH of about 7. The dispersed protein is spray dried to yield the soy protein isolate. This procedure produces about 33 kg isolate, 33 kg of spent flakes, and 33 kg of dissolved solutes in the whey from 100 kg of defatted soybean flakes. The dissolved solutes in the whey are comprised of the carbohydrates and other salts but include the valuable albumins, which represent a loss.

Our alternative soy protein isolate process utilizing the swellable gel technology is shown in Fig. 9. Instead of precipitating the protein, we subject the mother liquor to a series of five poly(N-isopropylacrylamide) gel treatments to concentrate and purify the protein stream. In the first gel stage, the mother liquor is contacted with collapsed gel at 5 °C. Approximately 40% of the water in the mother liquor is absorbed by the gel as it swells. The swollen gel is then separated from the retentate in a centrifuge. This retentate is diluted with water

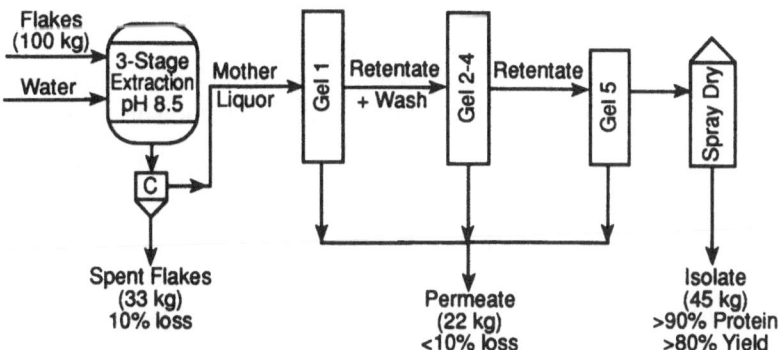

Fig. 9. Soy protein isolate process based on temperature sensitive gels. This proposed process, which will operate at approximately 5 °C, should produce more, higher quality protein than the conventional process in Fig. 8

to the original volume and then contacted with additional collapsed gel in stages 2–4. This separates the protein from low molecular weight impurities. In stage 5, retentate 4 is not diluted but is contacted solely with collapsed gel to increase the protein concentration (to 15% on dry basis) for spray drying. The retentate from stage 5 is then spray dried to yield a soy protein isolate containing albumins but not phytins, a product superior to the acid precipation product. This procedure can produce 45 kg isolate, 33 kg of spent flakes, and 22 kg of dissolved solutes in the whey from 100 kg of defatted soybean flakes.

The new process does have problems. Although the protein molecules are too large to be absorbed into the gel, they do absorb onto the gel particle surface. This absorbed protein would represent a significant loss if not re-covered. Recovery of the protein can be accomplished by two consecutive water washes, and subsequent separations of the gel from the wash water. The wash water recovered from the first wash contains most of the absorbed protein removed from the gel particles. This protein containing wash water is therefore used in the gel treatment stages and as make-up water in the flake extraction stage. The twice washed gel particles are then heated to 55 °C, at which point they have released virtually all of the absorbed solutes and water into a whey stream. The collapsed gel particles are then separated from the whey stream using centrifuges and recycled back to the gel treatment stages.

The use of the gel technology to produce soy isolates has several advantages over the conventional processes now being used [25]. These are:

1. One can work at temperatures near freezing, thus essentially eliminating spoilage.
2. No alkali is needed to neutralize the acid product before spray drying.
3. The new process has a greater yield of protein than current processes, since the albumins are recovered.
4. Since the separation process is physical rather than chemical, the proteins recovered by the gel process are more "native".
5. The protein may have better flavor.
6. Phytins are not present in the soy protein isolate made by the gel process.
7. The Protein Equivalence Ratio (PER), a criterion for protein nutritional value, is higher for the gel produced protein.

We have designed a commercially sized plant to evaluate the economic feasibi-lity of implementing the swollen gel technology for the production of soy protein isolate [28]. This design includes the overall capital cost, including equipment, installation, process piping, electrical service, instrumentation, building require-ments, and start-up expenses. The design also includes the total yearly operating expenses, consisting of the cost of ingredients, utilities, labor, and analytical services. This simple economic analysis showed that the profits balance the initial capital investment after 3.2 years of plant operation. This corresponds to a return on investment (ROI) of 30%. Therefore, the swollen gel technology appears to be an economically attractive alternative for production of soy protein isolate.

6 References

1. Tanaka T (1981) Sci Am 244:124
2. Cussler EL, Stokar MR, Varberg JE (1984) AIChE J 30:578
3. Stokar MR (1983) Gels as size selective extraction solvents. Thesis, University of Minnesota, Minnesota
4. Gehrke SH (1986) Kinetics of gel volume change and its interaction with solutes. Thesis, University of Minnesota, Minnesota
5. Gehrke SH, Cussler EL (1989) Chem Eng Sci 44:559
6. Freitas RFS (1986) Extraction with and phase behavior of temperature sensitive gels. Thesis, University of Minnesota, Minnesota
7. Freitas RFS, Cussler EL (1987) Chem Eng Sci 42:97
8. Freitas RFS, Cussler EL (1987) Sep Sci Tech 22:911
9. Lee KK, Cussler EL, Marchetti M, McHugh MA (1990) Chem Eng Sci 45:766
10. Marchetti M (1989) Phase equilibria of gels. Thesis, University of Minnesota, Minnesota
11. Marchetti M, Prager S, Cussler EL (1990) Macromol 23:1760
12. Marchetti M, Prager S, Cussler EL (1990) Macromol 23:3445
13. Trank SJ (1988) Design and application of temperature-sensitive hydrogels. Thesis, University of Minnesota, Minnesota
14. Trank SJ, Johnson DW, Cussler EL (1989) Food Tech 43:78
15. Flodin P, Gelotte B, Porath J (1960) Nature 188:493
16. Gehrke SH, Andrews GP, Cussler EL (1986) Chem Eng Sci 41:2153
17. Ilavsky M, Hrouz J, Ulbrich K (1982) Polym Bull 7:107
18. Dušek K, Patterson D (1968) J Polym Sci A2 6:1209
19. Albin GW, Horbett TA, Miller SR, Ricker NL (1987) J Contr Rel 6:267
20. Hoffman AS (1987) J Contr Rel 6:297
21. Marchetti M, Cussler EL (1989) Sep Purif Meth 18:177
22. Vasheghani-Farahani E, Cooper DG, Vera JH, Weber ME (1992) Chem Eng Sci 47:31
23. Badiger MV, Kulkarni MG, Mashelkar RA (1992) Chem Eng Sci 47:3
24. Cussler EL (1985) Method of size-selective extraction from solvents. U.S. Pat. 4,555,344
25. Cussler EL (1989) A temperature-sensitive method of size-selective extraction from solutions. U.S. Pat. 4,828,701
26. Cussler EL (1989) Isolation process using swellable poly(N-isopropylacrylamide) gels. U.S. Pat. 4,863,613
27. Trank SJ, Cussler EL (1987) Chem Eng Sci 42:381
28. Burban J (1992) Unpublished results

Received 17 August 1992

Synthesis, Equilibrium Swelling, Kinetics, Permeability and Applications of Environmentally Responsive Gels

Stevin H. Gehrke
Department of Chemical Engineering, Mail Location 171,
University of Cincinnati, Cincinnati, OH 45221-0171, USA

This article is a review of the research on environmentally responsive polymer gels done by Stevin H. Gehrke's research group at the University of Cincinnati. This group has studied a wide variety of responsive gels, including crosslinked cellulose ethers, poly(N-isopropylacrylamide), and radiation-crosslinked poly(vinyl methyl ether). Equilibrium swelling has been studied both experimentally and theoretically, with the general goal of learning how to control the nature of the volume transition. Conditions under which transition-inducing stimuli affect the rate of volume change have been identified and a diffusion analysis is shown to be broadly useful for correlating the kinetics. Polymer scaling theory provides a qualitative basis for understanding how polymer composition and stimulus interval influence the magnitude of the network diffusion coefficient. The response rate of gels can be several orders of magnitude greater than this if they are microporous and can absorb and expel solvent by a convective process. Diffusion coefficients of solutes in gels decline as the gel shrinks, in generally good agreement with free volume theories. Hydrophobic interactions between gels and drugs are often quite strong, especially in deswollen gels. The ability of these gels to dewater coal slurries has also been established.

Advances in Polymer Science, Vol, 110
© Springer-Verlag Berlin Heidelberg 1993

List of Symbols and Abbreviations

APS	ammonium persulfate
C	mass of crosslinking monomer per 100 g of monomers
CMC	carboxymethylcellulose
D	diffusion coefficient
D_o	diffusion coefficient in free solution
DSC	differential scanning calorimetry
DVS	divinyl sulfone
f	friction factor between network and solvent
G	shear modulus of gel
GRAS	generally recognized as safe
HPC	hydroxypropylcellulose
HPMC-E	hydroxypropylmethylcellulose, Type E
HPMC-K	hydroxypropylmethylcellulose, Type K
k	mass transfer coefficient
K	bulk modulus of gel
K	partition coefficient
$K_{o/w}$	octanol/water partition coefficient
LCST	lower critical solution temperatures
L_0	initial sample thickness
L_t/L_∞	normalized change in characteristic sample dimension
MC	methylcellulose
M_t/M_∞	normalized approach to equilibrium mass
N	number of polymer structural units in an effective chain segment
NIPAAm	N-isopropylacrylamide
P_g^*	polymer cohesive energy density
PNIPAAm	poly(N-isopropylacrylamide)
P_s^*	solvent cohesive energy density
PVME	poly(vinyl methyl ether)
Q	equilibrium swelling degree of gel
SDS	sodium dodecyl sulfate
SMB	sodium metabisulfite
T	mass of monomers (in grams) added to 100 ml of water
T	temperature
t	time
TEAC	tetraethylammonium chloride
TEMED	N,N,N',N'-tetramethylethylenediamine
T_e	equilibrium temperature
T_i	initial temperature
T_s	spinodal temperature
T_t/T_∞	normalized approach to equilibrium temperature

v	velocity
Z_{sg}	correction for deviation from a geometric mean mixing rule
ϕ	polymer volume fraction
η	solvent viscosity
ξ	correlation length

1 Background

1.1 Definition of Environmentally Responsive Gels

Although the term "gel" is a field-specific term, gels are most generally understood to be polymeric networks which absorb enough solvent to cause macroscopic changes in the sample dimension. Hydrophilic gels in aqueous solutions have been the most widely studied, but almost any polymer can be crosslinked to form a gel which will swell in a sufficiently good quality solvent. The three-dimensional network is stabilized by crosslinks which may be provided by covalent bonds, physical entanglements, crystallites, charge complexes, hydrogen bonding, van der Waals or hydrophobic interactions. Gels have many technologically important roles in chemical separations, biomedical devices and absorbent products, to name a few areas. The properties that make gels useful include their sorption capacities, swelling kinetics, permeabilities to dissolved solutes, surface properties (e.g. adhesiveness), mechanical characteristics and optical properties. The single most important property of a gel is its swelling degree, since most of the properties are directly influenced by this [1–3].

"Responsive" polymer gels are materials whose properties, most notably their solvent-swollen volumes, change in response to specific environmental stimuli including temperature, pH, electric field, solvent quality, light intensity and wavelength, pressure, ionic strength, ion identity and specific chemical triggers like glucose [4–13]. They have also been termed "actuated," "stimuli-sensitive", and "smart" gels. The property which often changes most dramatically is the swollen volume, which can increase by virtually any value in response to virtually any level of input stimulus, but the other properties can also change (usually as the result of the large volume change, but also at constant volume). These changes may occur discontinuously at a specific stimulus level (a phase transition) or gradually over a range of stimulus values. All of these changes are reversible with no inherent limit in lifetime. In fact, the cycles have been demonstrated to be repeatable upwards of 500 times with stable properties over many months for several types of responsive gels [14–16]. Perhaps no other class of materials can be made to respond to so many different signals.

Thus devices based on responsive polymer gels can be made to function in response to an incredibly broad range of stimuli. These gels are being considered for use in various devices, including switches, sensors, mechanochemical actuators, drug delivery devices, recyclable absorbents, specialized separation systems, bioreactors, bioassay systems, artificial muscles (e.g. sphincter replacements) and display systems. A product using responsive gels which is now reaching the marketplace is a screen or 'electric window' which makes use of a reversible opaque/transparent transition. Thus ideas for uses of responsive gels have come from many directions, as outlined in the following sections.

1.2 Historical Development of Responsive Gels

In 1949 Katchalsky created the first responsive polymer gels by crosslinking water-soluble polyelectrolytes to form gels which swelled and shrank in response to changes in solution pH [17]. He and his coworkers also created machines powered by gels responsive to pH and ionic strength [14, 18]. The gels studied by Katchalsky and those building upon his work responded gradually to changes in pH, electric field, ionic strength, or temperature. For example, a change of roughly two pH units was required to transform the swollen, permeable state to a shrunken, impermeable state and vice versa. However, in 1968 Dusek and Patterson postulated that a discontinuous phase transition in the swollen volume of the gel might exist [19]. That is, a swollen and a shrunken gel phase could coexist and the transition between the two states would occur at a fixed value of the surrounding environment, in a manner analogous to the vapor/liquid phase transition observed with pure fluids. Although their analysis indicated that such transitions would occur only in experimentally inaccessible regions, in 1978 Tanaka observed just such a phase transition in ionized polyacrylamide gels at specific concentrations of acetone in water [7]. Since that time, Tanaka's research group and others have demonstrated that this is a general phenomenon which should be observable in all gel/solvent systems where the appropriate region of the phase diagram is experimentally accessible. They have also demonstrated the diversity of the stimuli to which these gels could respond.

As a result of this ground-breaking work by Katchalsky, Tanaka and scientists at the Institute of Macromolecular Chemistry in Prague, the field of responsive polymer gels has expanded dramatically worldwide since the mid 1980s. In Japan, interest in responsive gels has accelerated in part due to the designation of responsive gels as a priority research area by the Ministry of International Trade and Industry [20]. While much of the literature in the area has been devoted to syntheses of new gels and the development of novel applications, broadly-based, fundamental work in the area has also appeared. Currently, a good qualitative understanding of responsive gels and their properties exists, though successful quantitative interpretations are fewer.

1.3 Research on Responsive Gels

Many different types of responsive gels have been synthesized by various techniques. Copolymerization/crosslinking of monofunctional and polyfunctional monomers has been extensively used to produce responsive gels of vinyl monomers, especially of a wide variety of acrylamide derivatives such as N-isopropylacrylamide, acrylic acid, and diethylacrylamide [4–11, 21–36]. Crosslinking of linear polymers by either chemical means or γ-radiation has also been used to produce a wide range of responsive gels. The precursor polymers used include poly(vinyl methyl ether), cellulose ethers, poly(vinyl alcohol/vinyl acetate) and biological polymers such as elastin [37–41]. Responsive gels have also been made from polymers which form complexes held together by non-covalent interactions, such as crystallites, hydrogen bonding or electrostatic interactions [42–44]. Gehrke et al. and Kabra et al. have determined that gel synthesis conditions such as reaction temperature and polymerization time can have a significant effect upon on the structure and properties of the resulting gels [45, 46]. In general, however, the relationships between the synthesis techniques and the resulting responsive gel properties have not been very carefully examined.

The thermodynamic theories of gel swelling assume that swelling is the result of several independent free energy changes which occur upon contacting the gel and the solvent. These changes are related to the inherent tendency of the polymer chains to dissolve in the solvent, the elastic resistance of the network to deformation caused by swelling, and, for polyelectrolyte gels, the osmotic pressure generated by the dissociation of ionic groups. Tanaka's, Ilavsky's and others' theories for swelling of ionic responsive gels have proven quite successful qualitatively but are less useful quantitatively [2, 7, 47, 48]. Work is ongoing to improve the theory for responsive gels, especially in terms of the observed non-ideal entropic effects which cause temperature sensitivity [34, 49–53]. Related topics that have also attracted recent attention include determination of the order of the volume phase transition and examination of surface patterns which can form on the gels as they swell or shrink [54–56].

The response kinetics of these gels will significantly influence the ultimate success of most of the applications envisioned for responsive gels. The rate of response can be controlled by either the stimulus rate itself or the rate of network response. Gehrke and Cussler outlined these concepts qualitatively and Grimshaw et al. developed a quantitative model for treating the kinetics of electrically- and pH-responsive gels [5, 57]. These papers show that swelling kinetics are frequently limited by ion exchange rates, although shrinking tends to be dominated by the network response. Furthermore, Siegel and Firestone and Chou et al. have explained that the kinetics of the pH response also depend strongly upon the type of buffer used in the solution [32, 58]. Huang et al. have developed a model for temperature-responsive gels when heat-transfer is rate-limiting, although Gehrke et al. have shown that the kinetics of the most commonly studied temperature-sensitive gels are not influenced by heat transfer [39, 59]. Much work to model the kinetics and to interpret the fundamentals of the network response has been done by Tanaka's and Gehrke's groups. The rate of volume change can be characterized using either the equations of motion [60] or Fick's law [61], except when the phase transition is crossed, where consistent deviations from these theories are observed. Tanaka and coworkers identified the diffusion coefficients obtained from macroscopic measurements of swelling and shrinking with the microscopic diffusion coefficients obtained from light-scattering experiments [60, 62–64]. Akhtar and Gehrke observed that the rate of volume change depends primarily upon the concentration of polymer in the gel.

Their explanations of these results adapted scaling law concepts originally developed for semi-dilute polymer solutions [61].

Recently, groups have reported that gels of heterogeneous, porous microstructures can shrink and swell in response to changes in temperature thousands of times faster than did the homogeneous, nonporous gels described above [12, 39, 46, 65]. An interconnected porous structure which allows water to be expelled from the pores by a convective process rather than a diffusive one results in this enhanced response rate. Developing a controllable technique for the production of such gels and gaining a clear picture of this enhanced response is important for future work in this area.

Although many of the proposed applications for these gels requires that they operate under an applied pressure or generate some kind of mechanical force, a detailed understanding of these relationships does not currently exist. There is data available on the effect of the load on the rate of work and stroke, or generated force vs time, for example, but this is often presented on an empirical basis. Furthermore, much of the work has been carried out under conditions where the stimulus is rate-limiting, rather than the polymer network [66, 67]. The development of a mathematical description of these phenomena using independently obtainable polymer parameters is needed.

Since the permeability of gels to dissolved solutes is normally related to the swelling degree, and since specific chemical stimuli can be used to control gel swelling, these stimuli can also alter a gel's permeability to solutes [28, 68, 69]. In general, solute permeability is expected to decrease as the gel contracts due to an accompanying decrease in the effective mesh size of the polymer network. However, Palasis and Gehrke have recently shown that this simple relationship between gel swelling and solute permeability can be altered by hydrophobic interactions [70]. In principle, responsive gels could function with either continuously adjustable permeability if swelling changes gradually or in an "on/off" mode if swelling changes abruptly. A variety of novel drug delivery and separation techniques have been proposed for responsive gels based on these phenomena, as described below.

1.4 Applications Development

The unique properties of responsive gels have resulted in substantial applications research, especially over the past decade. In general, these applications can be classified as mechanical devices, controlled solute delivery devices, or chemical separation techniques. A recent book edited by DeRossi et al. covers these applications in more detail [71].

An actuator is a motor or transducer that converts electrical, pneumatic, or hydraulic energy into power for motion or action. They are used in robotics for the movement of arms, wrists, and fingers to handle and lift loads, or as gradual switches and automatic valve controllers. Because responsive gels can generate force by swelling or shrinking in response a variety of signals, they have potential uses as actuators and sensors. In addition to the diversity of signals to which responsive gels respond, a key attraction to the use of responsive gels in such systems is the self-contained sensing and actuating mechanism of the gel – the "responsiveness." In other words, the sensing capability of a gel actuator is inherent in the device, so it would be very simple and virtually fail-safe. It could even be poured as a solution to form 'in place'. The materials also display promising strength and power, since responsive gels have been measured by Urry et al. to pick up weights 1000 times their dry weights, while Suzuki and Tatara showed that gels could generate power densities comparable to skeletal muscles [72–74]. Another advantage of using water-based responsive gels in biomedical applications is that they tend to be biocompatible, as they resemble human tissue due to their high water content and low interfacial tension. Suzuki and Tanaka have proposed the use of a photosensitive gel as an optically controlled sphincter [8, 75].

Several groups are working on the development of responsive gel actuators or "muscles". To date, much of the effort has gone into preparing gels which respond to the appropriate stimulus and into understanding the thermo-dynamics behind this response. Primarily qualitative interpretations have been provided for the kinetic data obtained (e.g. series of force or deformation vs time curves) [41, 71, 74, 76–78]. But Chiarelli and DeRossi have mathematically modeled the dimensional changes of electrically responsive gels, although the parameters had not been independently determined [79]. These mechanical applications are not currently ready for commercial exploitation, though Shiga et al. recently published an interesting demonstration of an electrically respon-sive gel which could lift and carry a quail's egg [80, 81]. Suzuki demonstrated a similar gel "elbow" driven by the addition of water or acetone to the gel [73].

The utility of responsive gels as sensors, independent of actuation, is also being examined. Sheppard et al. developed a conductimetric biosensor in which the bioactive layer is a responsive hydrogel [82]. Osada et al. have proposed the use of polyelectrolyte gels as switching devices with nerve-like properties, based on the repetitive oscillations of electric current generated by these gels under constant electrical potentials [83, 84]. Mechanical devices of a different sort are the light screens and display units being developed by companies based on the opaque/transparent transition which some responsive gels display in response to a given stimulus. For example, the SunTek Corporation has developed "responsive windows" which become opaque when a particular temperature is reached [85, 86]. These materials can be used as self-regulating sunscreens in skylights and greenhouses.

Many of the applications being developed for responsive gels are for controlled drug delivery, as reviewed by Gehrke and Lee, Hoffman, Tanaka and Urry [1, 28, 87, 88]. There are several ways in which these devices might respond to an appropriate stimulus. They could expand like pistons to pump drug from reservoirs, contract around reservoirs to squeeze out drug solutions, or function as membranes whose permeabilities to the drug can be turned on and off as needed [25–28, 32, 33, 57, 88–100]. As one example of such a responsive gel delivery system, consider the insulin delivery device for diabetes treatment proposed by Horbett et al. [13, 101–104]. In this system, a glucose-sensitive responsive gel membrane surrounds an insulin reservoir; at low blood glucose levels, the membrane is contracted and impermeable, blocking any insulin release. Above a critical glucose level, the membrane swells and becomes permeable to the insulin, which then diffuses out; when the glucose level falls back to the normal range, the membrane becomes impermeable again and insulin release stops.

The variable permeabilities of responsive gels can also be used to carry out chemical separations. The gels can be used in fashions analogous to conventional membranes, but will have permeabilities which can be adjusted by applying electric fields or by changing the solution pH or temperature [69, 90, 91, 105–107]. Cussler et al. developed a unique cyclical process driven by responsive gels for dewatering various solutions [29]. This system has been used to concentrate solutions of macromolecules and ionic solutes, and to dewater various slurries and sludges [4, 16, 30, 108, 109].

Applications of responsive gels in biotechnology have also been examined. For example, Dong and Hoffman proposed the use of temperature sensitive hydrogels for the immobilization of enzymes in bioreactors [110]. Enzyme activity can be turned on and off by causing the gel to swell or shrink; when shrunken, the enzyme activity is turned off, since the pores are blocked to the substrate. Different applications are possible for this system besides enhanced bioreactor performance, including bioassays.

2 Overview of Responsive Gel Research at the University of Cincinnati

I began research on responsive polymer gels while a Ph.D. candidate under E.L. Cussler at the University of Minnesota in 1983. My thesis work involved the development of the technique of gel extraction, which can be used to dewater a variety of macromolecular solutions [5, 30, 111]. After completing my Ph.D. and accepting a position as a faculty member at the University of Cincinnati, I continued research on responsive gels, as well as initiating research into the swelling of glassy polymer gels and the use of such materials for the controlled delivery of bioactive solutes [1, 112, 113]. I have also continued work on the development of novel gel-based techniques of biochemical separation and purification [114]. The general philosophy of my research program has been to learn how to control the properties of gels which influence their effectiveness in applications, both current and envisioned. Thus synthesis, swelling, kinetics and permeability have all been studied at UC.

Most of my research at UC has been a collaboration between myself and graduate students. Lii-Hurng Lyu received a M.S. degree for his demonstration of the feasibility of gel extraction as a method for dewatering fine coal slurries [59, 115–117]. His work provided the basis for more fundamental studies on the synthesis and response kinetics of poly(N-isopropylacrylamide) gel carried out by M. Kamal Akhtar and Maria Palasis [45, 118]. Akhtar studied the kinetic response of PNIPAAm gels as a function of gel composition and driving stimulus. He received a M.S. degree for his demonstration of the general utility of the diffusion analysis for the volume change kinetics of responsive gels, plus his use of the polymer scaling concepts of de Gennes to interpret the variations in the swelling and shrinking diffusion coefficients observed [46, 61, 118–121]. Palasis is currently working on her Ph.D. on the topic of "Permeability of Responsive Polymer Gels". To date, she has aided in the development of highly reproducible synthesis techniques for PNIPAAm gels, demonstrated the effectiveness of the Yasuda et al. free volume theory in predicting the relationship between responsive gels swelling and solute diffusivity and uncovered the existence of substantial hydrophobic interactions between responsive gels and drugs [45, 70, 122, 123]. David Harsh received a Ph.D. for his development and studies of a new class of responsive polymer gels based on cellulose ethers. These gels are made from polymers on the "Generally Recognized As Safe" list of the United States Food and Drug Administration, are available as commodity chemicals, and can be biodegradable. Thus they may have utility in food and drug use or large scale application in agriculture or disposable consumer products. Harsh established the connections between the properties of the precursor polymer and the resulting gel on both an experimental and theoretical basis [38, 52, 53, 124, 125]. For his Ph.D., Bhagwati G. Kabra has worked to create responsive polymer gels with enhanced response rates [46, 65, 118, 126]. To do this, he has developed techniques for creating microporous gels which can absorb and desorb solvent by convective, rather than diffusive, processes. His work has involved fundamental studies of the connections between synthesis conditions and microstructure development. He has also worked to establish the connection between the nature of the microstructure and the rate-limiting steps which determine the observed response. Ming-Chien Yang worked with several individuals in the group during a stint as a post-doctoral associate [45, 59, 118, 127]. He wrote many of the programs used to analyze the data and built much of the equipment used to synthesize and characterize the gels. Finally, numerous other graduate and undergraduate students carried out short term research projects for academic credit that aided the thesis work of the people mentioned here.

The goal of this article is to review our work on responsive gels. Rather than describe the work in chronological order, or in a thesis-by-thesis fashion, I have chosen to present the work in a topical order. We begin with a discussion of the synthesis principles of gels used in our lab and then move to an overview of our work on the equilibrium swelling of responsive gels. Next we present a review of our work on kinetics, followed by our permeability studies and applications work.

3 Synthesis of Responsive Gels

Responsive gels can be synthesized by any technique which can be used to create a polymeric network, including copolymerization/crosslinking of monomers, chemical or physical crosslinking of linear polymers and chemical conversion of one gel type to another. Copolymerization/crosslinking, especially the reaction of monovinyl monomers with divinyl monomers, has been the most widely used technique; we have used this technique to produce a variety of copolymer gels using N-isopropylacrylamide, acrylamide, acrylic acid, sodium methacrylate and the divinyl crosslinking monomer N,N'-methylenebisacrylamide. We have also created responsive gels by crosslinking linear polymers with chemical reagents or γ-radiation; specifically, we have produced a variety of responsive gels by crosslinking different cellulose ether polymers with divinyl sulfone and a thermally responsive gel by crosslinking poly(vinyl methyl ether) with γ-radiation. Chemical conversion was used to convert commercially available polyacrylamide gel beads to pH-sensitive responsive gels by basic hydrolysis of amide groups to carboxylic acid groups [30]. In this case, the ultimate composition is determined by the extent of hydrolysis allowed. In this review, we will focus on unique contributions that we have made to the literature in developing novel synthesis techniques and refining synthesis techniques originally developed by other groups.

3.1 Copolymerization/Crosslinking of Monomers

One of the most intensively studied responsive gels has been crosslinked poly (N-isopropylacrylamide) (PNIPAAm). This gel is produced in solution by using a free-radical initiation system to induce the copolymerization of N-isopropyl-acrylamide monomer with a relatively small amount of a crosslinking monomer, typically N,N'-methylenebisacrylamide. This solution polymerization technique was developed by several research groups in the mid-eighties; Hirokawa and Tanaka at the Massachusetts Institute of Technology published the first equilibrium swelling curves of this material in 1984 [22], while Freitas and Cussler at the University of Minnesota developed the material for use in chemical separations [4] and Hoffman et al. at the University of Washington created the material for solute delivery systems [27]. These papers had their roots in the earlier work by Taylor and Cerankowski at Polaroid Corp. [21] and Ilavsky, Hrouz, Ulbrich and Kopecek at the Institute of Macromolecular Chemistry in Prague [128, 129]. Since the mid-eighties, many different research groups the world over have studied this gel.

Monomer purity can be expected to have a significant effect on the properties of the gel synthesized because it has been shown by numerous investigators that copolymerizing NIPAAm monomer with even small amounts of different monomers can alter the transition temperature, swelling degree and sharpness of the volume change at the transition. For example, hydrophilic and ionic comonomers increase the swelling degree and the transition temperature, while hydrophobic comonomers reduce them [21, 130]. Furthermore, even a small percentage of ionic comonomer has been shown to induce discontinuous transitions in gels [130]. Thus hydrophilic, hydrophobic or ionic monomers present as impurities and copolymerizable with NIPAAm would alter the properties of the gel obtained.

In the course of our research on PNIPAAm gel, we purchased numerous lots of N-isopropylacrylamide from both Polysciences and Eastman Kodak; the NIPAAm received ranged from a free-flowing powder to a coarse granular material with large chunks. A typical assay of this material by high performance liquid chromatography (HPLC) indicated the presence of about 5% impurities by weight, including 0.7 wt % acrylamide and 0.5 wt % acrylic acid – enough to noticeably affect swelling properties. All of the impurities could be removed by the following recrystallization procedure [45]. First the monomer was dissolved in water in which the Cl⁻ form of an anion exchange resin was added to help absorb anionic impurities like acrylic acid. The solution was then extracted with a series of chloroform aliquots; the first one or two were usually yellowish and thus were discarded. The chloroform was removed by evaporation in a hood. The dry NIPAAm was then dispersed in warm heptane. Just enough acetone was added to the mixture to dissolve the monomer, after which the solution was cooled to 0 °C. Large needle-like crystals formed and were recovered by filtration and washing with cold heptane and dried in a vacuum desiccator. Yield depends upon how many of the initial chloroform aliquots must be discarded to eliminate discoloration but is typically between 50% and 75%.

We have studied the effects of the synthesis conditions on the resulting gel, which is described in Sect. 4.1.1. Here we will simply outline the procedure developed for the highly reproducible synthesis of transparent, homogeneous NIPAAm gels [45]. To describe the gel formulation used, we use the $T \times C$ notation system originated in the gel electrophoresis literature; T is the mass in grams of the combined monomers added to 100 ml of water and C is the mass of crosslinking monomer used per 100 g of monomers. Except for the NIPAAm monomer, which was purified as described above, all other reagents were reagent grade and used as received. The crosslinker used is N,N'-methylenebisacrylamide and the redox initiation couple is ammonium persulfate and N,N,N',N'-tetramethylethylenediamine (TEMED).

All of the synthesis operations are carried out in a glove box under a nitrogen atmosphere. The monomer and crosslinker are dissolved in distilled water, and the initiators are added to this solution from separate aqueous solutions in amounts which result in final solution concentrations of 30 mg APS and 15 mg TEMED per 100 ml of reaction mixture. All of the solutions are degassed under vacuum prior to mixing, and mixture is further degassed under vacuum while stirring on a magnetic stirrer kept within the glove box. Glass molds (either tubes or plates separated by a silicon rubber gasket) are then filled with solution, sealed and placed in an ice bath within the glove box. Gelation occurs within 1–2 h, depending upon the monomer and crosslinker concentrations. After gelation, the molds are removed from the ice bath and the glove box and put into a refrigerator at approximately 3 °C for at least 24 h to allow the reaction to approach completion. The gels are then removed from the mold and leached free from the sol fraction by immersion in a distilled water bath for several days, during which the water is changed several times. At this point, the gels are used in experiments.

3.2 Chemical Crosslinking of Linear Polymers

Most of the responsive gels which have been studied have been made by simultaneous copolymerization and crosslinking of monomers with one or two vinyl groups. The archetypal responsive gel, poly(acrylamide-*co*-acrylic acid) is a member of this class, as is the PNIPAAm gel discussed in the previous section. However, responsive gels can also be created by crosslinking linear polymers through physical interactions, as in the poly(vinyl alcohol)-poly(acrylic acid) or poly(ethylene glycol)-poly(methacrylic acid) systems [44, 77, 81], through γ-irradiation, as in the poly(vinyl methyl ether) or polypentapeptide systems [41, 46, 72, 88], or by chemical crosslinking, as in the cellulose ether-divinyl sulfone system to be discussed in this section [38, 52, 125].

There are several advantages to forming responsive gels by crosslinking linear precursor polymers over the more common copolymerization technique. These include the simpler synthetic methods, ready availability of a variety of well-characterized linear polymers, ability to form the gel in the presence of a solute with less concern for side reactions plus the potential for easier approval from the United States Food and Drug Administration (FDA) of gels made from approved precursor polymers. The latter two points are particularly important for biomedical and food-processing applications of responsive gels, which have driven much of the work on these gels [28, 88, 131]. Gels which form by physical phenomena such as crystallite formation or hydrophobic interactions can entrap a solute dissolved or dispersed in the solution as the gel forms [132]. Gels formed by chemical reaction typically cannot be loaded with the solute as the gel formation reaction occurs due to the possibility of side reactions between the reactive species and the bioactive solute. Furthermore, most of the vinyl monomers used to create gels by copolymerization/crosslinking are highly toxic, carcinogenic or teratogenic. Although the gels themselves are usually benign, FDA requirements for proof of purity could be difficult to meet. FDA approval could be easier to win for responsive gels made from polymers already on the Generally Recognized As Safe (GRAS) list.

These considerations sparked our investigation into the possibility of producing responsive polymer gels from cellulose ether polymers. Cellulose ethers are alkyl-substituted cellulose derivatives which are on the FDA's GRAS list for use in food and pharmaceutical formulations. Since most cellulose ethers display lower critical solution temperatures (LCST), as gels they should be temperature and solvent sensitive. With the incorporation of ionizable groups, they should become pH-, electric field- and salt-sensitive. Many different types of cellulose ethers – ionic, nonionic, very hydrophobic, very hydrophilic, low molecular weight, very high molecular weight, and so on – are commercially available in many different grades. Since they are available as bulk commodity chemicals, they are relatively inexpensive, and as materials of biological origin, they can also possess favorable biodegradation properties. Their low cost could improve commercial viability in large scale, cost-sensitive applications, and biodegradability could enhance their acceptance in agricultural or disposable applications. Thus the goals of this work were to demonstrate that it was in fact possible to create responsive polymer gels from commercially available cellulose ethers and to relate the properties of the resulting gels to the solubility properties of the precursor polymers. The latter point would enable ready synthesis of responsive gels with the desired properties from polymers of known characteristics.

Gels cannot be made directly from native cellulose (a polymer of anhydroglucose), since it is water-insoluble. However, each glucose unit of the cellulose polymer has 3 hydroxyl groups which may be chemically derivatized. Substitution disrupts the crystallinity of the native cellulose, allowing the polymer to be solubilized by water, unless the substituent groups themselves are significantly hydrophobic. Thus the properties of the derivative are strong functions of both the type of substituent as well as the extent of substitution. The properties of these semi-synthetic polymers have been extensively studied due to their importance in pharmaceutical applications [133–135]. We have examined the synthesis of gels from methylcellulose, hydroxypropylmethylcellulose, hydroxypropylcellulose and carboxymethylcellulose, all crosslinked with divinyl sulfone.

The synthesis of these gels is straightforward, in contrast to the complexities of PNIPAAm gel synthesis discussed earlier. First, the dry polymer is dissolved in pH 12 NaOH solution and is given time to hydrate fully. A typical solution concentration is 10 wt % polymer. After full hydration is achieved, the crosslinker divinyl sulfone (DVS) is thoroughly mixed into the solution (about 10^{-4} mol crosslinker/g polymer). This mixture is then poured into the desired mold; gelation typically occurs in an hour or less. 24 h is typically allowed for reaction, after which the gel is neutralized with HCl and leached free of the sol fraction with distilled water. Ionizable gels can be made by mixing carboxymethyl cellulose with a nonionic cellulose ether, and crosslinking the polymers together with DVS.

3.3 Radiation Crosslinking of Linear Polymers

While chemical crosslinking is the most common means of synthesizing gels from linear polymers, as discussed in the previous section, linear polymers can also be crosslinked by non-chemical means such as irradiation. High energy radiation splits covalent bonds into unpaired radicals which then recombine randomly. Depending upon the relative rates of scission and recombination, an irradiated polymer can either degrade into low molecular weight fragments or wind up crosslinked into a network [3].

In this section, we describe the synthesis of gels made by crosslinking poly(vinyl methyl ether) (PVME) with γ-irradiation. Our interest in studying this material was sparked by the report by Huang et al. of a radiochemically crosslinked PVME gel that appeared to swell and shrink by orders of magnitude faster than any other responsive gel of comparable dimensions [39]. Because of the technological value of increasing the response rate of these gels so dramatically, we decided to characterize these gels as a function of radiation dose and intensity.

PVME was obtained as a 30 wt % solution (average molecular weight approximately 90 000) from the Aldrich Chemical Co. Ferric oxide (Fe_3O_4) was mixed into this solution to increase the thermal conductivity of the resulting gels, as suggested by Huang et al. [39, 108]. Gels can be formed without the addition of the ferric oxide, however. The solutions were degassed and poured into glass molds – either cylinders or glass plates – and sealed. The molds were irradiated with γ-rays emitted by the ^{60}Co source in the College of Engineering at University of Cincinnati. The dose rate was varied from 0.24 to 0.85 Mrad/h, and the total dose was varied from 5.0 to 16.9 Mrad. Solutions irradiated with total doses less than 5.0 Mrad did not gel. While the gels formed were macroscopically homogenous, they normally contained gas bubbles, sometimes as large as 1 mm in diameter. Presumably, these voids were formed by the gases generated by the irradiation. Prior to characterization, the gels were leached by repeatedly swelling and shrinking them in 23 °C and 50 °C water. After several such cycles over a week's time, the gels were characterized, as described in Sect. 4.1.3.

4 Equilibrium Swelling of Responsive Gels

The most important property of responsive gels is their degree of swelling at equilibrium in solvents. The equilibrium swelling degree, or swelling ratio, is usually denoted by the variable Q or q, and is defined as the ratio of the solvent-swollen weight of the gel to the dry weight of the gel. For any given gel, the amount of solvent uptake is dependent upon the chemical nature of the gel and its network structure plus the nature of its environment-solvent composition, pH, temperature and so on. In this section, we will review our studies of the equilibrium swelling properties of different classes of responsive gels as a function of gel type, synthesis conditions, and solvent composition. We will also discuss use of a thermodynamic model to describe the temperature dependence of the swelling degree of temperature-sensitive gels.

4.1 Dependence of Swelling upon Synthesis Variables

The equilibrium degree of swelling of a responsive gel is a strong function of the synthesis variables, particularly chemical composition and crosslink density. Chemical composition can usually be fixed with precision though the choice of reactants. Crosslink density is more difficult to pin down. First of all, crosslink formation is a random process, and many crosslinks do not function as elastically effective junctions; instead, they are incorporated into loops, dangling ends, densely crosslinked clusters, and so on. Specifying the amount of cross-linking achieved is particularly difficult with non-chemical crosslinking means like irradiation, complex or crystallite formation. Finally, the microstructure of responsive gels which is "locked-in" by crosslinking can also be strongly influenced by the reaction conditions; reactions near a phase separation condition can yield a heterogeneous gel while reactions carried out well within the stable single phase region tend to yield gels with relatively homogeneous structures. Thus network properties tend to be significantly influenced by reaction conditions. This fact can lead to significant variability in the properties of the gels unless reaction conditions are specifically defined and carefully controlled.

In this section, we will focus on how synthesis conditions other than chemical composition can affect the equilibrium swelling properties of the resulting gels. However, we will also discuss the swelling properties of cellulose ether gels as a function of their chemical composition, since these are a novel class of responsive gels which we developed. Finally, we will also review our use of the thermodynamic model of Marchetti et al. to correlate the swelling behavior of the cellulose ether gels [50, 51, 53].

4.1.1 Poly(N-isopropylacrylamide) Gels

When we first began to work with this gel in 1988, we found reproducible synthesis of this gel difficult to achieve as the properties were influenced by many subtle variables such as monomer purity, initiator freshness, dissolved oxygen concentration and so on. This sensitivity of PNIPAAm gel properties to details of synthesis procedure is also reflected in the variations in properties of the gel as reported by different investigators. For example, gels with approximately the same nominal compositions had noticeably different swelling degrees and transition temperatures. Furthermore, the temperature-induced volume transition of this gel has been described as both continuous [27, 34, 45, 92] and discontinuous [4, 22]. Thus we undertook a detailed study of the variables which can influence the properties of PNIPAAm gel of given composition, which we will review here [45]. We found that monomer purity, temperature variations of the solution during the polymerization, initiator type and concentration and total reaction time all had significant effects on the properties of the gels obtained.

Reaction conditions such as reaction temperature, initiator concentration, dissolved oxygen concentration and polymerization time strongly affect the properties of the PNIPAAm gel by altering the microstructure and the effective crosslink density of the gel. Microstructure could not be directly measured, but was implied from the degree of transparency of the gel; increases in turbidity indicate increasing heterogeneity or domain sizes at a microscopic level. Effective crosslink density can be calculated from the shear modulus obtained using a uniaxial compression device [38]. Also useful for interpretation of results is the crosslink efficiency, the ratio of the effective crosslink density to the theoretical crosslink density (that which would be obtained if each crosslinking molecule formed an elastically effective crosslink junction).

The facts that the LCST of PNIPAAm is only 32 °C and the polymerization reaction is highly exothermic can also lead to problems with reproducibility of the gel synthesis. If the solution is deliberately warmed above the LCST during polymerization, an opaque gel with a permanently phase separated microstructure is formed [65]. Depending upon the initial solution temperature, initiator concentration and type, and heat transfer characteristics of the gel mold, the solution temperature can rise substantially during polymerization and alter the gel's properties.

We determined that the sodium metabisulfite (SMB)/ammonium persulfate (APS) initiation system induced a much faster temperature rise than did the TEMED/APS initiation system. In fact, the temperature in the SMB/APS system rose as much as 8 °C in ten minutes, with a corresponding increase in turbidity. For solutions initially at room temperature, this is enough to cause phase separation of the polymer solution. This yields gels with greater turbidities and equilibrium swelling degrees (at all temperatures) and lower shear moduli than are obtained with less active initiator systems or lower initial solution temperatures. Incremental increases in SMB/APS initiator concentrations caused swelling degrees to rise incrementally, up to 24%. Decreasing the initial solution temperature from 21 °C to 0 °C, with other variables kept constant, reduced the swelling degree 10–30% while eliminating any turbidity in the gels.

Because of the importance of solution temperature on the resulting gel, the type of mold and its heat transfer characteristics can also affect the properties of PNIPAAm gels. For example, gels made in 4 mm outside diameter cylindrical tubes with 0.1 mm thick walls have lower degrees of swelling at all temperatures than do gels made from the same starting solution but polymerized between 3 mm thick glass plates separated by a 2 mm thick gasket. This appears to be a result of the fact that the cylinders have much better heat transfer characteristics than the sheets and since the glass walls were much thinner for the cylinder than the sheets. Most likely, this led to a greater temperature rise in the sheets than the tubes, which causes an increase in the equilibrium swelling degree at all temperatures, as noted above. This also indicates that gels made by suspension polymerization would also differ in properties from gels made by solution polymerization, even if the same initial solution was used.

Polymerization time affects the extent of reaction and thus the properties of the gels. Procedures which have been published for PNIPAAm gels allow vastly different polymerization times; some quench the reaction shortly after gelation, while others allow the reaction to continue for extended periods of time and approach completion of reaction. We determined that the greater the concentration of the monomer and/or crosslinker in the solution, the faster the gelation. However, if the reaction is quenched at the gel point, all gels have approximately the same shear moduli and swelling degrees, indicating that at the gel point, comparable amounts of network are formed, regardless of the starting composition (Table 1a). However, significant property differences develop if the reaction is allowed to continue well past the gel point (Table 1b). For dilute, lightly crosslinked gels, the properties are little changed with the additional reaction time. However, for more concentrated solutions with larger proportions of crosslinker, the properties change significantly with increased reaction time: the shear modulus can rise an order of magnitude and the equilibrium swelling degree can be cut in half. Then swelling degree is observed to fall and the

Table 1a. Properties of PNIPAAm gels at 25 °C when the polymerization reaction was quenched shortly after the gel point. Reprinted from Polymer International (1992) 28: 29 by permission of the copyright holder, the Society of Chemical Industry [45]

Gel type	Approximate gelation time (min)	Shear modulus $(g\,cm^{-2})$	Equilibrium degree of swelling	Effective crosslinking density $(mol\,cm^{-3})$	Crosslinking efficiency (%)
10×1	60	30 ± 5	25.7	3.6×10^{-5}	22
16×1	60	44 ± 5	19.7	3.6×10^{-5}	23
10×4	45	28 ± 4	18.2	3.0×10^{-5}	5
16×4	40	25 ± 4	17.6	2.0×10^{-5}	3

Table 1b. Properties of PNIPAAm and poly(NIPAAm-co-sodium acrylate) gels at 25 °C when the polymerization reaction approached completion (24 h). Note: the percentage indicated in the gel type notation is the molar substitution of sodium acrylate for NIPAAm in the pre-gel solution. Reprinted from Polymer International (1992) 28: 29 by permission of the copyright holder, the Society of Chemical Industry [45]

Gel type	Shear modulus $(g\,cm^{-2})$	Equilibrium degree of swelling	Effective crosslinking density $(mol\,cm^{-3})$	Crosslinking efficiency (%)
10×1	33 ± 5	19.6	3.6×10^{-5}	23
16×1	87 ± 6	14.8	6.5×10^{-5}	39
16×1–3%	69 ± 5	20.2	5.7×10^{-5}	40
10×4	124 ± 7	11.1	1.1×10^{-4}	18
10×4–1%	115 ± 7	12.2	1.1×10^{-4}	19
10×4–3%	106 ± 7	15.4	1.1×10^{-4}	19
16×4	220 ± 5	7.8	1.3×10^{-4}	20
16×4–3%	165 ± 6	12.3	1.2×10^{-4}	18

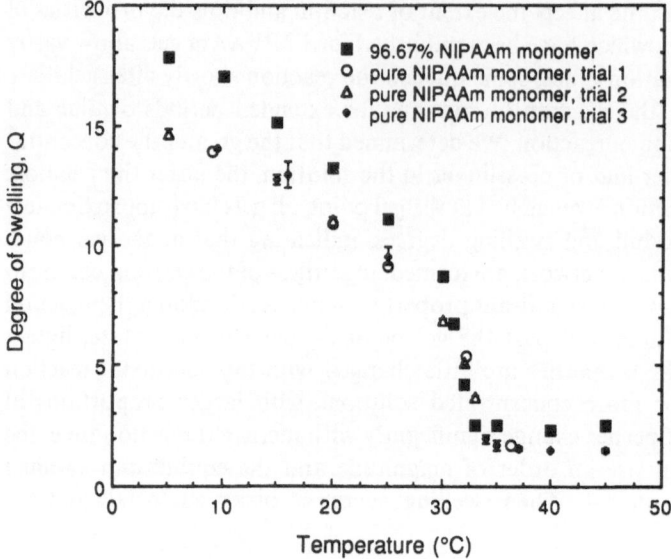

Fig. 1. The equilibrium swelling degree of different batches of NIPAAm gels as a function of temperature. The use of identical molds and controlled synthesis conditions allows good reproducibility of swelling properties. Thicker glass molds resulted in PNIPAAm gels with greater swelling degrees at all temperatures. Legend: ■ 3 mm glass plates with a 2 mm spacing; ○ 0.1 mm thick glass tubes with 4 mm O.D., Trial 1; △ 0.1 mm thick glass tubes with 4 mm O.D., Trial 2; ◇ 0.1 mm thick glass tubes with 4 mm O.D., Trial 3. Error bars on Trials 1–3 are the' standard deviations of triplicates (most error bars are smaller than the symbols). Reprinted from Polymer International (1992) 28: 29 by permission of the copyright holder, the Society of Chemical Industry [45]

modulus to rise with increasing monomer concentration and crosslinking percentage, as intuitively expected.

Because different investigators invariably choose somewhat different reaction conditions and use monomers of varying purity, it is not surprising that a variety of swelling characteristics have been published for PNIPAAm gel. However, with careful control and precise specification of reaction conditions, good reproducibility in gel properties can be achieved, as shown in Fig. 1. The swelling degrees of several different batches of PNIPAAm gel at different temperatures vary by no more than 4.7%, and in most cases the variability is much less. The figure also demonstrates that using different molds to noticeable differences in the equilibrium swelling curve of the gel (see Ref. [45] for further details).

4.1.2 Cellulose Ether Gels

Polymer type, crosslink density and ionic content were varied to determine their influence on the swelling degree of the cellulose ether gels as a function of

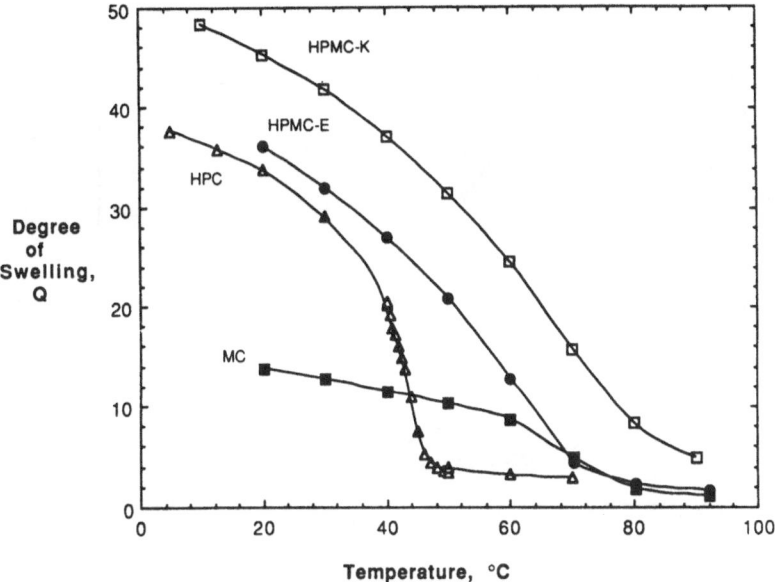

Fig. 2. Swelling behavior of cellulose ether gels synthesized from different polymer types. The gel transition temperature is near the linear polymer LCST; a lower LCST indicates a more hydrophobic polymer. At comparable crosslink densities, this means a reduction in swelling degree (HPC is less crosslinked than the other gels). Lines added are to guide the eye. Reprinted from the Journal of Controlled Release (1991) 17: 175 by permission of the publishers, Elsevier Science Publishers BV [52]

temperature. We were particularly interested in determining the relationship between the LCST of the precursor linear polymer and the transition temperature of the gel as well as the sharpness of transition. As described in the original paper on this topic, the transition temperature was defined as the inflection point on a plot of the swelling degree as a function of temperature, while the sharpness of transition was the slope at this point, in terms of percent of total volume change per degree Celsius. This can be determined either graphically [52] or from a numerical curve fit of the data [125]; the difference between methods is minor.

Figure 2 illustrates the temperature dependence of the swelling degree as a function of precursor polymer type. Methylcellulose (MC), hydroxypropyl-methylcellulose, type E (HPMC-E) and hydroxypropylmethylcellulose, type K (HPMC-K) gels have comparable effective crosslink densities of about 2×10^{-5} mol/cm^3 (as determined from uniaxial compression testing), while the crosslink density of the hydroxypropylcellulose (HPC) gel is about half this [52]. The transition temperature for each gel is within several degrees of the precursor polymer lower critical solution temperature (LCST), except for the MC gel, which has a transition temperature 9 °C higher than the LCST. The sharpness of the transition was about 3%/°C, except for the HPC gel transition, which was much sharper – about 8%/°C.

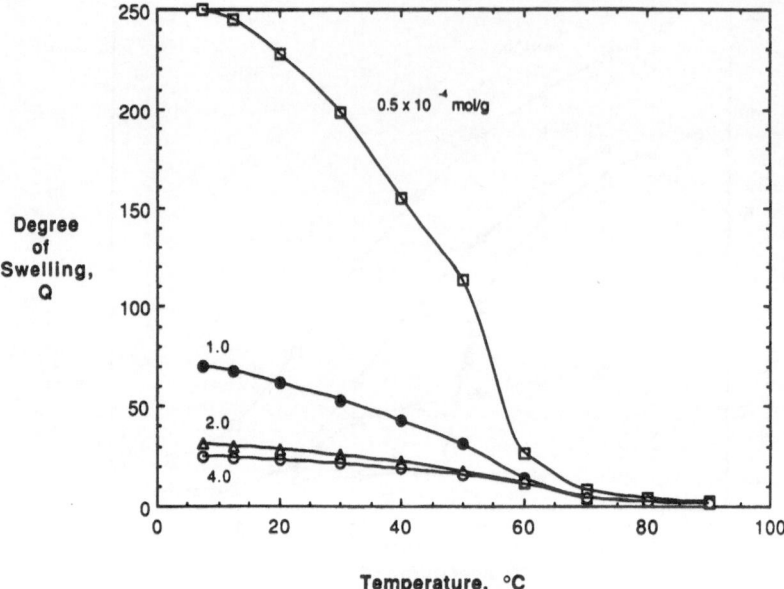

Temperature, °C

Fig. 3. Effect of varied crosslink density on swelling of HPMC-E gels. Increased crosslinking decreases swelling below the transition temperature but has little effect on gel transition temperature, sharpness of transition, or degree of swelling above the transition. Lines added are to guide the eye. Reprinted from the Journal of Controlled Release (1991) 17: 175 by permission of the publishers, Elsevier Science Publishers BV [52]

The dependence of the swelling degree on crosslinker dosage for HPMC-E gels is shown in Fig. 3. As is expected, increasing the crosslinker dosage decreases the swelling degree of the gels, especially since the crosslinking efficiency rises from 5% for the lowest dosage to 12% at the higher dosages. The increase in crosslinking causes, at best, a slight increase in the transition temperature, but has no effect on the sharpness of transition; about 3%/°C for all gels.

Crosslinking a solution of a nonionic cellulose ether and carboxymethyl cellulose (CMC) produced polyelectrolyte gels. Figure 4 shows that the incorporating CMC into an HPMC-E gel increases the swelling degree at all temperatures, due to the increased ionic swelling pressure [52]. While each incremental increase in ionic content increases the transition temperature several degrees, the sharpness of the transition is virtually constant.

To summarize the experimental work on cellulose ether gels, we have examined the effects of polymer type, crosslink density and ionic content on the degree of swelling as a function of temperature, the volume transition temperature and the sharpness of the volume transition. The swelling degree can be increased by using more hydrophilic polymers, reducing the crosslink density, or by increasing the ionic content of the polymer. Ionic content increases the swelling degree at all temperatures, while hydrophilicity and crosslink density

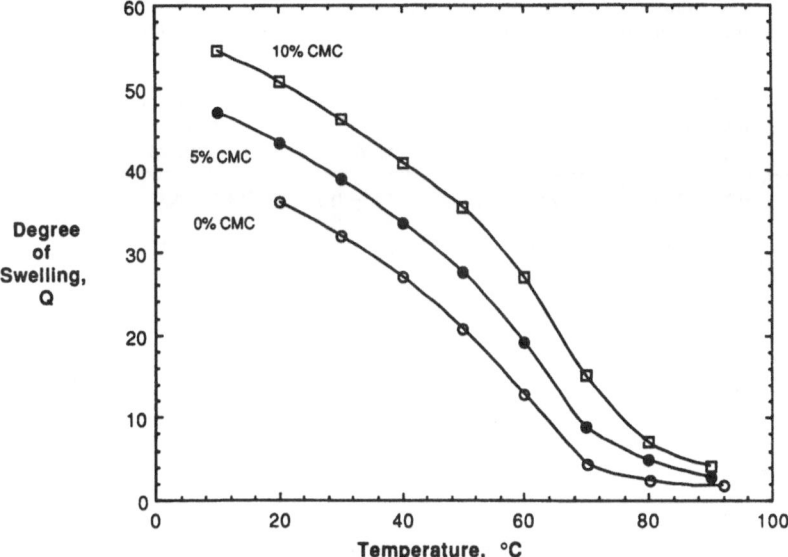

Fig. 4. Swelling of HPMC-E/CMC gels, showing the effects of added ionic content. Ionic content increases the degree of swelling at all temperatures; the transition temperature is increased slightly as ionic content increases but the sharpness of transition is little changed. Lines added are to guide the eye. Reprinted from the Journal of Controlled Release (1991) 17: 175 by permission of publishers, Elsevier Science Publishers BV [52]

have relatively little effect on swelling above the transition temperature. The transition temperature is usually about that of the linear precursor polymer, although it rises a few degrees Celsius with substantial increases in crosslink density and ionic content. The sharpness of the volume transition depends primarily upon the precursor polymer and is not significantly influenced by any other variable; HPC has a significantly sharper transition than the other gels.

The results of this experimental work clearly demonstrate the success of the concept of using cellulose ethers to produce responsive gels. Explanations for the details of the observed behavior is more difficult, though in the original manuscript describing this work, we provided qualitative interpretations of the observed trends [52]. We also examined the results in light of the thermodynamic model of Marchetti et al. [53]. In Sect. 4.2, we will summarize the modeling results.

4.1.3 Poly(vinyl methyl ether) Gels

The equilibrium swelling degrees of PVME gels were determined by a simple gravimetric "blot and weigh" technique, after equilibration in distilled water kept to $\pm 0.1\,°C$ in constant temperature water bath. The swelling degrees at $23\,°C$ as a function of radiation dose and intensity are given in Table 2. This

Table 2. Synthesis conditions and equilibrium swelling degrees of PVME gels made by γ-irradiation of aqueous solutions containing 30 wt % PVME and 15 wt % ferric oxide. Reprinted from Polymer (1991) 33: 990 by permission of the publishers, Butterworth Heinemann [46]

Trial number	Radiation intensity, Mrad h^{-1}	Radiation time, h	Total dose, Mrad	Swelling degree at 23 °C, (swollen wt) · (dry wt)$^{-1}$
1	0.24 ± 0.01	24.6	5.8 ± 0.2	13.3 ± 0.3
2	0.24 ± 0.01	34.2	8.2 ± 0.4	5.7 ± 0.1
3	0.24 ± 0.01	49.8	12 ± 0.5	5.4 ± 0.1
4	0.85 ± 0.115	13.0	11.1 ± 1.5	5.4 ± 0.1

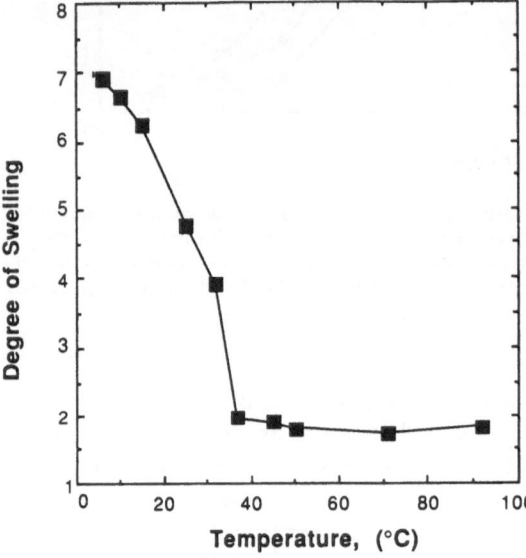

Fig. 5. Equilibrium degree of swelling of radiation-crosslinked PVME gel (Trial 4) at different temperatures. The line added is to guide the eye. Reprinted from Polymer (1991) 33: 990 by permission of the publishers, Butterworth Heinemann [46]

table shows that as the radiation dose is increased at constant intensity, the swelling degree decreases. This observation is consistent with the concept of increased crosslink density with increased dose. Intensity appeared to have little effect on the swelling degrees, however, as gels irradiated at different intensities for the same total dose had the same swelling properties at 23 °C. The entire swelling curve with temperature for trial 4 gels is shown in Fig. 5. The shape of the curve is qualitatively similar to that of the PNIPAAm and cellulose ether gels discussed earlier; the volume transition occurs at 37 °C and is sharp but continuous.

Thus in terms of equilibrium swelling, radiation crosslinked PVME appears to follow the same trends as the gels made of other polymers by other techniques. However, the kinetic properties of these gels could be quite different, as will be discussed in Sect. 5.4.1.

4.2 Thermodynamic Modeling of Equilibrium Swelling

Development of an application for a responsive gel will require control of the swelling degree as a function of the stimulus level, plus the value of the stimulus at which the volume transition occurs. Transitions due to changes in ionization and ionic strength have been treated by several authors with excellent qualitative success, but less quantitative success [7, 38, 47, 48, 125, 136, 137]. Modeling gel systems which display lower critical solution temperatures is more difficult due to the inability of classical Flory–Huggins polymer–solvent interaction theory to treat this phenomenon and the system-specificity of the interactions which drive many of these transitions. Nonetheless a variety of theories are being developed to model this class of gels [49–51, 54]. We chose to modify the model developed by Marchetti et al. to fit the swelling curves obtained for cellulose ether gels [50, 51]. We used their expressions for polymer–solvent interactions unchanged, but modified the elastic term to account for non-Gaussian behavior. We also added a term to account for network ionization.

The model of Marchetti et al. is based on the compressible lattice theory which Sanchez and Lacombe developed to apply to polymer–solvent systems which have variable levels of free volume [138–141]. This theory is a ternary version of classic Flory–Huggins theory, with the third component in the polymer–solvent system being vacant lattice sites or "holes". The key parameters in this theory which affect the polymer–solvent phase diagram are:

P_g^* = polymer cohesive energy density;
P_s^* = solvent cohesive energy density;
Z_{sg} = correction for deviation from a geometric mean mixing rule.

In general, the closer the values of P_g^* and P_s^*, the better the quality of the solvent for the polymer. Theoretically, narrowing the difference between these values causes an increase in the swelling degree and a rise in the transition temperature of the gel. When the other parameter, Z_{sg}, is negative, the polymer–solvent interactions are more favorable than expected from the geometric mean mixing rule; in short, the more negative the value of Z_{sg}, the better the quality of the solvent for the polymer. For responsive hydrogels it is helpful to recast these concepts in the more familiar terms of hydrophilicity and hydrophobicity [21]. The value of P_s^* for water is a constant at 642.2 cal/cm^3; P_g^* values for polymers are generally much less than this. Thus increasing P_g^* for a polymer is equivalent to increasing the hydrophilicity. Of course Z_{sg} is also a significant parameter, and is not directly related to the cohesive energy densities. If modifying a polymer enhanced specific interactions between the polymer and the solvent (such as hydrogen bonding), this would cause a more negative value of Z_{sg}. Although all three of these parameters can be obtained from independent measurements, the theory proves to be highly sensitive to the specific values of each parameter. Small variations (< 1%) in these parameters cause substantial shifts in the predictions of the degree of swelling, transition temperature and

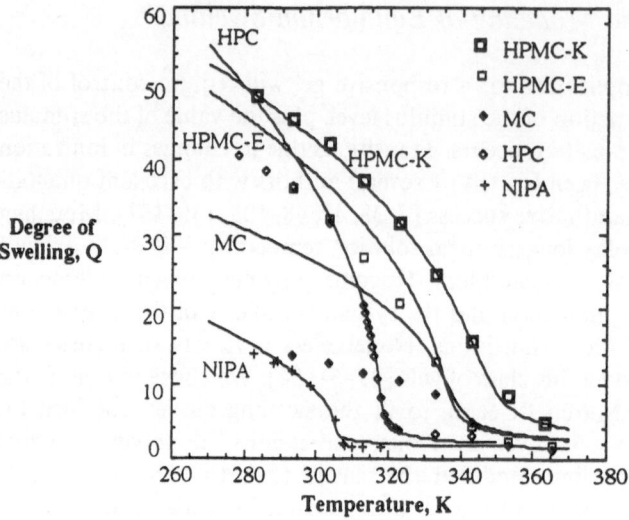

Fig. 6. Least squares fit of the modified theory of Marchetti et al. to swelling data for gels of different polymer types. Values of fit parameters are given in Table 3. Reprinted with permission from [53]. Copyright [1993] American Chemical Society

sharpness of transition [53, 125]. Thus, it is difficult to move beyond qualitative interpretations of these parameters.

To aid our interpretation of cellulose ether gel swelling, however, we fit the theory to our swelling data for the cellulose ether gels and one type of PNIPAAm gel. We treated the values of P_g^* and Z_{sg} as adjustable parameters and looked to see if the trends were consistent with the qualitative concepts outlined in the preceding paragraph. In general, the model could not fit the entire swelling curve accurately. However, at high swelling degrees (over 10), the polymer–solvent interactions are very strong and weaknesses in the elastic expression dominate the swelling predictions (Use of a non-Gaussian elastic term improves the curve fit for $Q > 10$, but does not improve the fit in the region of the transition. But its use introduces a new parameter, N, which is the number of polymer structural units in an effective chain segment). In contrast, the solubility theory has the dominant influence at lower degrees of swelling and in the vicinity of the volume transition. Since we are primarily interested in the volume transition, we optimized our curve fit in the region of the sharpest transition, as described in references 53 and 125. After optimizing the fit of P_g^* and Z_{sg} using the experimentally measured values of effective crosslinking density, we re-optimized the fit treating crosslink density as an adjustable parameter. In general, allowing the crosslink density to function as an adjustable parameter did not cause much shift in its value.

The results of this work for gels made from different cellulose ethers at comparable levels of crosslinking are given in Fig. 6. The parameters obtained from the curve fit are given in Table 3. The quality of the curve fit depends

Table 3. Parameters from the modified Marchetti et al. model, extracted from the curve fits for various temperature-sensitive gels as shown in Fig. 6. Reprinted with permission from [53]. Copyright [1993] American Chemical Society

Polymer gel	Polymer cohesive energy density P_g^* $(cal\,cm^{-3})$	Deviation from geometric mean mixing rule parameter Z_{sg}	Non-Gaussian elasticity parameter, N	Curve fit crosslink density $(10^5\,mol\,cm^{-3})$	Experimental crosslink density $(10^5\,mol\,cm^{-3})$
PNIPAAm	306.7	− 0.0648	30	12.0	13.0
HPC	314.0	− 0.0598	8	0.97	1.1
HPMC-E	322.0	− 0.0547	6	2.9	2.5
HPMC-K	324.8	− 0.0531	4	5.1	2.0
MC	319.9	− 0.0562	3	1.5	2.1

significantly upon the type of polymer and ranges from excellent to poor. The gels based on HPMC-K and PNIPAAm are well-correlated over the entire temperature range, HPC is fit very well except at very high swelling degrees, HPMC-E data are fit well only at low swelling degrees near the transition, while the methylcellulose gel could not be fit well at all. The reasons for the disparity in the curve fit quality are unclear. It may be that the least hydrophilic polymers like MC do not hydrate well thus do not crosslink uniformly, leading to the formation of highly non-ideal networks.

The parameters obtained from the fit are consistent with theoretical expectations. Based on the LCST's of the linear precursor polymers; we would order these polymers in terms of increasing hydrophilicity as PNIPAAm > HPC > MC ≈ HPMC-E > HPMC-K [50, 125]. This is the same order in which the values of P_g^* are ordered. We also see that the sharper volume transitions correspond to more negative values of Z_{sg} (which is the theoretical prediction [53, 125]), which occur for the more hydrophobic polymers. This makes physical sense from the standpoint that for the more hydrophobic polymers to swell substantially, as observed, they must have significant favorable interactions with the solvent; in other words, Z_{sg} must be significantly negative (hydrogen bonding is the important specific interaction known to exist between these polymers and water).

5 Swelling and Shrinking Kinetics of Responsive Gels

Applications being developed for responsive gels include drug delivery systems, novel separation systems, artificial muscles and the like, switches and sensors. The rate of response to the environmental changes may directly influence the system performance, as in switches and sensors, or indirectly, as in the cases of recyclable absorbents, where cycling times must be minimized to ensure the economic feasibility of the application. Much of the research on responsive gels has been to demonstrate the possibilities of applications of the phase transition phenomenon. Thus the bulk of the research has been concerned with gel synthesis, thermodynamics and application development, which establish the potential of these materials. However, kinetics may well determine which applications can move from the lab to the marketplace. Thus we have invested significant effort into determining how to correlate swelling and shrinking data, what factors determine the rate of volume change, and how to maximize the response rate.

5.1 Rate Limiting Steps

The rates of swelling or shrinking can be influenced by a number of different processes, including stimulus rate, diffusion and polymer relaxation. If a single step is rate limiting, successfully modeling that step alone will be sufficient to correlate data and predict rates. If the different phenomena occur at comparable rates, their influences will be coupled and modeling will be more complex. Thus we first sought to identify the relative rates of the different kinetic phenomena, to determine the conditions under which each can become the rate-limiting step, and to model those systems which have a single rate-influencing step.

Before a responsive gel can respond to a stimulus, that stimulus must permeate the gel itself. For example, a temperature-sensitive gel will not undergo a volume transition in response to a change in its environmental temperature until the temperature within the gel has changed. In other words, the rate of heat transfer may influence the swelling or shrinking rate of such a gel. Similarly, a pH-sensitive gel cannot respond to a change in the solution pH until an ion exchange process causes a shift of the pH within the gel itself. Once the stimulus has altered the gel, then the mass transfer process which allows the movement of solvent into or out of the gel will occur. This could involve the mutual diffusion of the polymer network and the solvent or the rate of the polymer network relaxation to a more expanded or contracted configuration. Since we have worked with pH-, temperature- and solvent-sensitive systems, we have determined what conditions could cause the rates of ion exchange, heat transfer or solvent mass transfer to influence the rates of gel swelling or shrinking. We have also studied the volume change kinetics of different gels under conditions when these stimuli are not rate influencing.

5.1.1 Rates of Heat Transfer

Increasing the temperature of the solution surrounding a thermally responsive gel based on a polymer with a LCST in the solution will cause it to shrink, while decreasing the temperature will cause it to swell. The rate limiting step for this could be either heat transfer or mass transfer. Since convection does not occur within conventional non-porous gels (microporous gels will be discussed in

Sect. 5.4), the heat transfer process can be modeled using classical unsteady state heat conduction theory [142–144]. From the mathematical solutions to heat conduction problems, a thermal diffusivity can be extracted from measurements of temperatures vs. time at a position inside a gel sample of well-defined geometry.

To identify the rate limiting steps, we carried out a series of parallel experiments; one in which we determined the rate of thermal equilibration of cylindrical temperature-sensitive gels in response to different temperature driving forces and the other in which we determined the rate of swelling or shrinking of identical gel samples under identical conditions.

The thermal experiment began by inserting a K-type thermocouple along the centerline of a 1.5 cm diameter 16×1 poly(N-isopropylacrylamide) gel. The gel was then equilibrated in a water bath at a given temperature. When the sample reached thermal and swelling equilibrium, the gel was immersed into another temperature bath kept at a different temperature. The change in the temperature measured at the centerline was recorded until thermal equilibrium was reestablished. An identical experiment was also run, but in this case, the mass was measured by a simple "blot-and-weigh" procedure. Since this involved removal of the sample from the constant temperature bath, this might not have been a valid experiment, except that it was soon noted that the rate of heat transfer was very much faster than the rate of mass transfer, as seen in Fig. 7 (note that T_t/T_∞ is the normalized approach to the equilibrium centerline

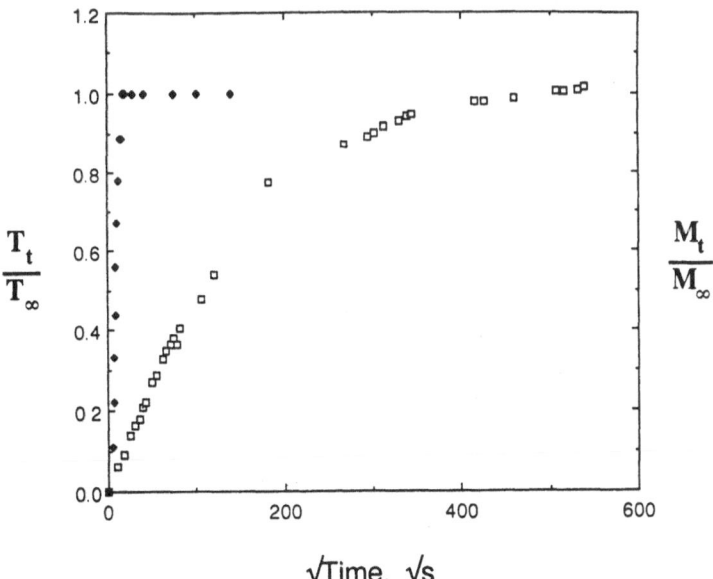

Fig. 7. Heat transfer occurs much faster than mass transfer in temperature-sensitive gels. This example compares the rate of temperature change with the rate of shrinking for a 16×1 PNIPAAm gel taken from 27 °C to 35 °C [59]

temperature, and is defined as the ratio of $(T - T_i)$ to $(T_e - T_i)$, where T is the temperature at a given time, T_e is the equilibrium temperature and T_i is the initial temperature of the gel. M_t/M_∞ is the normalized approach to equilibrium mass, and is defined analogously to T_t/T_∞. The plot is made against the square root of time to linearize the short time data and to compress the data at long times).

Thus Fig. 7 shows that thermal equilibrium is reached well before any significant change in the mass of the gel. Furthermore, the experimental time-temperature profiles match the theoretical conduction equations quite well. The thermal diffusivities obtained increased gradually with temperature from 2.3 $\times 10^{-3}$ cm^2/s at 35 °C to 3.0×10^{-3} cm^2/s at 47 °C [116,117]. This is somewhat greater than the thermal diffusivity of water, which is 1.6×10^{-3} cm^2/s over this temperature range. Thus the thermal diffusivities are about four orders of magnitude greater than the mass diffusivities of these gels (note that in these experiments, the gels do not swell and shrink by purely diffusive processes, since the transition temperature is crossed; see the discussion in Sect. 5.2).

Although the values of these diffusivities depend somewhat upon the type of gel and the specifics of the temperature interval, it can be safely assumed that the thermal diffusivity will always be so much greater than the mass diffusivity of nonporous gels that heat transfer effects can be neglected for such temperature-sensitive gels. The exceptions to this statement are microporous gels in which mass transfer occurs by a convective process rather than a usual diffusive process. The discussion of microporous gels is given in Sect. 5.4.

5.1.2 Rates of Ion Exchange

Gels which are sensitive to changes in the pH of a solution usually have ionizable groups, typically carboxylic acid or basic aminoalkyl moieties. When these groups become ionized, a substantial osmotic swelling pressure is generated inside the gel, causing the swelling degree of the gel to increase. If ionized groups are converted to nonionic forms, this swelling pressure is lost and the gel shrinks. The process of ionization and deionization is an ion exchange process, and thus the rates of ion exchange may influence the overall swelling or shrinking kinetics. For example, a gel with carboxylic groups in the un-dissociated hydrogen form (RCOOH) will become ionized upon reaction with a base such as sodium hydroxide, converting the carboxylic group to the disso-ciated sodium carboxylate form (RCOO$^-$ Na$^+$). Thus, upon ionization, the swelling pressure in the gel increases, and it will absorb water to dilute the increase in sodium ion concentration within the gel, since the sodium ions cannot leave the gel due to electroneutrality constraints. The overall rate of swelling of the gel will be determined by the relative rates of sodium hydroxide transport into the gel and the rate at which the network can reconfigure to allow the influx of additional water into the gel.

The goal of this section is to examine under what circumstances ion exchange can influence or control the kinetics of volume change in a pH-sensitive gel. The relevant principles of ion exchange kinetics were developed by Helfferich in the 1960s [145, 146], and these were applied to pH-sensitive gels by Gehrke and Cussler [5]. Gehrke and Cussler showed that under some circumstances (such as swelling in dilute, unstirred base), ion exchange kinetics dominate the volume change kinetics, while under others they have little influence (such as shrinking in acid) [5]. The data and explanations of this phenomenon are provided by this book in the chapter by Wang, Burban and Cussler. Rather than repeat this discussion, we will review some of our recent work at Cincinnati which quantifies certain aspects of ion exchange kinetics in responsive gels [127].

Since the ion exchange reactions themselves are very fast, the rate limiting steps become either diffusion of ions across the unstirred boundary film surrounding the gel ("film diffusion") or within the gel itself ("particle diffusion"). Film diffusion becomes dominant when the gel is highly swollen (minimizing particle diffusion) with a high concentration of ionizable groups relative to the solution concentration (depleting ions in the boundary film) in a poorly agitated solution (increasing the boundary layer thickness). Particle diffusion becomes the rate-limiting step under the opposite extremes: low swelling degrees (so that diffusion of ions within the gel is retarded), low concentrations of ionizable groups within the gel relative to the solution (so that the concentration gradient in the boundary layer is minimal) in a well-stirred solution (minimizing boundary layer thickness).

Typical pH-sensitive gels are weak acids or weak bases, and the ionization or neutralization of the acidic or basic groups involves a reaction which consumes a pair of counterions, such as the reaction of a hydrogen ion with a hydroxide ion to form water. It has been demonstrated that in analogous ion exchange resins, a sharp reaction front develops that moves toward the center of the particle with time [147]. These moving ion exchange fronts can be modeled with a shrinking core model similar to that used to analyze fluid-solid reactions [127, 147, 148].

For transparent pH-sensitive gels, this ion exchange front can be visually monitored as a sharp refractive index gradient. We modified the shrinking core model of Dana and Wheelock to apply to a cylindrical pH-sensitive gel in the acid form being converted to the ionized sodium form [127, 147]. The model includes both film and particle diffusion resistances, but neglects the volume change. This does not introduce serious error in the model, since the total change in the diffusion path length is relatively small (since dimension scales only with the cube root of the volume change) and since the concentration gradient is shallow in the portion of the gel which is ionized and undergoing volume change (behind the front). The model does not contain any parameters which cannot be independently determined or estimated.

In this experiment, crosslinked poly(acrylamide-co-sodium methacrylate) (0.5–5.0 mol% sodium methacrylate) gel cylinders of different radii in the unionized hydrogen form at pH 4 are immersed in an unstirred solution of pH 12 sodium hydroxide solution. The moving fronts, clearly observable within the gels, are monitored through a microscope with a calibrated eyepiece. A least squares fit of the theory to the data for one set of gels is given in Fig. 8; the curve fit quality is excellent when the mass transfer coefficient in the boundary film (k) and the diffusion coefficient (D) in the gel of the sodium hydroxide are treated as adjustable parameters. From the curve fits, good internal consistency of the values of D and k is obtained. To use the model in a predictive capacity (as opposed to correlative use as in Fig. 8), well-defined flow fields past the gel particle must exist so that accurate estimates of the mass transfer coefficient can be made using literature correlations.

The main point of this discussion is that while ion exchange kinetics can influence, even control, the rate at which pH-sensitive gels swell and shrink, under other conditions ion exchange kinetics can be ignored. But the theory of ion exchange kinetics has been developed to the point that qualitative predictions can be made about which systems are likely to be limited by ion exchange kinetics, and which are not. Furthermore, kinetic ion exchange phenomena can be modeled with good results, as seen in Fig. 8. More extensive modeling of ion

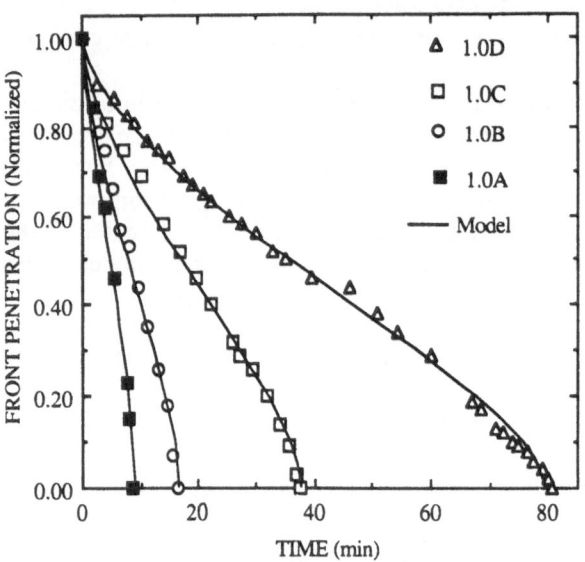

Fig. 8. Examples of a least squares fit of a shrinking core model to measurements of moving ion exchange fronts which develop within the acidic form of poly(acrylamide-*co*-sodium methacrylate) gel cylinders of different radii when immersed in pH 12 NaOH solution. Legend: All gels were made from a solution containing 1 mol % of solution methacrylate; $A = 1.8$ mm radius; $B = 2.8$ mm; $C = 4.2$ mm; $D = 6.4$ mm. Reprinted with permission from [127]. Copyright [1992] American Chemical Society

exchange phenomena and its effect on the swelling and shrinking kinetics of pH-sensitive gels can be found in the work of Grimshaw et al. [57, 90, 91].

5.1.3 Rates of Solvent Transport

Immersing a gel in a solvent with a different composition that the one with which it had reached equilibrium will cause the gel to swell or shrink, depending upon whether the new environment is a better or poorer solvent for the polymer. As with ion exchange and heat transfer, however, the gel volume will not change until the solvent composition within the gel changes. The question is whether the gel volume can change as quickly as the solvent molecules can exchange between the interior of the gel and the solution surrounding the gel. The distinction made here is between the movement of individual solvent molecules and the movement of the network through the solvent. The expansion or contraction of the network to accommodate a macroscopic volume change requires a net displacement of the polymer chains through the solvent; this process will be defined by a mutual or cooperative diffusion coefficient, as discussed in the following section. However, solvent molecules can also diffuse into and out of the network independently of any volume change; this process would be akin to the diffusion of a dissolved solute into or out of a network (as discussed in Sect. 6).

To clearly distinguish between these two modes of solvent penetration of the gel, we immersed poly(acrylamide-co-sodium methacrylate) gels swollen with water and equilibrated with either pH 4.0 HCl or pH 9.2 NaOH solution into limited volumes of solutions of 10 wt % deuterium oxide (D_2O) in water at the same pHs. By measuring the decline in density of the solution with time using a densitometer, we extracted the diffusion coefficient of D_2O into the gel using a least squares curve fit of the exact solution for this diffusion problem to the data [121, 149]. The curve fit in each case was excellent, and the diffusion coefficients obtained were $2.3 \times 10^{-5} cm^2/s$ into the ionized pH 9.2 gel and $2.4 \times 10^{-5} cm^2/s$ into the nonionized pH 4.0 gel. These compare favorably with the self diffusion coefficient of D_2O, which is $2.6 \times 10^{-5} cm^2/s$, since the presence of the polymer can be expected to reduce the diffusion coefficient about 10% in these cases [150]. In short, these experiments show that individual solvent molecules can rapidly redistribute between the solution and the gel by a Fickian diffusion process with diffusion coefficients slightly less than in the free solution.

In the next section, we will see that the volume change kinetics of the gels themselves typically occur with diffusion coefficients on the order of $10^{-7} cm^2/s$. Thus the rate at which a volume change-inducing solvent would penetrate a gel is likely to occur so much faster than the volume change that solvent diffusion would influence the kinetics little. This should hold true not just for solvents, but for any dissolved low molecular weight solute, such as a salt, which induces a volume change simply by diffusing into the gel, since diffusion coefficients of such solutes are more than an order of magnitude greater than the network diffusion coefficients.

5.2 Diffusion Analyses of Swelling and Shrinking Kinetics

In the previous section, we saw that the rate of stimulus change is not often a rate-limiting step. Although there are exceptions to this observation due to specific interactions between volume change-inducing solutes and gels which occur at rates slower than simple diffusion (such as certain ion exchange processes), in most cases, the rate of network motion is generally rate-limiting. This is true because volume-change inducing solutes have diffusion coefficients in gels one to two orders of magnitude greater, and the thermal diffusivity of the gel is four orders of magnitude greater, than network diffusion coefficients. Thus it is the rate of network motion that will typically determine the rate of gel volume change, and this is the subject of this section.

In the preceding discussion we have already stated that the rate of volume change can often be correlated with a diffusion coefficient characteristic of the network's motion in the solvent. There are two ways of obtaining such a diffusion coefficient from kinetic volume change data. One is to analyze the volume change using Fick's Law in a polymer-fixed reference frame for the particular geometry of the gel [5, 46, 61, 111, 112, 118–121, 149]. The other approach is to use the equations of motion to describe the changes in dimension as a function of time [60, 62–64].

In Fick's law, the diffusion coefficient is a proportionality constant between flux and concentration gradient. Thus developing theoretical insight into the significance of the values of the diffusion coefficients obtained from different experiments can be difficult. In contrast, a more comprehensive theoretical background has been developed for the diffusion coefficients obtained from the perspective of the equations of motion. Tanaka et al. defined the diffusion coefficient of the gel network as the ratio of the longitudinal modulus of the gel to the friction coefficient between the gel network and the fluid [60, 151]. While the friction coefficient is a proportionality constant comparable in nature to the Fickian diffusion coefficient, this approach does have the advantage of dove-tailing neatly into the framework developed to describe the fluctuations of polymer chains in solution around their equilibrium configurations, as measured by quasi-elastic light scattering. Tanaka has explicitly shown that the "microscopic" diffusion coefficients obtained from light scattering experiments match the values obtained from macroscopic swelling experiments [60, 151]. Since de Gennes and others have done a great deal of fundamental work on the microscopic diffusion coefficients, these ideas can aid in the interpretation of the macroscopic values [120, 121, 152].

Both of these approaches have identical mathematical solutions, except that the variable correlated by the equations of motion is the normalized change is characteristic sample dimension L_t/L_∞, while Fick's law correlates the normalized concentration changes, or equivalently, mass and volume changes, M_t/M_∞ [46, 60, 61, 119, 121, 149]. Thus these two approaches will yield the same diffusion coefficient only in the limit of no volume change, and these values diverge as the magnitude of the volume change increases. However, both approaches prove equally successful in correlating swelling and shrinking data. Examples of the success of these correlations are given in Figs. 9 and 10 (they are plotted against the square root of time to linearize the initial portion of the curve and compress the long time data). We have used both models successfully to fit a broad set of swelling and shrinking data [119, 121]. They have worked for gels made of PNIPAAm and polyacrylamide synthesized by solution copolymerization/crosslinking [5, 59, 111, 121], poly(2-hydroxyethyl methacrylate) made by bulk polymerization [112, 113], poly(vinyl methyl ether) crosslinked by γ-radiation [46], and hydroxypropylcellulose crosslinked by divinyl sulfone [119, 125]. They have worked for volume changes induced by a number of stimuli, including pH, ionic strength, solvent composition, and temperature [61, 119, 121]. They correlate both swelling and shrinking experiments, and although each was derived under conditions strictly valid for small deformations, they have worked for volume changes as large as factors of 7. They have worked for gels with swelling degrees less than 3 and over 80. They can be applied to planar, cylindrical and spherical geometries. Despite the diverse systems studied, the values of the diffusion coefficients obtained usually fall in the range of $0.8-8 \times 10^{-7}$ cm^2/s.

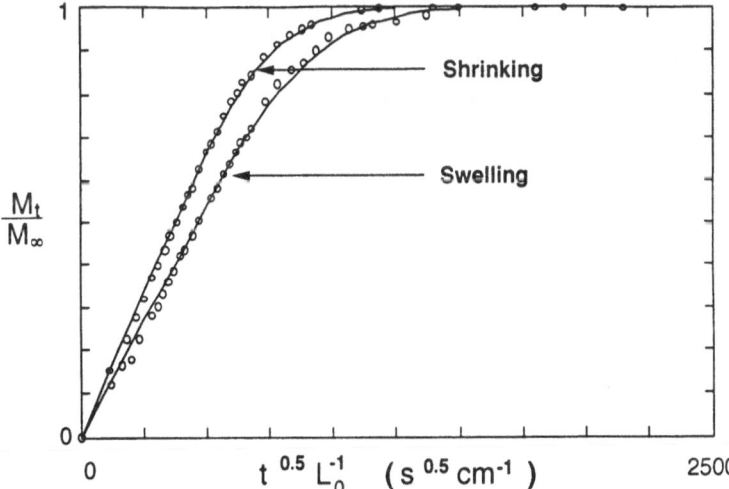

Fig. 9. Plot of normalized approach to equilibrium mass against the square root of time for a temperature-sensitive 10×4 PNIPAAm gel sheet swelling and shrinking between 10 and 25 °C. Shown are the curve fits to the kinetic data of theory developed from Fick's law of diffusion in a polymer-fixed reference frame [149]. The equilibrium degree of swelling is 17.0 at 10 °C and 11.1 at 25 °C; the diffusion coefficients obtained from the curve fits are 2.3×10^{-7} cm^2/s for swelling and 3.6×10^{-7} cm^2/s for shrinking [121]

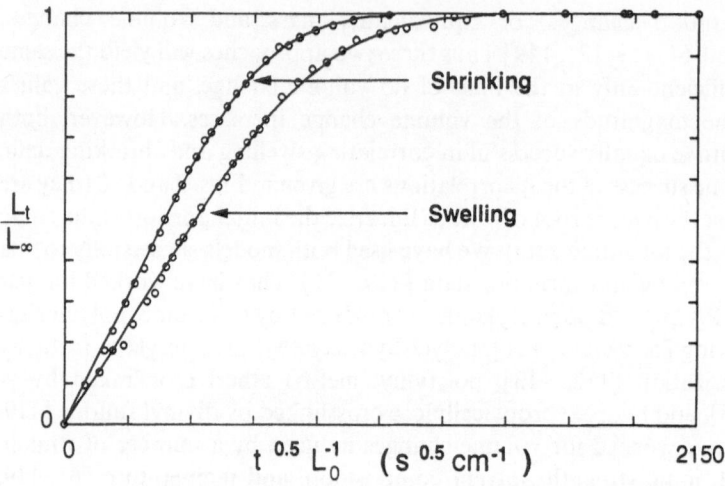

Fig. 10. Plot of normalized approach to equilibrium thickness against the square root of time for a temperature-sensitive 10×4 PNIPAAm gel sheet swelling and shrinking between 10 and 25 °C. Shown are the curve fits to the kinetic data of theory developed from equations of motion (Tanaka and Fillmore theory) [60]. The equilibrium degree of swelling is 17.0 at 10 °C and 11.1 at 25 °C; the diffusion coefficients obtained from the curve fits are 2.3×10^{-7} cm^2/s for swelling and 3.6 $\times 10^{-7}$ cm^2/s for shrinking [121]

Whether the volume transition occurs with respect to temperature, salt or solvent composition, each model works well. Furthermore, neither model outperforms the other in terms of statistical quality of fit of data to theory. Also, the diffusion coefficients obtained from either model are normally comparable, even if a correction for the difference in reference frames is applied [119, 121, 153]. Theoretically, the values of D obtained from the different models differ significantly only for very large volume changes. Thus, if the desire is to correlate different experiments or to reduce kinetic data to a single parameter either model can be used satisfactorily.

However, the models do fail when the volume transition is approached or crossed. This failure occurs in a qualitatively consistent manner for different systems [119, 121]. Upon swelling, the data appear sigmoidal when plotted against the square root of time, although equilibrium is approached at about the same rate as simple Fickian processes. This reason for this deviation is uncertain, but could be the result of a finite surface relaxation time. Upon shrinking across the volume transition, the rate of volume change starts out at a typical rate and progresses with the square root of time. However, once 70 to 90% of equilibrium is approached, the rate of volume change slows dramatically. Matsuo and Tanaka have suggested that this might be due to the formation of a low permeability skin on the surface of the gel which blocks transport of solvent out of the gel [56].

5.3 Dependences of Diffusion Coefficients

In the previous section, we stated that a diffusion analysis successfully correlates the volume change kinetics of a wide variety of gels under different circumstances. The only consistent failure was kinetic response across the volume transition, whether induced by changes temperature, solvent or ionic strength. Thus swelling and shrinking experiments can be reduced to a single parameter characterizing the rates, the diffusion coefficient. The question addressed in this section is "what determines the magnitude of the diffusion coefficient for a particular experiment?"

To examine this question, we made a set of PNIPAAm gels in planar geometry with different compositions and allowed them to reach equilibrium at 10 °C in distilled water [120, 121]. Then they were immersed in slightly warmer water, and the rate at which they shrank was measured by a simple gravimetric "blot-and-weigh" procedure. Using a least squares curve fit to the data of the exact solution to the diffusion equation, the diffusion coefficient was determined for the interval for each gel. Since the rate of heat transfer is so much faster than the rate of mass transfer, the shrinking occurs at virtually constant temperature (at the bath temperature). This process of increasing the temperature of the gel a few degrees and then determining the diffusion coefficient of the process was continued until the gels reached equilibrium at 32 °C, just short of the transition temperature (above which the diffusion analysis will no longer work well). The process was then reversed, lowering the temperature in increments, and measuring the diffusion coefficients for the resulting swelling, until the gels had been returned to equilibrium at the original temperature of 10 °C [61, 120, 121].

The diffusion coefficients obtained from these experiments are given in Fig. 11. Complex behavior is observed, but certain patterns do arise. First of all, shrinking diffusion coefficients are invariably greater than swelling diffusion coefficients at the same temperature (this is generally observed in many different responsive gel systems). While the shrinking diffusion coefficients have little dependence on the gel composition, for swelling at a given temperature, the 16 × 4 gel has the greatest diffusion coefficient, followed by the 10 × 4 gel and then the 16 × 1 gel. This pattern is the opposite of the swelling degrees at any given temperature: the 16 × 4 gel always swells the least and the 16 × 1 gel swells the most. Finally, for both swelling and shrinking, the diffusion coefficients rise toward a maximum around 20–25 °C, and then decline sharply as the transition temperature is approached. As they decline, the swelling and shrinking diffusion coefficients for all gels also appear to converge. We have also determined the diffusion coefficients for all gels also appear to converge. We have also determined the diffusion coefficients for these and other gels over larger temperature intervals, and these same patterns persist, except that over larger temperature intervals, the shrinking diffusion coefficients tend to fall into the same patterns of dependence on gel composition as are seen here for swelling – the less a gel swells at a given temperature, the larger its diffusion coefficient. Furthermore, although

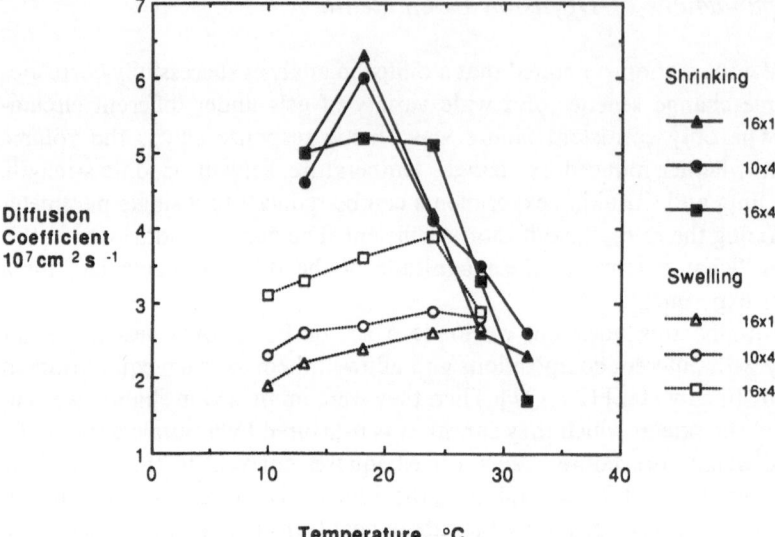

Fig. 11. Diffusion coefficients for PNIPAAm gels obtained from swelling and shrinking experiments over small temperature intervals. The gels were initially equilibrated at 10 °C, then warmed in increments to 32 °C, where the process was reversed and the gels were cooled until they returned to 10 °C. The temperature axis is the temperature at which the volume change occurred (bath temperature). The trends are interpreted using polymer scaling concepts [121]. The lines drawn are to guide the eye

the complex patterns suggest large experimental error, the data are actually quite reproducible [121]. Also, comparable trends are observed with gels made to swell and shrink in response to changes in solvent composition or ionic strength of the solution, although our data are less extensive in these cases [121].

It is clear that diffusion coefficient dependences are complex and will be difficult to quantify. However, we have been able to make headway in interpreting this behavior using polymer scaling concepts. Scaling concepts developed for semi-dilute solutions can also be applied to swollen gels [152]. These scaling relationships allow different network parameters such as crosslink density, polymer concentration and polymer–solvent interaction parameter to be expressed in terms of a common parameter, the correlation length, ξ. In semi-dilute concentrations in good solvents, the correlation length is the average distance between two neighboring crosslinks (thus the greater the crosslink density, the lower the correlation length).

Tanaka's identification of the diffusion coefficients obtained from macroscopic swelling and shrinking experiments with the collective or cooperative diffusion coefficients used in scaling theory is the key concept here [60]. During swelling and shrinking, all the network chains of the gel interact and move together [152]. Since the solvent molecules move much faster than the network (see Sect. 5.1.3), the motion of the solvent molecules can be neglected apart from viscous effects. The relationship between this diffusion coefficient and the swollen gel properties can be arrived as follows [154]. First of all, the diffusion coefficient is related to gel properties by the following relationship [151]:

$$D = \frac{(K + \frac{4}{3}G)}{f} \tag{1}$$

where: K = bulk modulus of gel;
         G = shear modulus of gel;
         f = friction factor between network and solvent.
For nonionic gels in pure, good solvents, the following relationships between K, G, f and the polymer volume fraction ϕ can be developed [152, 154]:

$$K, G \propto \phi^{9/4} \tag{2}$$

$$f \propto \phi^{3/2} \tag{3}$$

Combining relationships (1)–(3) yields the following scaling law between the diffusion coefficient and the polymer volume fraction:

$$D \propto \phi^{3/4} \tag{4}$$

Thus the prediction is that the diffusion coefficient should increase as the polymer concentration increases, or equivalently, as the swelling degree of the gel declines. From constant volume light scattering experiments with poly-acrylamide gels, the exponent actually observed lies between 0.66 and 0.77 [152, 154]. The variation in the exponent has been interpreted as the result of dangling ends and other imperfections in the network which distort the scaling relations. Also, as the solvent quality becomes poorer, the correlation length varies over several network chains.

First of all, we examined the validity of the shear modulus scaling law – Eq. (2) – by performing uniaxial compression experiments on a series of PNIPAAm gels with different compositions to obtain their shear moduli. The scaling exponent obtained was 2.0, close to the theoretical value of 2.25 [45]. Secondly, we extracted scaling exponents for different temperatures from the swelling and shrinking data from Fig. 11; the results are given in Table 4. For the swelling experiments, we see excellent agreement with the scaling law, and the scaling exponents, while somewhat smaller than the theoretical values, match experimental values obtained from light scattering experiments. But there is no apparent dependence of the diffusion coefficient on the gel composition for the shrinking experiments. This might be due to small number of diffusion coeffic-ients obtained in the low temperature, good solvent region, where the scaling laws are expected to hold.

To examine some of the other features of Fig. 11 in light of scaling concepts, it is useful to present the scaling law Eq. (4) written explicitly in terms of the correlation length ξ [152, 154]:

$$D \propto T/\eta\xi \tag{5}$$

Table 4. Scaling exponents from the scaling law $D \propto \phi^n$ for PNIPAAm gels of various compositions caused to swell and shrink over a series of temperature intervals, as shown in Fig. 11 [121]

Temperature interval (°C)	Average ratio of final to initial swelling degree	Scaling exponent n	Regression coefficient r^2
13–10	1.08	0.65	0.97
18–13	1.17	0.62	0.98
24–18	1.25	0.67	0.99
28–24	1.24	0.68	0.97
32–28	1.48	0.43	0.92
10–13	0.93	0.04	0.04
13–18	0.87	− 0.50	0.94
18–24	0.81	0.44	0.50
24–28	0.79	0.54	0.71
28–32	0.66	− 0.40	0.26

where: T = absolute temperature
           η = solvent viscosity.

This relationship indicates that this diffusion coefficient has conventional temperature and viscosity dependences: increasing in proportion with absolute temperature and inversely with viscosity.

With this relationship, we can make some headway into interpreting the temperature dependences observed in Fig. 11. First of all, the steady increase in the swelling and shrinking diffusion coefficients with increases in temperature in the good solvent region (at low temperature) is consistent with Eq. (5). The proportionality with absolute temperature, the decreases in solvent viscosity and ξ (since the gel shrinks) with temperature all act to increase the value of D. The magnitudes of the temperature dependences observed in Fig. 11 are less than the prediction in Eq. (5), however. One problem with applying this relationship here is that it only applies in the good solvent regime, which lies well below the phase separation temperature. The magnitude of concentration fluctuations (and thus ξ) increase as the system nears phase separation, reducing D. In fact, the magnitude of the correlation length increases rapidly as the phase separation temperature is approached; specifically, $\xi \to \infty$ as $T \to T_s$, where T_s is the spinodal temperature [152]. This explains the rapid drop in the diffusion coefficients as the volume transition is approached. An equivalent way of viewing the drop in the diffusion coefficient near the phase separation temperature is that randomly occurring concentration fluctuations tend to persist as the phase separation temperature is neared; fluctuations of polymer chains slow dramatically at this point, and the diffusion coefficient approaches zero. Tanaka et al. have demonstrated the validity of this concept with quasielastic light scattering experiments on PNIPAAm gels [62]. Furthermore, since the gels are not highly crosslinked, the average molecular weight between crosslinks is quite large, and thus it is likely that incipient phase separation leads to values of ξ which are not well-correlated with effective crosslink density. This could explain the insensitivity of the diffusion coefficient to the gel composition near the transition temperature.

A final feature of Fig. 11 that is seen in many different systems is the observation that shrinking invariably occurs faster than swelling at a given temperature or over a given interval. Again de Gennes' scaling concepts can prove helpful. There is friction between the polymer chains and the solvent, and the amount of energy dissipated per unit volume will scale with $\eta v^2/\xi^2$, where v is the relative velocity. Thus the greatest frictional resistances and energy losses occur for the highest velocities and smallest "pores" (visualized as spaces between chains with diameters roughly equivalent to ξ). We have already identified swelling and shrinking as diffusional processes, and these scale with the square root of time. Thus the rate of volume change is at a maximum at time zero and declines steadily as equilibrium is approached. Now consider the difference between swelling and shrinking processes occurring between the same states. A shrinking gel has its largest "pores" at the beginning of the process, when the relative velocity between solvent and polymer is greatest. In contrast, a swelling gel has its smallest "pores" at early times, when relative velocities are high. Thus frictional resistance and dissipated energy will be higher for a swelling gel than a shrinking gel, and this will cause shrinking processes to be slower than swelling.

Admittedly, much of our interpretation of the kinetic behavior of responsive gels can be criticized as "hand-waving". More work is required to prove that the interpretations we have presented here are valid. However, the analogies between semi-dilute polymer solutions and gels have been well established by others. The step we have made here has been to extend these concepts from perturbations from equilibrium to the macroscopic shifts of practical interest in responsive gel applications. While this extension is fraught with hazards, the qualitative connections which seem to persist remain helpful and point the way for future work on kinetics.

5.4 Enhancing Response Rates of Gels

We have established that the volume change kinetics of responsive gels are usually diffusion-controlled processes. Even when the diffusion analysis failed, the rates were comparable to or slower than a classical diffusive process. The implications of this for practical applications are quite negative, since diffusive processes are quite slow. A gel slab 1 mm thick with a diffusion coefficient of 10^{-7} cm^2/s will take over an hour to reach 50% of equilibrium and more than six hours to reach 90% of equilibrium in response to a stimulus. This is far too slow for almost all potential applications of these materials. Since diffusion times scale with the square of dimension, decreasing the characteristic dimension of a sample will increase the rates dramatically. Thus if an application can make use of submillimeter size gels, millisecond response times become possible. Unfortunately, it may not always practical to use gels of such small dimension.

However, a report by Huang et al. described gels of radiation-crosslinked PVME of macroscopic dimension that swelled and shrank at rates far greater than a diffusional mechanism could allow [39]. Later on, they noted that these gels were microporous, and that this microporosity was the result of thermally-induced phase separation of the polymer solution caused by radiative heating. This was the key to the synthesis of gels with greatly enhanced rates of swelling and shrinking. If responsive gels are created with microporous, open-celled structures, the crosslinked cell walls will respond by a diffusive mechanism in milliseconds or faster, since they will have characteristic dimensions in microns, while the solvent could be absorbed or expelled by convection through the micropores in the gel. Thus we began a line of research to create microporous gels and examine the rates of volume change as a function of the microstructure. We were able to create gels of PVME, PNIPAAm and HPC as samples with multi-millimeter thicknesses which could respond to stimuli in seconds, rather than the hours required by conventional nonporous gels.

5.4.1 Fast Response PVME Gels

The synthesis technique for the production of PVME gels by radiation cross-linking was described in Sect. 3.3, and the equilibrium swelling properties were discussed in Sect. 4.1.3. Here we will describe their swelling and shrinking kinetics. We began work with PVME gel before we had a clear understanding of how to create a microporous gel and how this would affect the swelling kinetics. We began this work because we were intrigued by a figure in the paper by Huang et al. that indicated that a 1 cm diameter sphere of PVME gel could swell to equilibrium in about 10 minutes, far faster than the PNIPAAm gel we were working with at the time [12]. Thus we worked to reproduce their work and learn the origin of this fast response [46].

Most of the PVME gels that we synthesized with varying intensities and doses displayed kinetic behavior quite similar to the PNIPAAm gels described in Sect. 5.2 [46]. Swelling and shrinking processes occurred by Fickian processes when the temperature changes stayed below the transition temperature of the gel, with diffusion coefficients comparable in magnitude to PNIPAAm results. When the transition temperature was crossed, the shrinking curves had a late-stage slowdown in rate and the swelling curves appeared sigmoidal when plotted against the square root of time [46]. However, one trial stood out – the Trial 4 gels from Table 2. For this gel type, shrinking occurred with an apparent diffusion coefficient of 1×10^{-5} cm^2/s, almost 100 times greater than we had observed for any other gel. Surprisingly, this gel swelled at about the same slow rate as others, as shown in Fig. 12. Since this shrinking "diffusion coefficient" was of the same magnitude as the self-diffusion coefficient of water, it was obviously not a diffusion controlled process, but must have involved convective flow. The fact that the shrinking curve nearly follows Fickian mathematics is probably a result of the fact that unsteady state flow through porous media

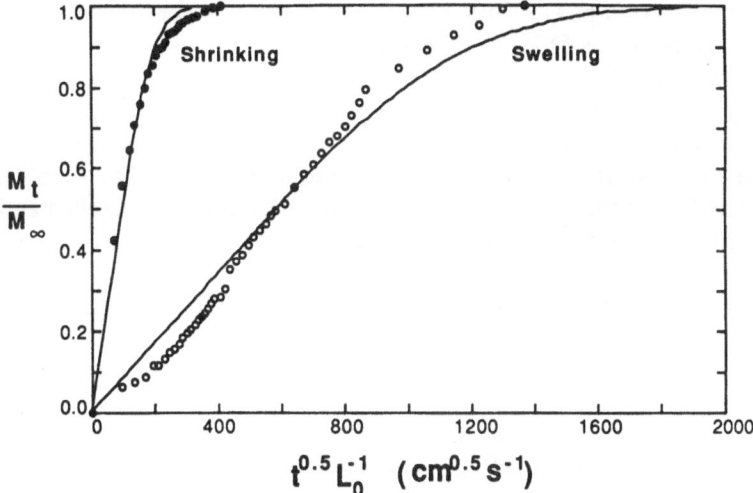

Fig. 12. Volume change kinetics of PVME gel (Trial 4) between states above and below the volume transition of 37 °C. Non-Fickian behaviour is observed for both swelling and shrinking. For comparison, lines are calculated which provide the closest fit of Fickian theory to the data. ○: Swelling (from 50 to 24 °C), $D = 4.0 \times 10^{-7}$ cm^2/s; ●: Shrinking (from 23 to 50 °C), $D = 1 \times 10^{-5}$ cm^2/s. Reprinted from Polymer (1991) 33: 990 by permission of the publishers, Butterworth Heinemann [46]

(such as this gel) has been successfully modeled with an equation mathematically identical to Fick's law [155]. Scanning electron micrographs of this gel confirmed its porous microstructure. However, the other, apparently diffusion controlled, gels also appeared to be microporous, so microporosity in and of itself was not sufficient to guarantee a fast response; pore interconnectivity, pore size, and so on apparently are also important.

We originally believed that the porosity was caused primarily by the gas bubbles generated by polymer degradation during irradiation. However, Huang et al. published further work on this gel that indicated that the microporous structure was the result of phase separation of the solution due to radiative heating [108]. Since the greatest heating should occur for the highest radiation dose rate, this could explain why Trial 4 gels had the highest rates. The difficulty of controlling the temperature rise in this system also explained why we could not reproduce their results; undoubtedly the heat transfer characteristics of our system did not match theirs. Thus we changed research directions to examine systems where we had better control of both gel formation and microstructure development.

5.4.2 Fast Response Poly(N-isopropylacrylamide) Gels

Our first attempt to produce fast response, microporous gels of PNIPAAm was to polymerize an emulsion with a monomer-rich continuous phase and a monomer-free dispersed phase. This concept had been used to produce micro-porous polystyrene foams [156–158]. However, we were unable to develop a successful emulsion system due to NIPAAm's good solubility in both water and common organic solvents. Thus we switched to the idea of using deliberate heating of the monomer solution *during* the reaction to "lock-in" a microporous structure via chemical crosslinking as phase separation occurred. With this idea, we were successful in creating a microporous PNIPAAm gel which swelled and shrank several thousand times faster than nonporous gels of the same composi-tion and dimension [65].

The synthesis of the microporous gel started out in the same fashion as preparing a conventional nonporous gel, as described in Sect. 3.2. The only difference is that after the polymerization has begun, but before the solution reaches the gel point, the mold containing the solution is immersed into a water bath at a temperature above the LCST of the polymer. This causes phase separation to occur, but if the sol has achieved a sufficient molecular weight, gelation can still occur (a latex can be formed if the reaction is carried out entirely above the LCST). The gel produced by this technique is milky white, unlike conventional gels, which are transparent. This is the result of light scattering from the heterogeneous structure "locked-in" by the reaction in the phase separated state.

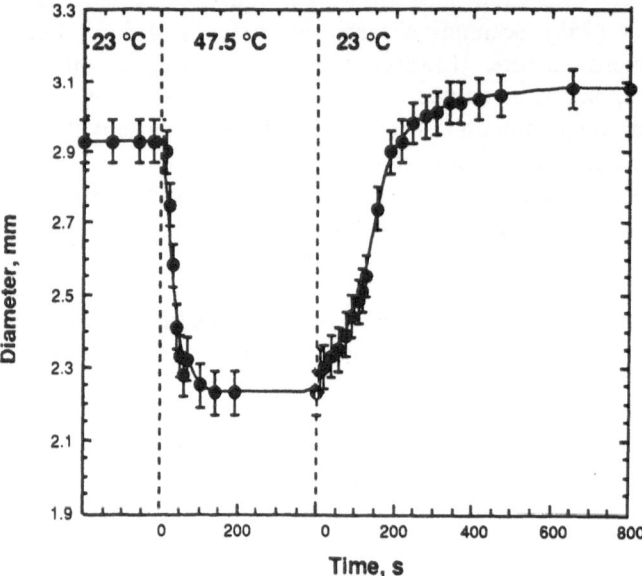

Fig. 13. A swelling/shrinking cycle of PNIPAAm gel crosslinked in a phase-separated state, across the volume transition temperature of 33 °C. The time to complete one cycle is several thousand times less than required for a homogeneous PNIPAAm gel of identical chemical composition and dimensions. The lines are to guide the eye. Reprinted from Polymer Communications (1991) 32: 322, by permission of the publishers, Butterworth Heinemann [65]

Figure 13 is an example of the swelling and shrinking kinetics for a microporous PNIPAAm gel. Although the diameter of the swollen gel is 3 mm, it shrinks to equilibrium is 100 s, and reswells in 400 s. This rate is so fast that it could not be determined by simple blot and weigh techniques. Instead we had to make use of a videomicroscopy and image analysis system [46, 65]. These swelling and shrinking times are 120 and 3000 times shorter, respectively, than for a homogeneous, nonporous gel of the same overall composition and dimensions.

5.4.3 Fast Response Hydroxypropylcellulose Gels

We sought to prove the generality of the technique of locking in a microporous structure by crosslinking during phase separation by extending the idea to chemically crosslinked linear polymers. The test system was the hydroxypropyl-cellulose-divinyl sulfone gel described in Sect. 3.2. Again, crosslinking of the solution while phase separation is occurring can lead to the formation of a microporous gel which swells and shrinks in response to changes in temperature at rates comparable to the PNIPAAm gel illustrated in Fig.13. Furthermore, we have shown that such cellulose ether gels also respond very rapidly in response to changes in solvent and salt concentration. For these gels, unlike nonporous gels (as in Figs. 2–4), the stimuli are often the rate limiting steps. Even heat transfer proves to be the rate limiting step for the fastest gels. However, we have also demonstrated that the relationships between synthesis conditions, micro-structures, response rate and equilibrium swelling degree are quite complex. A continuum of rates are seen, from very fast, stimuli-limited gels, to microporous but apparently diffusion-controlled gels. For intermediate systems, the shrinking rate may be very fast and appear to occur by convection, while the swelling rate is slow and apparently diffusion-controlled, as was seen in Fig. 12 for PVME gel. Discovering these relationships and interpreting these phenomena is the focus of B. Kabra's Ph.D. thesis; publications detailing these results will be forthcoming [126]

6 Permeability of Responsive Gels to Solutes

The literature contains a number of examples of responsive gels that can function as valves which turn solute flux on and off in response to changes in environment-turning "on" when the gel swells, and "off" when the gel shrinks [25, 57, 69, 90, 107]. There is generally a direct relationship between the swelling degree of a gel and its permeability to solutes; as the water content of the gel declines, so does its average pore size, and thus obstruction of solutes increases. Thus the permeability of responsive gels to solutes is expected to decline as the gels shrink in response to an appropriate stimulus. However, this simple picture will hold only if the network is inert with respect to the solute; i.e., no attractive or repulsive interactions exist. However, general inertness of gels had not been established, and neither had the magnitude of declines in permeability which could be expected. Thus we have examined these issues.

6.1 Theoretical Predictions of Permeability

To be specific, permeability is the product of the partition coefficient K and diffusion coefficient D of a solute within a material. The partition coefficient is usually defined as the ratio of solute concentration in the material to the concentration in the solution; it is a measure of the interaction between the polymer and the solute. The partition coefficient should decline as a gel shrinks and the average pore size declines, if other interactions between the solute and the gel are unchanged. The diffusion coefficient is also expected to decline as water content falls and obstruction to movement increases. While K is an equilibrium property, D is a transport property, and so these parameters can be evaluated independently.

The most commonly used theory for diffusion in swollen gels is that by Yasuda et al. [159]. It assumes that solutes diffuse only through the free volume of the solvent held within the gel, and thus predicts that D should decline as the amount of solvent held within the gel declines. If the theory holds, a plot of the logarithm of the diffusion coefficient against $(Q - 1)^{-1}$ will be a straight line with an intercept equal to the free solution diffusion coefficient and a slope proportional to the ratio of the solute volume to the free volume of the solvent. While this theory has been successful for a number of solute/hydrogel systems, the limits of its applicability have not been established [70]. Based on the assumptions behind its development, however, it is expected to hold only for homogenous gels of relatively high water contents.

The partition coefficient of a solute between a gel and a solution is generally the result of a number of interactions between the gel and the solute. These interactions include size exclusion, hydrophobic interactions, electrostatic interactions and biospecific interactions. Biospecific interactions include phenomena like antibody-antigen complexation, and will not be present unless appropriate ligands are attached to the network. Electrostatic interactions can be very important if the network and the solute carry electrical charge. If they carry the same charge, the solute will tend not to permeate the gel due to Donnan ion exclusion [5, 146]. If they are oppositely charged, the gel may strongly absorb the solute via ion exchange [146]. Hydrophobic interactions may exist if the solute and gel include hydrocarbon regions. Size exclusion is a phenomenon which arises when a molecule's dimension comes within an order or magnitude or so of the pore dimensions. Because of the restrictions on the orientational configuration of such a solute within the pore, entropy favors a solute's presence in the solution over permeation of the gel. Thus as a gel shrinks, the partition coefficients of solutes will approach zero; this is the basis of the "on-off" concept behind the use of responsive gels as chemical valves. The other interactions are not directly correlated with the swelling degree of the gels, however.

If the gel is inert with respect to the solute – obstructing its movement, but having no net attraction or repulsion – size exclusion will be the only significant interaction between a gel and a solute. In fact, this has been observed in a number of hydrogel systems, including polyacrylamide gel, so it might be expected to be true in the case of responsive gels also [70]. A number of scientists have developed theories to describe size exclusion; the most recent theoretical advance is by Schnitzer [160]. This theory predicts sharp declines in the partition coefficients of smaller solutes as a gel shrinks, but more gradual declines for larger solutes [114]. The maximum value of K for size exclusion is 1 (concentration of solute in the gel equals that in the solution) and the minimum value is zero (concentration of solute in the gel is zero).

Both the Yasuda et al. and Schnitzer theories indicate that the key variables defining permeability in a gel are the swelling degree of the gel and the average dimension of the solute. Thus the dependence of permeability on swelling degree for a responsive gel inert to solutes can be predicted using the theory of Yasuda et al. to obtain the ratio D/D_o (the ratio of the diffusion coefficient of a solute in the gel to its value in free solution) and the theory of Schnitzer for ideal size exclusion to obtain K. A dimensionless permeability can be defined as the

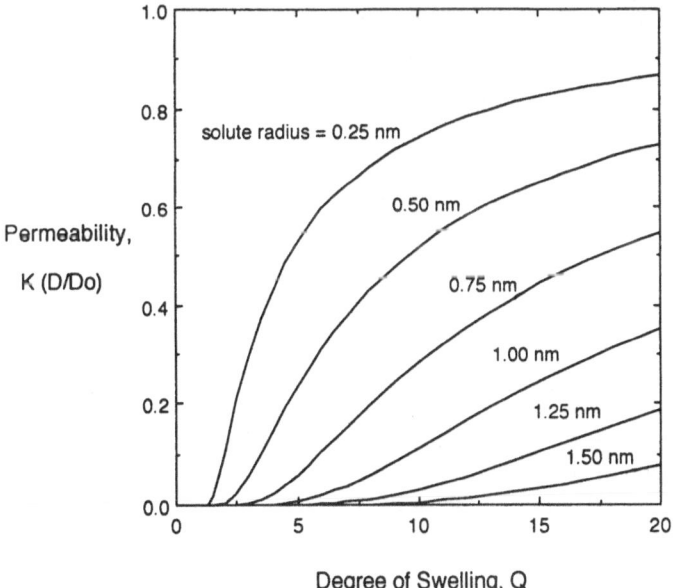

Fig. 14. Prediction of solute permeability for solutes of different hydrodynamic radii in swollen gels. Permeability is calculated as the product of the partition coefficient, K, using the size exclusion theory of Schnitzer and the ratio of the solute diffusion coefficient in the gel, D, to its value in solution, D_o, using the theory of Yasuda et al. [123, 159, 160]

Table 5. Selected hydrodynamic radii of various solutes for comparison with the predictions of Fig. 14 [161]

Solute	Molecular weight	Hydrodynamic radius (nm)
Urea	60	0.25
Creatine	131	0.31
Uric acid	168	0.33
Theophylline	180	0.37
Sucrose	342	0.47
Vitamin B-12	1355	0.84
Myoglobin	17,000	1.75
Ovalbumin	45,000	1.76
Bovine serum albumin	66,200	2.25

Data taken from [161]

product KD/D_o and calculated as a function of solute dimension and swelling degree. Figure 14 shows the result of this calculation (assuming a fiber radius of 0.4 nm and a value of 1.8 nm^{-3} for the free volume overlap factor divided by the free volume of water [113, 159]). Taking note of the hydrodynamic radius data for solutes given in Table 5 [161], we see that Fig. 14 predicts that for solutes the sizes of common drugs, large drops in permeability should be observed over easily achieved swelling degree changes at the volume transitions of responsive gels. However, the dependence of permeability on swelling is more gradual for very large solutes like proteins. Thus, if these theories apply, they indicate that delivery or separation systems using on-off permeability may be more easily created for lower molecular weight species, where the magnitude of volume change required is less than for large solutes. Our next goal was to examine the experimental validity of these equations in responsive gel systems.

6.2 Solute Diffusion Coefficients

Previously, the Yasuda et al. theory had been tested either by using a series of solutes of different sizes with a single swollen gel, or else diffusion of a single solute was measured in a series of different polymer gels with different swelling degrees. With responsive gels, it is possible to use a single solute and a single gel; the water content is varied by modulating the external environment. By measuring the diffusion of acetaminophen in the PNIPAAm gel over a relatively small temperature interval, and of vitamin B-12 in a pH-sensitive polyacrylamide copolymer gel, we established the applicability of this theory to responsive gels [5, 70].

The diffusion coefficients of acetaminophen in PNIPAAm gel at different temperatures were measured by a simple sorption experiment. PNIPAAm gel with a 10×4 composition (see Fig. 1) was synthesized in cylindrical form by the procedure described in Sect. 4.1.1. After equilibration of a gel sample in water at

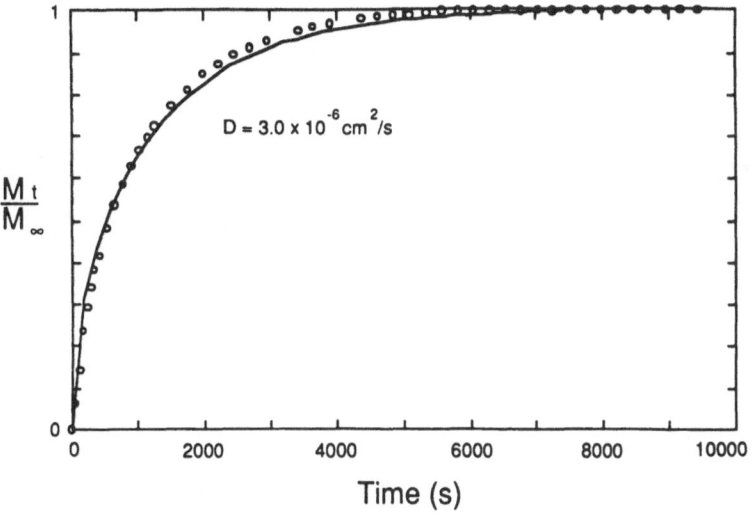

Fig. 15. Diffusion of dilute, aqueous acetaminophen into a long, swollen cylinder of 10×4 PNIPAAm gel at 25 °C. The diffusion coefficient is extracted from a nonlinear least squares curve fit of the exact solution for diffusion into a cylinder of infinite length immersed in a well-stirred solution of finite volume to the data [123, 149]

a given temperature, it was immersed in a tube containing solution of acetaminophen kept within the same temperature bath. The solution to gel volume ratio was kept less than 10 so that the concentration in the solution (continuously monitored by a UV/VIS spectrophotometer) would fall measurably over the course of the experiment as the solute diffused into the gel. The solution was kept mixed by a submersible magnetic stirrer and by the steady flow of the solution through the system caused by pumping the fluid through the spectrophotometer with a HPLC pump. The solute diffusion coefficient can be extracted from a least-squares curve fit to the exact mathematical solution of this problem to the concentration vs. time data obtained [149]. An example of this curve fit is shown in Fig. 15.

The plot of the diffusion coefficients obtained at different temperatures is given in Fig. 16. The linearity of the data as plotted matches the predictions of the Yasuda et al. theory. The intercept is about 25% below the free solution volume of 6.0×10^{-6} cm²/s, however. This may be due to boundary layer resistance which could not be eliminated even by increased the stirring rate. It might also be due to sieving effects which have not been considered, since these are not expected to influence solutes as small as acetaminophen [159, 161, 162]. A more significant deviation from the theory is seen at 35 °C, above the transition temperature of the gel. The value of D at this temperature was 35 times greater than predicted by extrapolation of the Yasuda et al. theory as indicated in the figure. This is particularly surprising, since D might have been expected to be substantially *less* than predicted by free volume theory since the water content of the gel at this temperature is low and essentially all bound

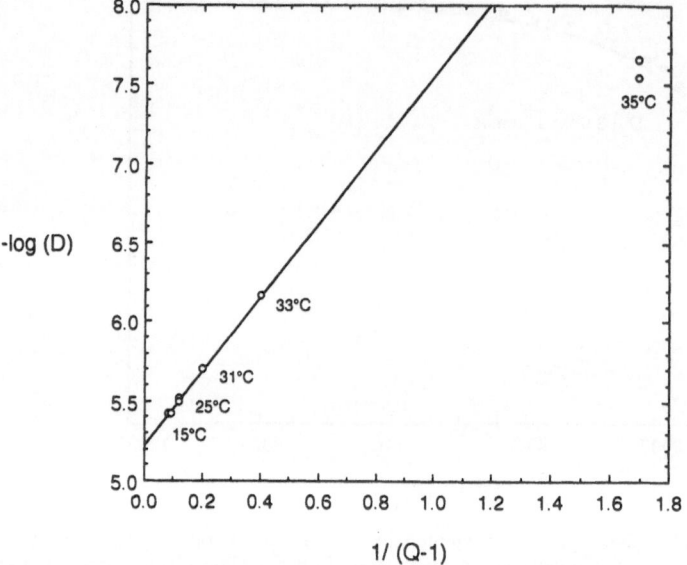

Fig. 16. The diffusion coefficient of acetaminophen in 10×4 PNIPAAm gels falls as the swelling degree (Q) of the gel decreases due to increasing temperature. Below the transition temperature of the gel, the linear relationship between log D and $(Q - 1)^{-1}$ predicted by the free volume theory of Yasuda et al. [10] is observed. Above the transition temperature, the theory underestimates D by 35 times. Reprinted from the Journal of Controlled Release (1992) 18: 1, by permission of the publishers, Elsevier Science Publishers BV [70]

water (as indicated by DSC measurements) [94]. Diffusivities of solutes through bound water regions are expected to be very low. However, we observe that the collapsed gel is quite opaque, indicating heterogeneity of the gel's micro-structure. Thus, the network mesh size is likely substantially greater in the collapsed gel than would be assumed by the Yasuda et al. theory. However, this point needs more study.

The same type of experiment was also performed with a series of pH-sensitive gels, crosslinked poly(acrylamide-*co*-sodium methacrylate), and the solute vitamin B-12 [5]. In this case, the swelling was modulated both by changing the solution and by altering the degree of crosslinking of the gel. Even when combining both types of points, a straight line is obtained when the data are plotted as indicated by the Yasuda et al. theory as shown in Fig. 17. For this system, the intercept yields a value of the diffusion coefficient that is within 3% of the measured free solution value of 3.8×10^{-6} cm^2/s, agreement well within the experimental error. Since vitamin B-12 is a much larger solute than acetaminophen, and no sieving effects by the mesh were observed with B-12, this indicates that sieving likely does not explain the low value of the intercept seen with acetaminophen. However, since the radii of the pH-sensitive gel cylinders used in the vitamin B-12 experiments were almost ten times larger than those used in the acetaminophen experiments, boundary layer resistances would likely

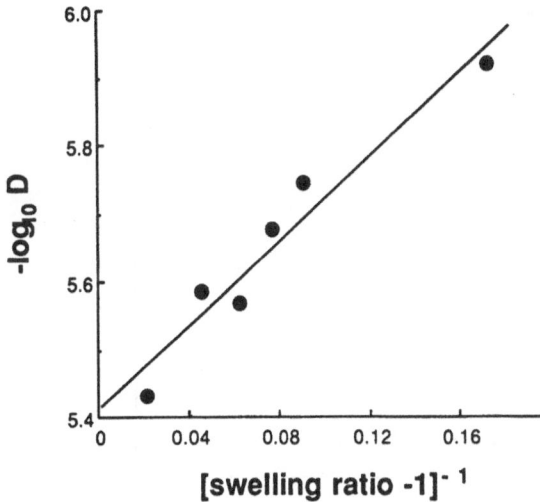

Fig. 17. Vitamin B12 diffusion coefficients vs poly(acrylamide-*co*-sodium methacrylate) gel swelling. The variation is that expected from free volume theories. Reprinted with permission from Chemical Engineering Science, 44, S.H. Gehrke and E.L. Cussler, Mass Transfer in pH-sensitive Gels, Copyright (1989), Pergamon Press [5]

to be more significant in the latter cases. No deviation from the Yasuda et al. theory is seen at the lower degrees of swelling in this figure, because the swelling degrees of the pH-sensitive gels were all very high (between 6.8 and 47) and all were transparent. Thus they appeared to be homogeneous, and the percentage of water held within the gel which was bound rather than free was likely to be quite small. Thus, no deviation from the theory would be expected on these grounds.

6.3 Solute Partition Coefficients

The experiments which yielded the diffusion coefficients for acetaminophen in PNIPAAm gel in Fig. 16 also yielded the corresponding partition coefficients. While the diffusion coefficients fit theory, the partition coefficients as plotted in Fig. 18 do not at all. In fact, a trend opposite to theory is observed as the partition coefficients are seen to *increase* as the gel swelling decreases. In fact, above the transition temperature of the gel, at 35 °C, the partition coefficient is seven times the maximum possible size exclusion coefficient, 1. This implies the dominance of hydrophobic effects over steric effects, since acetaminophen is a relatively small, nonionic but hydrophobic solute, and while the gel mesh size shrinks with increasing temperature, its level of hydrophobicity increases with temperature.

To test this, we determined the partition coefficients of two different solute extremes – vitamin B-12 and norethindrone. Vitamin B-12 is a relatively large, hydrophilic solute, while norethindrone is a relatively small, hydrophobic solute. Thus we would expect size exclusion to dominate partitioning with vitamin B-12, while hydrophobic interactions should dominate with norethindrone. This behavior is in fact observed in Figs. 19 and 20. The values of K

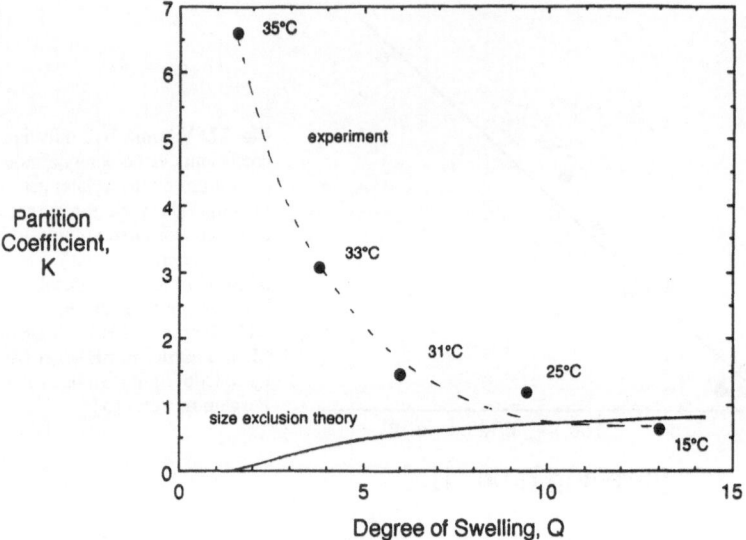

Fig. 18. The partition coefficients of acetaminophen in 10×4 PNIPAAm gels as a function of the swelling degree of the gel. The trend is opposite to that estimated by the ideal size exclusion theory of Schnitzer, shown as a *solid line* [17]. The dashed line is to guide the eye. Reprinted from the Journal of Controlled Release (1992) 18: 1, by permission of the publishers, Elsevier Science Publishers BV [70]

predicted by the Schnitzer theory are lower than what is actually observed, but the trends are comparable, including the sharp decline across the transition. The slight increase in K with temperature below 33 °C may be an indication of slight hydrophobic interaction between vitamin B-12 and the gel, since these interactions are known to strengthen with temperature. The data for norethindrone stand in sharp contrast to that for B-12, however: instead of a sharp drop in K across the transition temperature, a dramatic rise is observed, to values of K over 35. Instead of blocking hydrophobic solutes upon gel collapse, the gel absorbs them! Further evidence of the connection between a sharp increase in partition coefficient when the temperature is increased above the transition value and solute hydrophobicity is given in Table 6 [123, 163, 164]. The magnitude of the partition coefficient above the transition temperature is seen to increase as the drug's octanol/water partition coefficient ($K_{o/w}$) increases ($K_{o/w}$ is a commonly used measure of drug hydrophobicity).

As further evidence of the importance of hydrophobic interactions in these systems, we examined the partition coefficient of methyl orange in the presence of water structure-forming and water structure-breaking salts above and below the transition temperature [70]. Methyl orange is an easily detected, hydrophobic dye which has been sulfonated to improve water solubility. Water structure-breaking salts like tetraethylammonium chloride (TEAC) are known to minimize hydrophobic interactions while water structure-forming salts like ammonium sulfate are known to increase hydrophobic interactions [165, 166].

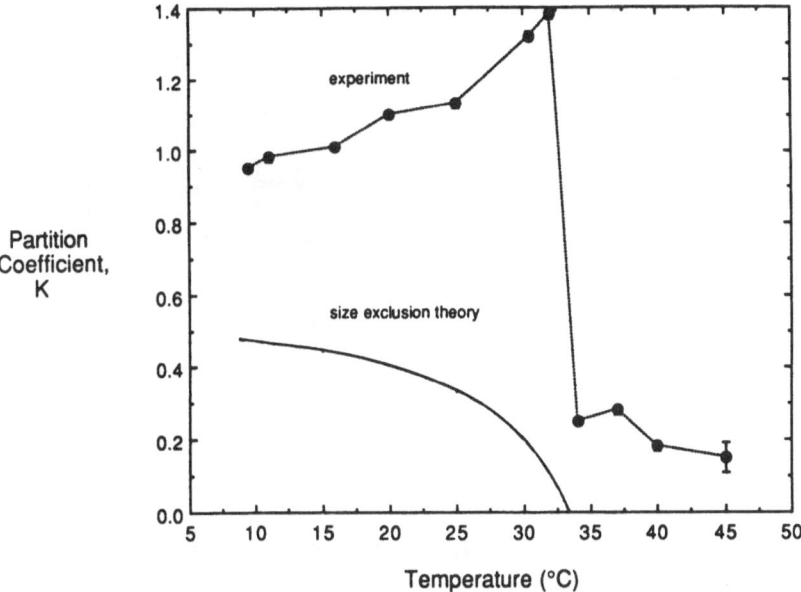

Fig. 19. Experimental partition coefficients and those predicted by size exclusion theory (Eq. (4), shown as a *solid line*) for vitamin B12 (a relatively large, hydrophilic solute) in 10×4 PNIPAAm gel. Although size exclusion theory consistently underestimates the value of K, the sharp drop predicted at the transition temperature is observed. Error bars (standard deviation of three samples) not shown are smaller than the symbol. The dotted line is to guide the eye. Reprinted from the Journal of Controlled Release (1992) 18: 1, by permission of the publishers, Elsevier Science Publishers BV [70]

Table 6. Comparison of octanol/water and gel/solution partition coefficients for hydrophobic drugs [123, 163, 164]

Drug	Octanol/water partition coefficient $K_{o/w}$	Partition coefficient between water and 10×4 PNIPAAm gel at 25 °C	Partition coefficient between water and 10×4 PNIPAAm gel at 40 °C
Acetaminophen	0.8[b]	1.2	4.1
Phenacetin	1.57[a]	1.1	6.3
Ibuprophen	3.51[a]	1.0	12.1
Norethindrone	[c]	1.6	28.0

[a] Taken from [163]
[b] Taken from [164]
[c] $K_{o/w}$ not available, but this has the lowest water solubility among these drugs

The TEAC had little effect on K at 5 °C, but caused K to decline steadily with increasing TEAC concentration at 40 °C, above the transition temperature. This is as expected since at 5 °C, the gel is hydrophilic and hydrophobic interactions are insignificant even in the absence of TEAC. However, at 40 °C, hydrophobic interactions are strong, so the adding TEAC reduces them. In contrast, the

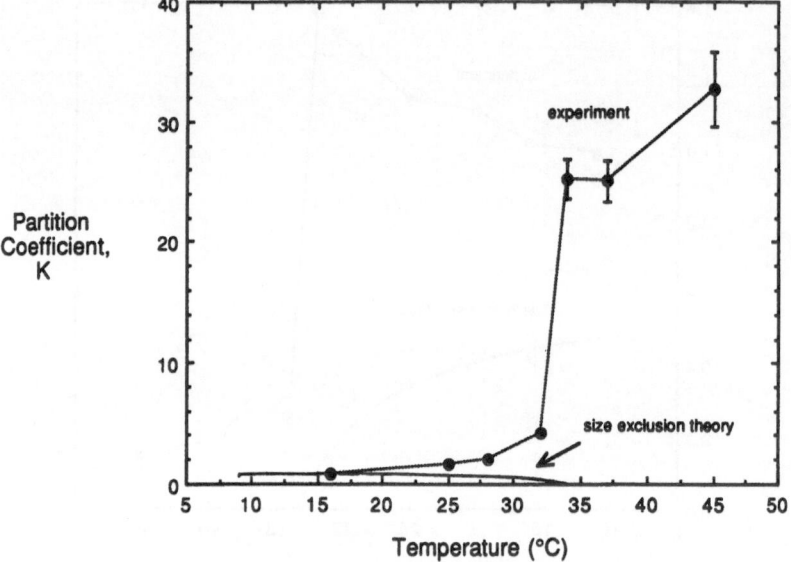

Fig. 20. Experimental partition coefficients and those predicted by size exclusion theory (Eq. (4), shown as a *solid line*) of norethindrone (a small, hydrophobic solute) in 10×4 PNIPAAm gel. Norethindrone is absorbed by the collapsed gel above the transition temperature, rather than being excluded from it as predicted by Eq. (4). Error bars (standard deviation of three samples) not shown are smaller than the symbol. The dotted line is to guide the eye. Reprinted from the Journal of Controlled Release (1992) 18: 1, by permission of the publishers, Elsevier Science Publishers BV [70]

addition of ammonium sulfate to the solution causes the value of K at 5 °C to rise dramatically, from 0.60 ± 0.2 in pure water to 195 ± 30 in 1.5 M ammonium sulfate solution. This is consistent with the concept that the addition of ammonium sulfate to the solution should promote hydrophobic interactions. This phenomenon is complex, however, as K drops at sufficiently high values of ammonium sulfate concentration (to 7.9 ± 1.1 in 2.0 M ammonium sulfate). It is conceivable that this phenomenon could be used to load solute into a gel used as a drug delivery device.

In conclusion, we see that while the Yasuda et al. theory works well below the transition temperature, it predicts values which are much smaller than observed in the low water content collapsed state. Furthermore, while a sudden decline in K for solutes as the gel is pushed into the collapsed region can be observed for relatively large, hydrophilic solutes (though the drop is not nearly as sharp as predicted), very large *increases* in K may be observed for hydrophobic solutes. Taking together the larger than expected values of D in the collapsed gel and the possibly steep increase in K in the collapsed gel, "on-off" permeability may sometimes be difficult to achieve with responsive gels. In fact, the permeability could even increase when a gel collapses rather than decrease!

7 Applications Research Using Responsive Gels

Our research on responsive gels has been carried out with applications in separations and drug delivery in mind. Except for the slurry dewatering process described in the next section, however, we have not focused on the development of specific applications. One novel idea which we pursued, however, was the use of temperature-sensitive PNIPAAm for the preparative gel electrophoresis, discussed in Sect. 7.2.

7.1 Dewatering Fine Coal Slurries

The main factor limiting the use of coal for the production of electricity is the costliness of necessary pollution control measures. High levels of sulfur commonly found in many coals, including that mined in Ohio, cause the most significant problems. A substantial percentage of the sulfur found in coal is in the form of inorganic pyrites. This form of sulfur can be removed from coal prior to combustion by physical separation techniques which exploit pyrite's significant differences in density and surface properties from the carbonaceous portion of the coal. However, these techniques require that the coal be ground into particulate form and immersed in water; thus the clean coal product obtained from these cleaning processes are coal–water slurries. The more finely the coal is ground, the greater the percentage of pyritic sulfur which can be removed. However, mechanical dewatering techniques are no longer effective for such deep-cleaned slurries of powdered coal. Thus we examined the potential of using the gel extraction process originally developed by Cussler et al. for the concentration of macromolecular solutions to dewater such coal slurries [29].

The slurry dewatering process using temperature-sensitive gels works as follows. The gel in its shrunken state, above LCST, is contacted with coal slurry at a temperature below the LCST. Thus the gel swells by absorbing water from the slurry. Next, the dewatered slurry is separated from the swollen gel. The gel is then regenerated and the water absorbed is released by warming it above its LCST. Once shrunken, the gel can be reused in another dewatering cycle. Since the gel is recyclable and the dewatering process can be driven by low grade heat, its operating costs should be low. In this project, we examined the effectiveness of the gels in removing water from various coal slurries, the impact of the coal on the equilibrium and kinetic swelling properties of the gels, the service lifetime of the gels in this application, and evaluated the economic competitiveness of the technique with respect to currently used dewatering techniques [16, 115–117]. Here we will present the highlights of this work.

The slurry dewatering experiments were simple in principle but required great care to complete with accuracy. 10×4 PNIPAAm gels were synthesized as flat sheets 1 mm thick and cut into square slabs approximately 30 mm on an edge. The slabs were equilibrated in water at temperatures above the LCST, and then immersed in various coal slurries of different concentrations and particle size ranges. After equilibration at room temperature (below the LCST), the gel slabs were removed from the slurries, dipped into a beaker of room temperature water to dislodge coal particles clinging to the surface of the gel, and then shrunken in warm water. By rigorously accounting for water and coal mass at each stage in the process, the amount of water removed from the slurry by the gel could be determined. Slurry concentration were determined by evaporation to surface dryness.

One expects that the more dilute the initial slurry and the more concentrated the final slurry desired, the greater the amount of gel required relative to slurry. Thus Fig. 21 shows that for a slurry of 40 wt % solids, the greater the amount of gel, the higher the final concentration of solids in the slurry – up to plateaus whose values depend upon the particle size in the slurry. The maximum solids concentration achievable was about 60% for the finest particles (mud-like slurries), but 80% for the coarser slurries (more granular slurries). The reason that complete dewatering is not possible, and that the extent of dewatering achieved drops as particle size is reduced is the result of the fact that the swelling degree of the gels declines sharply in the range of 50 wt % for the finest particulate slurries and drops in the range of 70 wt % for the more coarse slurries. In fact, gels *deswell* when placed in slurries of concentrations greater than these values – diluting the slurries rather than dewatering them! The reason for this decline in swelling was traced to the decline in the chemical potential of

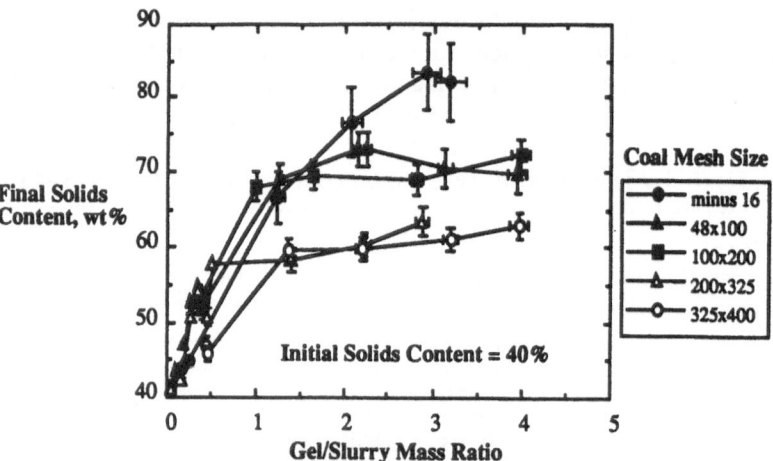

Fig. 21. Coal slurry dewatering capability by gel extraction using thermally responsive 10×4 PNIPAAm gel [117]. The final solids contents rises as the ratio of gel to slurry increases, but plateaus due to reduced sorption capacities of gels in high solids content slurries. The maximum extent of dewatering declines as the particles become finer

water in high solids content slurries. In these slurries, most of the water coats the surface of the coal particles, and its vapor pressure – directly related to chemical potential – was measured to decline sharply in the same ranges that the gels deswelled. The increase in surface area per unit mass with the decrease in particle size is the reason that the decline in swelling occurs at lower solids contents with the finer particles. Nonetheless, the solids contents achieved for the fine particle slurries with gel extraction are better than are achievable with conventional mechanical dewatering techniques, except for centrifugation (a costly technique).

We also examined the swelling and shrinking kinetics of the gels in the slurries in comparison to kinetics in pure water. The rates were reduced in slurry somewhat – swelling and shrinking diffusion coefficients dropped by 60–70%. But the rates were fairly independent of slurry concentration over the range of 20 to 70 wt % solids.

Economic viability of the process will be strongly affected by the number of dewatering cycles in which the gels can be used without deterioration. We found that over a two-month, 21-cycle test period, gel properties remained stable. Specifically, the swollen and shrunken swelling degrees remained constant, as did the swelling and shrinking rates. Furthermore, no physical deterioration of the gel could be observed under a light microscope. Nor was abrasive degradation of the gel slabs immersed in flowing slurry observed.

An economic analysis of the commercial feasibility of the gel extraction process showed that the process could be economically competitive with current techniques (centrifuges, dewatering screens, disc filters) if the gel cycle times were short – minutes rather than hours. This fast cycling time requires that conventional diffusion-controlled gels be of dimension well under a millimeter in diameter. If gels greater than a millimeter are to be used, the process will be economically competitive only if the gels with enhanced sorption rates described in Sect. 5.4 are used. The problem with using particulate gels of submillimeter dimension is the difficulty of separating such gels from the dewatered slurry. However, particulate gels might be usable if a hydrodynamic technique for separating the gel from the slurry particles can be developed, since the gels themselves are almost neutrally buoyant in water.

7.2 Bioseparations

Gel electrophoresis is a powerful, widely-used technique for the separation of complex mixtures of high molecular weight species. There are many variations of the technique, but all are based on the fact that charged species migrate at different rates under the influence of an electric field. If the species are forced to migrate through a gel, convective mixing is eliminated, and the solutes separate into bands as a function of their mobility in the gel. Mobility and thus separation is a complex function of molecular size, shape and charge. Separation of proteins occurs primarily on the basis of molecular size if the detergent sodium dodecyl sulfate (SDS) is added to the solution, since SDS binds with proteins and largely masks any other property differences [167].

Once the biomolecules have been resolved into distinct bands within the gel they can be isolated from each other simply by cutting the gel between the bands. There are several methods for recovering the isolated biomolecules from these gel strips [167]. The first is simply to cut the gel up into smaller pieces, immerse those pieces into solution and wait for the solute to diffuse into the free solution; then the solution is filtered to remove the stripped gel particles. While simple, this method is slow and has a yield of only about 60%. It also dilutes the biomolecule, which then requires additional steps for reconcentration, which is not a trivial problem with biomolecules [167]. Another method of recovering the solute from the gel is to destroy the crosslinks that hold the gel together by chemical treatment. There are several drawbacks to this method. First of all, conditions that can solubilize the gel may denature the biomolecular product. Secondly, the resulting solution is a mixture of the product and the polymer which made up the gel. Thus another separation step is required to produce the final, purified product. A third method of extracting the product from the gel is to use electrophoresis again; this time, to pull the biomolecule out of the gel strip and into free solution. Because electrophoresis is a slow, relatively expensive technique, using it for the simple task of removing a solute from a gel is inefficient.

Our concept was to develop a simpler method for collecting concentrated, purified protein fractions from the gel strips by carrying out the original electrophoresis step within a temperature-sensitive PNIPAAm gel, rather than conventional polyacrylamide gel. When the biomolecules are separated from one another by cutting the gel into strips after electrophoresis, the gel would be warmed several degrees, above the LCST, causing it to shrink and expel a solution containing the product. Since PNIPAAm is a chemical cousin of polyacrylamide gel (the only difference between these gels is the substitution of an isopropyl group for a hydrogen on the amine group of acrylamide monomer), the primary material used in gel elctrophoresis, it was expected to have comparable resolving power. Therefore, users of the standard technique might be able to switch to this new type of gel for product separation and recovery without acquiring new techniques and equipment.

To test this idea, a Bio-Rad Protean II electrophoresis cell and Bio-Rad Model 3000xi computer-controlled power supply were used to carry out the electrophoretic separation and recovery of pre-stained and unstained protein calibration standards (lysozyme, soybean trypsin inhibitor, carbonic anhydrase, ovalbumin, bovine serum albumin and phosphorylase B) obtained from Bio-Rad Laboratories. Standard SDS-gel electrophoresis techniques were used [167].

For use in the gel electrophoresis system, the PNIPAAm gel must be formed in a buffer solution containing sodium dodecyl sulfate (SDS). These additional solutes present in the reaction mixture altered the polymerization of the gel; most significantly, the initiator levels required to form the gel successfully in these circumstances are different from the values required for polymerization in pure water. Thus successful PNIPAAm gel formation within the required mold and solution was a challenge. The most successful gel formulation was a 12×2 formulation with SMB and APS concentrations of 14 mg/100 ml each; the solution was a pH 8 Tris-HCl buffer containing 0.1 wt % SDS. To carry out the electrophoretic separation, the gel (inside its glass mold) was inserted into the Protean II cell. Aliquots of 10–20 µl of the protein standard solutions were applied to the sample wells molded into the top of the gels. The separation was run for 5 h at a constant current of 16 milliamps. The cooling water flowing through the cell was set to keep the gel at 15 °C (effective cooling of the entire gel was a major technical problem, since temperature gradients caused by resistance heating caused the gel to swell and shrink in irregular fashions within the molds).

Several trials of separation of the pre-stained protein standards were marginally successful [168]. The proteins were separated on the PNIPAAm gel, but only as broad, smeared out bands. The separation was inferior to that achieved with standard polyacrylamide gel. The poor separation was probably related to non-uniform gel formation and microscopic heterogeneity. The pre-stained proteins could be leached out of the gels by warming them; however, we did not develop analytical techniques necessary to quantify recovery. We also tried separation of unstained proteins by carrying out the electrophoresis in the same mode as the pre-stained proteins. However, the staining process to fix the bands and make them visible after separation required immersion of the gel in dye solution containing organic solvent. When this was attempted, the dye stained the gel itself, and the solvent used caused the gel to shrink and crack.

Thus electrophoretic separation of proteins on PNIPAAm gel can be achieved, and the proteins can be recovered from the gel. However, the substantial technical difficulties we encountered made our vision of a simple swap of PNIPAAm gel for polyacrylamide gel in standard procedures impossible to realize. Thus, we discontinued further work on this project. However, the concept might still find use in cases of routine preparative gel electrophoresis where the problems and expense of solute recovery as described earlier were significant enough to warrant the developmental work required to make the PNIPAAm process practical.

8 Concluding Remarks

At this point in time, the principles for the synthesis of polymer gels which respond to a great diversity of chemical and physical signals have been widely demonstrated. A good qualitative understanding of how key parameters influence the important properties of responsive polymer gels exists among researchers in the field. Although useful models exist for making predictions of how properties like swelling degree or permeability will vary with gel properties like crosslink density or with environmental conditions like solution composition, qualitative predictive success is elusive on most counts. Some of the areas in which we feel further work is needed include:

1) Examination of how synthesis conditions can alter the physical structure of the gel network and how microstructure affects responsive gel properties.

2) Modeling and measuring the generation of mechanical force (rate and magnitude) by gels in terms of accessible parameters like modulus and solubility parameters.

3) Development of responsive gels with improved mechanical properties like increased tear strength.

4) Modeling of the interactions of solutes with gels; even a model which simply predicts trends would be a significant advance.

5) Continued fundamental work on equilibrium swelling thermodynamics and kinetics to allow systematic creation of responsive gels to match application requirements.

6) Continued development of novel applications for these materials. Devices whose performance depends upon low-precision control of a single property, such as transparency (as in a sun screen) or swelling degree (as in a sensor whose signal is a visible change in volume) should be fairly easy to develop if appropriate applications for the phenomena are found. Development of devices requiring simultaneous control of multiple gel properties or precise characteristics will take much more effort. For example, creation of bioresponsive drug delivery systems requires precise, accurate and reproducible control of the transition point, magnitude of the transition, response kinetics and more. And these characteristics must be designed into a material which also meets stringent requirements for biocompatibility!

Acknowledgments: M.K. Akhtar, K.D. Gehrke, D.C. Harsh, B.G. Kabra and M. Palasis provided invaluable assistance in the preparation of this chapter by reviewing text, preparing figures, compiling the reference list, and so forth. All of the work described in this chapter was carried out in collaboration with undergraduate (E. Balam, E. Cunningham, S. Fisher, P. Gwynne, T. Hashman, J.F. Johnson, R.M. Mason, J. Robeson, C. Stauss) and graduate students (M.K. Akhtar, D. Biren, D.C. Harsh, B.G. Kabra, L.H. Lyu, M. Palasis, N. Vaid), plus a post-doctoral assistant (M.C. Yang). In addition to the papers cited here (and many others not cited), discussions with fellow academic scientists (especially E. Cussler, C. Durning, R. Freitas, A. Hoffman, M. Ilavsky, M. Marchetti, N. Peppas, R. Siegel, M. Tirrell) have helped in the development of the ideas presented here. Financial support of this work has been provided by the Mobis Corp., 3M Co. (Non-tenured Faculty Grants), National Science Foundation (Grants CBT-8809271 and EID-9000779), National Institutes of Health (Grant BRSG-RR07075), Ohio Coal Development Office (Grant OCDO/R-87-57), and the University of Cincinnati Research Council.

9 References

1. Gehrke SH, Lee PI (1990) In: Tyle P (ed) Specialized drug delivery systems: manufacturing and production technology. Marcel Dekker, New York, p 333
2. Tanaka T (1981) Sci Am 244: 124

3. Peppas NA, Barr-Howell BD (1986) In: Peppas NA (ed) Hydrogels in medicine and pharmacy, vol I fundamentals. CRC Press, Boca Raton, FL, p 27
4. Freitas RFS, Cussler EL (1987) Chem Eng Sci 42: 97
5. Gehrke SH, Cussler EL (1989) Chem Eng Sci 44: 559
6. Tanaka T, Nishio I, Sun S, Ueno-Nishio S (1982) Science 218: 467
7. Tanaka T (1978) Phys Rev Lett 40: 820
8. Suzuki A, Tanaka T (1990) Nature 346: 345
9. Mamada A, Tanaka T, Kungwatchakun D, Irie M (1990) Macromolecules 23: 1517
10. Lee KK, Cussler EL, Marchetti M, McHugh MA (1990) Chem Eng Sci 45: 766
11. Ohmine I, Tanaka T (1982) J Chem Phys 11: 5725
12. Huang X, Unno H, Akehata T, Hirasa O (1988) J Chem Eng Jpn 21: 10
13. Horbett TA, Kost J, Ratner BD (1984) In: Shalaby SW, Hoffman AS, Ratner BD, Horbett TA (eds) Polymers as biomaterials. Plenum, New York, p 193
14. Sussmann MV, Katchalsky A (1970) Science 167: 45
15. Tatara Y (1978) Polym J 10: 563
16. Lyu LH, Gehrke SH (1989) AIChE Annual Meeting, Paper 98j, San Francisco, CA
17. Katchalsky A (1949) Experientia 5: 319
18. Katchalsky A, Sussmann MV, Oplatka A, Steinberg I (1967) US Patent No. 3,321,908
19. Dusek, K, Patterson D (1968) J Polym Sci A-2 6: 521
20. Research Institute for Polymers and Textiles Bulletin (1989) Ibaraki, Japan p 12
21. Taylor LD, Cerankowski LD (1975) J Polym Sci, Polym Chem Edn 13: 2551
22. Hirokawa Y, Tanaka T (1984) J Chem Phys 81: 6379
23. Katayama S, Hirokawa Y, Tanaka T (1984) Macromolecules 17: 2641
24. Bae YH, Okano T, Kim SW (1989) J Controlled Release 9: 271
25. Okano T, Bae YH, Jacobs H, Kim SW (1990) J Controlled Release 11: 255
26. Mukae K, Bae YH, Okano T, Kim SW (1990) Polym J 22: 250
27. Hoffman AS, Afrassiabi A, Dong LC (1986) J Controlled Release 4: 213
28. Hoffman AS (1987) J Controlled Release 6: 297
29. Cussler EL, Stokar MR, Varberg JE (1984) AIChE J 30: 578
30. Gehrke SH, Andrews GP, Cussler EL (1986) Chem Eng Sci 41: 2153
31. Trank SJ, Cussler EL (1987) Chem Eng Sci 42: 381
32. Siegel RA, Firestone BA (1988) Macromolecules 21: 3254
33. Firestone BA, Siegel RA (1988) Polym Commun 29: 204
34. Beltran S, Hooper HH, Blanch HW, Prausnitz JM (1990) J Chem Phys 92: 2061
35. Hooper HH, Baker JP, Blanch HW, Prausnitz JM (1990) Macromolecules 23: 1096
36. Yoshio N, Hirohito N, Matsuhiko M (1986) J Chem Eng Jpn 19: 274
37. Amiya T, Tanaka T (1987) Macromolecules 20: 1162
38. Harsh DC, Gehrke SH (1990) In: Brannon-Peppas L, Harland R (eds) Absorbent polymer technology. Elsevier, Amsterdam, p 103
39. Huang X, Unno H, Akehata T, Hirasa O (1987) J Chem Eng Jpn 20: 123
40. Peppas NA (1987) In: Hydrogels in medicine and pharmacy, vol. 2 polymers. CRC Press, Boca Raton FL, p 1
41. Urry DW, Harris D, Long MM, Prasad KU (1986) J Peptide Protein Res 28: 649
42. Klier J, Scranton AB, Peppas NA (1990) Macromolecules 23: 4944
43. Scranton AB, Klier J, Peppas NA (1991) J Polym Sci, Polym Phys Ed 29: 211
44. Scranton AB, Klier J, Aronson CL (1992) In: Harland RS, Prud'homme, RK (eds) Polyelectrolyte gels, ACS symposium series 480. American Chemical Society, Washington, DC
45. Gehrke SH, Palasis M, Akhtar MK (1992) Polymer Int 28: 29
46. Kabra BG, Akhtar MK, Gehrke SH (1992) Polymer 33: 990
47. Ilavsky M (1981) Polymer 22: 1687
48. Ilavsky M (1982) Macromolecules 15: 782
49. Prange MM, Hooper HH, Prausnitz JM (1989) AIChE J 35: 803
50. Marchetti M, Prager S, Cussler EL (1990) Macromolecules 23: 1760
51. Marchetti M, Prager S, Cussler EL (1990) Macromolecules 23: 3445
52. Harsh D, Gehrke SH (1991) J Controlled Release 17: 175
53. Harsh D, Gehrke SH (1993) In: El-Nokaly M, Piatt D, Charpentier B (eds) Polymeric delivery systems. ACS symposium series 520. American Chemical Society, Washington, DC, p 105
54. Otake K, Inomata H, Konno M, Saito S (1990) Macromolecules 23: 283
55. Tanaka T, Sun ST, Hirokawa Y, Katayama S, Kucera J, Hirose Y, Amiya T (1987) Nature 325: 796

56. Matsuo ES, Tanaka T (1988) J Chem Phys 89: 1695
57. Grimshaw PE, Grodzinsky AJ, Yarmush ML, Yarmush DM (1990) Chem Eng Sci 45: 2917
58. Chou LY, Blanch HW, Prausnitz JM (1992) J Appl Poly Sci 45: 1411
59. Gehrke SH, Lyu LH, Yang MC (1989) Polym Prepr 30: 482
60. Tanaka T, Fillmore DJ (1979) J Chem Phys 70: 1214
61. Gehrke SH, Akhtar MK (1990) In: 33rd IUPAC international symposium on macromolecules .
    July 8–13, Montreal, Canada
62. Tanaka T, Sato E, Hirokawa Y, Hirotsu S, Peetermans J (1985) Phys Rev Lett 55: 2455
63. Li Y, Tanaka T (1990) J Chem Phys 92: 1365
64. Li Y, Tanaka T (1991) In: De Rossi D, Kajiwara K, Osada Y, Yamauchi A (eds) Polymer gels:
    Fundamentals and biomedical applications. Plenum Press, New York, p 41
65. Kabra B, Gehrke S (1991) Polym Commun 32: 322
66. Tatara Y, Mori M (1972) Trans Jpn Soc Mech Eng 38: 2807
67. Chiarelli P, Umezawa K, De Rossi D (1991) In: De Rossi D, Kajiwara K, Osada Y, Yamauchi
    A (eds) Polymer gels: Fundamentals and biomedical applications. Plenum, New York, p 195
68. Leung BK, Robinson J (1990) J Membr Sci 52: 1
69. Feil H, Bae YH, Feijen J, Kim SW (1991) J Membr Sci 64: 283
70. Palasis M, Gehrke SH (1992) J Controlled Release 18: 1
71. De Rossi D, Kajiwara K, Osada Y, Yamauchi A (eds) (1991) Polymer gels: Fundamentals and
    biomedical applications. Plenum, New York
72. Urry DW, Harris RD, Prasad KU (1988) J Am Chem Soc 110: 3303
73. Suzuki M (1989) Kobunshi Ronbunshu 46: 603
74. Tatara Y (1987) Adv Robotics 2: 69
75. Bishop JE (1990) In: The Wall Street Journal, August 13, p B1
76. De Rossi DE, Galletti PM, Dario P, Richardson PD (1983) ASAIO J 6: 1
77. De Rossi DE, Parrini P, Chiarelli P, Buzzigoli G (1985) Trans Am Soc Artif Intern Organs
    31: 60
78. De Rossi DE, Chiarelli P, Buzzigoli G, Domenici C, Lazzeri L (1986) Trans Am Soc Artif
    Intern Organs 32: 157
79. Chiarelli P, De Rossi D (1988) Progr Colloid Polym Sci 78: 4
80. Stinson S (1990) Chem Eng News, January 8, p 30
81. Shiga T, Hirose Y, Okada A, Kurauchi T (1990) Polym Prepr 30: 310
82. Sheppard NF, Lee HL, Madrid Y, Melcher JR, Langer RS (1988) Annual AIChE Meeting,
    paper 120i, Washington DC
83. Osada Y, Umezawa K, Yamauchi A (1988) Makromol Chem 189: 597
84. Umezawa K, Osada Y (1987) Chem Lett 1795
85. Gillmore VE (1989) Popular Sci 235: 24
86. Wald ML (1992) In: The New York Times, August 16, p F9
87. Tanaka T (1992) In: Harland RS, Prud'homme RK (eds) Polyelectrolyte gels, ACS symposium
    series 480. American Chemical Society, Washington, DC, p 1
88. Urry DW (1990) Polymeric Materials Eng Sci 63: 329
89. Cornejo-Bravo JM, Siegel RA (1990) Proc Intern Symp Contr Rel Bioact Mat 17: 174
90. Grimshaw PE, Grodzinsky AJ, Yarmush ML, Yarmush DM (1989) Chem Eng Sci 44: 827
91. Grodzinsky AJ, Grimshaw PE, Yarmush ML (1990) Proc Intern Symp Contr Rel Bioact Mat
    17: 108
92. Afrassiabi A, Hoffman AS, Cadwell LA (1987) J Membr Sci 33: 191
93. Dong LC, Hoffman AS (1992) J Controlled Release 19: 171
94. Dong LC, Hoffman AS (1990) J Controlled Release 13: 21
95. Mukae K, Bae YH, Okano T, Kim SW (1990) Polym. J 22: 206
96. Bae YH, Okano T, Kim SW (1989) J Controlled Release 9: 271
97. Bae YH, Okano T, Kim SW (1991) Pharm Res 8: 531
98. Bae YH, Okano T, Kim SW (1991) Pharm Res 8: 624
99. Kwon IC, Bae YH, Kim SW (1991) Nature 28: 291
100. Kwon IC, Bae YH, Okano T, Kim SW (1991) J Controlled Release 17: 149
101. Horbett TA, Ratner BD, Kost J, Singh M (1984) In: Anderson JM, Kim SW (eds) Recent
     advances in drug delivery systems. Plenum, New York, p 209
102. Horbett TA, Ratner BD (1988) Proceedings Pharm Tech Conference '88. East Rutherford, NJ,
     p 356
103. Kost J, Horbett TA, Ratner BD (1985) J Biomed Mater Res 19: 1117
104. Albin GW, Horbett TA, Miller SR, Ricker NL (1987) J Controlled Release 6: 267

105. Osada Y, Takeuchi Y (1981) J Polym Sci, Poly Lett Ed 19: 303
106. Osada Y, Takeuchi Y (1983) Polym J 15: 279
107. Osada Y, Hasebe M (1985) Chem Lett 12: 85
108. Huang X, Akehata T, Unno H, Hirasa O (1989) Biotech Bioeng 34: 102
109. Hirasa O, Ito, S, Yamauchi A, Fujishige S, Ichijo H (1991) In: DeRossi D, Kajiwara K, Osada Y, Yamauchi A (eds) Polymer gels: Fundamentals and biomedical applications. Plenum Press, New York, p 241
110. Dong LC, Hoffman AS (1986) J Controlled Release 4: 223
111. Gehrke SH (1986) Kinetics of gel volume change and its interaction with solutes. Thesis, University of Minnesota, Minneapolis, MN
112. Kabra BG, Gehrke SH, Hwang ST, Ritschel W (1991) J Appl Polym Sci 42: 2409
113. Gehrke SH (1992) to appear in: Lee PI (ed) Fundamental aspects of polymers in pharmaceutics. Elsevier, Amsterdam
114. Gehrke SH, Johnson JF, Robeson J, Vaid N (1991) Biotech Prog 7: 355
115. Gehrke SH, Lyu, LH, Barnthouse K (1992) submitted
116. Lyu LH (1990) Dewatering fine coal slurries by gel extraction. Thesis, University of Cincinnati, Cincinnati, OH
117. Gehrke SH, Lyu LH (1990) Final project report: Dewatering fine coal slurries by gel extraction. Ohio Coal Development Office, Columbus, Ohio
118. Gehrke SH, Yang MC, Kabra BG, Lyu LH, Akhtar MK (1989) Proceed Intern Symp Control Rel Bioact Mater 16: 209
119. Gehrke SH, Akhtar MK (1992) submitted
120. Akhtar MK, Gehrke SH (1992) submitted
121. Akhtar MK (1990) Volume change kinetics in near critical gels. Thesis, University of Cincinnati, Cincinnati OH
122. Palasis M, Gehrke SH (1990) Proceed Intern Symp Contr Rel Bioact Mater 17: 385
123. Palasis M (expected, 1993) Permeability of responsive polymer gels. Thesis, University of Cincinnati, Cincinnati OH
124. Harsh DC, Gehrke SH (1989) Proceed Intern Symp Contr Rel Bioact Mater 16: 383
125. Harsh DC (1992) Controlling swelling characteristics of novel cellulose ether gels. Thesis, University of Cincinnati, Cincinnati OH
126. Kabra BG (expected, 1993) Relationships between synthesis, structure, and swelling properties of responsive polymer gels. Thesis, University of Cincinnati, Cincinnati OH
127. Gehrke SH, Agrawal G, Yang MC (1992) In: Harland RS, Prud'homme, RK (eds) Polyelectrolyte gels, ACS symposium series 480. American Chemical Society, Washington, DC, p 211
128. Hrouz J, Ilavsky M, Ulbrich K, Kopecek J (1981) Eur Polym J 17: 361
129. Ilavsky M, Hrouz J, Ulbrich K (1982) Polym Bull 7: 107
130. Hirotsu S, Hirokawa Y, Tanaka T (1987) J Chem Phys 87: 1392
131. Trank SJ, Johnson DW, Cussler EL (1989) Food Tech. 43: 78
132. Dualeh AJ, Steiner CA (1992) In: Harland RS, Prud'homme, RK (eds) Polyelectrolyte gels, ACS symposium series 480. American Chemical Society, Washington, DC, p 42
133. Doelker E (1990) In: Brannon-Peppas L, Harland RS (eds) Absorbent polymer technology. Elsevier, Amsterdam, p 125
134. Doelker E (1986) In: Peppas NA (ed) Hydrogels in medicine and pharmacy, vol. II polymers. CRC Press, Boca Raton, FL, p 115
135. Butler RW, Klug ED (1980) In: Davidson RL (ed) Handbook of water-soluble gums and resins. McGraw-Hill, New York, p 1
136. Brannon-Peppas L (1990) In: Brannon-Peppas L, Harland RS (eds) Absorbent polymer technology. Elsevier, Amsterdam, p 67
137. Brannon-Peppas L, Peppas NA (1988) Polym Bull 20: 285
138. Sanchez IC, Lacombe RH (1976) J Phys Chem 80: 2352
139. Lacombe RH, Sanchez IC (1976) J Phys Chem 80: 2568
140. Sanchez IC, Lacombe RH (1978) Macromolecules 11: 1145
141. Sanchez IC, Balazs AC (1989) Macromolecules 22: 2325
142. Welty JR, Wicks CE, Wilson RE (1984) Fundamentals of momentum, heat, and mass transfer, 3rd edn. John Wiley, New York
143. Perry RH, Green D (1984) Chemical engineering handbook, 6th edn. McGraw-Hill, New York
144. Carslaw AS, Jaeger JC (1959) The conduction of heat in solids, 2nd edn. Oxford Clarendon Press, Oxford

145. Helfferich F (1965) J Phys Chem 69: 1178
146. Helfferich F (1962) Ion Exchange. McGraw-Hill, New York
147. Dana PR, Wheelock TD (1974) Ind Eng Chem Fund 13: 20
148. Levenspiel O (1972) Chemical Reaction Engineering, 2nd edn. John Wiley, New York
149. Crank J (1975) The mathematics of diffusion, 2nd edn. Oxford University Press, London
150. Meares P (1968) In: Crank J, Park G (eds) Diffusion in polymers. Academic, London, p 373
151. Tanaka T, Hocker LO, Benedek GB (1973) J Chem Phys 59: 5151
152. de Gennes. PG (1979) Scaling concepts in polymer physics. Cornell Unoversity Press, Ithaca, New York
153. Billovits GF, Durning CJ (1989) Chem Eng Commum 82: 21
154. Peters A, Schosseler F, Candau SJ (1989) In: Glass J (ed) Polymer in aqueous media: performance through association. Advances in chemistry series 223. American Chemical Society, Washington, DC, p 45
155. Chatterjee PK, Nguyen HV (1985) In: Chatterjee PK (ed) Absorbency. Textile science and technology, vol. 7. Elsevier Science Publishers, Amsterdam, p 29
156. Willams JM, Wrobleski DA (1989) Langmuir 4: 656
157. Barby D, Haq Z (1982) European Patent 0,060,138
158. Williams JM, Wilkerson KR (1990) Polymer 31: 2162
159. Yasuda, H, Peterlin A, Colton CK, Smith KA, Merrill EW (1969) Die Makromol Chem 126: 177
160. Schnitzer JE (1988) Biophys J 54: 1065
161. Peppas NA, Reinhart (1983) J Membr Sci 15: 275
162. Lustig SR, Peppas NA (1988) J Appl Polym Sci 36: 735
163. Dearden JC, Tomlinson E (1971) J Pharm Pharmac 73: 735
164. Hansch C, Anderson S (1967) J Org Chem 32: 2583
165. Ruckenstein E, Lesins V (1988) In: Mizrahi A (ed) Downstream processes: equipment and techniques. Alan R. Liss, New York, p 246
166. Luck WA (1980) In: Rowland SP (ed) Water in polymers, ACS symposium series 127. American Chemical Society, Washington DC, p 43
167. Scopes RK (1987) Protein purification: principles and practice, 2nd edn. Springer, Berlin Heidelberg New York
168. Hashman T, Palasis M (1988) unpublished research

Received October 14, 1992

Polymer Gel Phase Transition in Condensation-Decondensation of Secretory Products

Pedro Verdugo
Center for Bioengineering, University of Washington WD-12, Seattle, Washington, 98195, USA

The molecular mechanisms of product storage and release in secretion remain unknown. Mucins stored in secretory granules form a highly condensed polymer network. An important condition for mucin condensation is the shielding of their polyionic charges, which depends on the presence of large amounts of Ca^{2+} inside the secretory granule. Experiments conducted on rabbit respiratory goblet cells, and in isolated giant mucin granules of the slug *Ariolimax columbianus*, indicate that upon exocytosis the mucin network undergoes dramatic swelling, expanding its volume as much as 250 to 1000 fold, respectively. Swelling of the granular matrix is governed by Donnan equilibrium, and follows a typical first-order kinetics. Depending upon the composition of the medium, the exocytosed mucin network can be recondensed. The observed volume changes are reversible and exhibit the characteristic discontinuity, as well as the temperature and pH dependence found in polymer gel phase transitions.

The evidence reviewed here is consistent with the idea that the condensed conformation of secretory products during storage in the cell, and their hydrated conformation upon release from the cell, reflect the corresponding condensed and decondensed phases of a polymer gel. Product release in exocytosis would result from a polymer gel phase transition that is probably triggered by a polycation $Z^{\geq +2}/Na^+$ ion exchange via the secretory pore.

Advances in Polymer Science, Vol. 110
© Springer Verlag Berlin Heidelberg 1993

1 Introduction

All higher forms of life occur in a gel phase. Cells and the environment they live in are made of complex polymer networks forming a broad range of hydrogels. Although the overall biological role of many of these gels has been well defined, their mechanisms of operation remain poorly understood. A striking paradox is that in spite of the rich and powerful body of polymer physics developed since the time of Flory [1] to explain the dynamics of polymer networks, the utilization of these principles to investigate the functioning of biological gels have been virtually ignored.

Here, we report an example in which the application of theoretical principles and methods to study polymer networks is changing drastically our view of the mechanism of storage and release in secretion.

First, we will describe briefly the biology of secretory cells in general and goblet cells in particular. Next, we will outline our earlier studies on the conformation of mucin networks using dynamic laser scattering. Short discussions on the Donnan swelling properties of the mucin network will bring us to the application of the theory of polymer gel phase transition to explain condensation and decondensation in secretion.

1.1 Secretory Cells

Cells that specialize in the production and export of materials are known as secretory cells. There are two modes by which cells export newly synthesized materials, known as regulated and non-regulated secretion. In the former, which is the subject of this review, secreted products follow a well-defined assembly line that begins with the transcription of nucleic acid into peptide chains. The resulting peptides enter a second reactor – the Golgi apparatus – where they undergo further chemical processing, including methylation and glycosilation. Secretory materials are then sorted and packed in membrane-bound vesicles. These vesicles bud from the Golgi apparatus to follow a complex and still obscure process of "maturation" whereby secretions are further refined and drastically condensed. The end result is the formation of smaller vesicles called secretory granules, in which densely packed secretions remain stored, awaiting release from the cell. Upon stimulation, the granules dock to the cell wall; a pore is formed connecting the interior of the granule with the extracellular space, and the content of the granule undergoes quick decondensation, driving the final release of secretory products in a process called exocytosis (see Fig. 4).

Together with the primary secretory products (enzymes, hormones, neurotransmitters, etc.) secretory granules contain a pair of countercharged chemical moieties: a polyanionic polymer matrix, and a cation or polycation. These countercharged pairs, which are found ubiquitously inside different types of secretory granules, gave us the first clue that a polymer gel phase transition –

similar to the one found in synthetic polymer gels – might be at work in the condensation of secretory products, while stored in the granule, and in decondensation of secretory products, upon their release from the cell.

Results obtained in mucin-secreting goblet cells, which will be reviewed here, and in histamine-secreting mast cells, which have been presented elsewhere [2], are prompting us to abandon some long-established ideas about packing and release in secretion and to sketch a new hypothesis based on current polymer-gel theory.

1.1.1 The Goblet Cell

The lumens of the respiratory, reproductive and digestive tracts are lined with a layer of cobblestone-shaped cells forming a pavement-like tissue called the mucosa. Normally coated by mucus, the mucosa has a critical role in the protection of adjacent tissues. This protective function results from a complex and still poorly understood interaction among the various cell types that form the mucosa, including the goblet cells. The specific role of goblet cells is to produce mucins, the large structural polymers that form the mucus gel matrix. Other cell types contribute with their corresponding functions, i.e. moving the mucus layer by ciliary action, absorbing nutrients, releasing anti-microbial agents or digestive enzymes, or controlling the movement of ions and water to and from the mucosal surface.

1.2 Mucins and the Mucus Gel

Mucus is one of the most ubiquitous polymer gels found in nature. In a broad range of animal species, mucus functions as a barrier that protects the skin or internal mucosa from chemical, physical, or bacterial injury. The polymer matrix of mucus is made of a heterogeneous group of polymers called mucins, which are the main secretory product of goblet cells. Mucins are glycoproteins of gigantic dimensions that can reach up to several microns in length. Their primary structure consists of a linear peptide backbone, or apomucin, to which are attached short polysaccharide side branches forming a bottle-brush structure. In these polymers, highly glycosilated, rigid hydrophilic domains, composing about 75% of the polymer length, alternate with non-glycosilated, flexible hydrophobic domains that account for the remaining 25%. Recent studies have shown that mucins, like other polymers containing periodic flexible/rigid domains, can self-associate, forming nematic liquid crystalline structures [3]. Approximately 80% of the mucin composition by weight is carbohydrate, the remaining 20% the protein backbone. The sugar side chains are two to twenty monosaccharides long and vary in composition, except for the first monosaccharide, which is always N-acetylgalactosamine. The glycosidic side chains are connected to the apomucin core by O-glycosidic bonds between

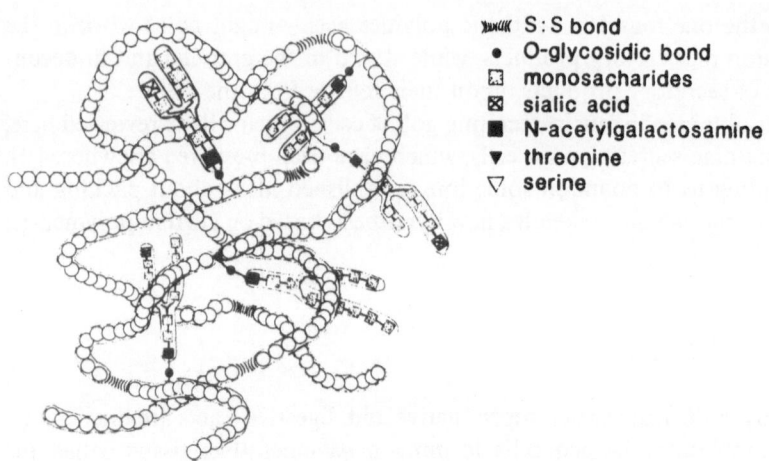

)wwwwwww S:S bond
• O-glycosidic bond
▣ monosacharides
⊠ sialic acid
■ N-acetylgalactosamine
▼ threonine
▽ serine

Fig. 1. Schematic representation of the mucin polymer network. Notice the tangled topology of the network, the linear conformation of the mucins and the presence of S:S bonds in the apomucin backbone

N-acetylgalactosamine and threonine or serine. Owing to the presence of cysteine residues, the non-glycosilated regions of the mucin chains are susceptible to cleavage [4–9]. The presence of sulfate and sialic acid terminals in the sugar side chains give mucins their characteristically strong polyanionic properties [10–12] (see Fig. 1).

1.2.1 Tangle Network Model for the Mucin Matrix

For many years, the mucin matrix was thought to be composed of branched mucin chains interconnected by cystein disulfide bonds. It was further believed that changes in the degree of S:S bonding among mucin chains were responsible for controlling the mucus rheological properties, a critical feature for the function of this gel [13, 14]. Regulatory messages from hormones or other chemical transmitters would prompt the secretory cells to produce a more or less S:S crosslinked mucin matrix, modulating in this way the viscoelastic properties of the mucus gel. A major weakness of this view was that it was based on indirect evidence derived from analytical methods that degrade precisely the level of molecular organization they attempt to define, namely the macromolecular conformation of the mucin-polymer network.

Non-destructive methods revealed a very different picture. Studies using dynamic laser scattering rendered the presence of two distinct diffusional components in the native undegraded mucus gel, a local diffusional fluctuation $(2 \times 10^{-9} \text{ cm}^2/\text{s})$ and a slow translational (reptational) diffusion $(8 \times 10^{-10} \text{ cm}^2/\text{s})$. This finding was the first compelling indication that mucins inside the mucus matrix must be held together by tangles and low energy bonds rather than by S:S interchain crosslinks. It also implied that mucins should be linear rather than branched, and that swelling, rather than the degree of S:S bonding, must control the rheological properties of mucus [15, 16].

1.2.1.1 Linear Conformation of Mucins

The linear conformation of mucins, proposed by Carlstedt and Sheehan [6], could be inferred from the tangle network model [15, 16]. It was later confirmed directly by several groups [9, 17, 18]. Electron microscopic images of mucin chains obtained by rotary shadowing demonstrated unequivocally the characteristic linear conformation of these large polymers. Light scattering measurements rendered a radius of approximately 270 nm for respiratory mucins, which compared to the micron-plus dimensions observed by electron microscopy suggest that, in solution, mucins must have a very flexible random coil conformation [6]. This is consistent with the observation that mucins exhibit the typical proportional log/log relationship between the radius of gyration and molecular weight that characterize a linear random coil conformation. Estimations of the Kuhn statistical elements suggest a mucin subunit of 150 nm, with a radius of gyration of 40 nm, and a molecular weight of 500 kDa. Longer chains would result from the association of these subunits [19].

1.2.1.2 Annealing of Mucus

Another illuminating implication of the proposed tangled network model was that it explained an old riddle, namely: how could mucin networks, which are normally released from the granules in small discrete packages, interpenetrate each other, annealing into a continuum gel mass? While the annealing property of mucus remained unexplained by the notion of an interchain S:S bonded matrix, it was readily explained by the idea of a tangled mucin polymer network.

1.2.1.3 Donnan Swelling of Mucus

The most substantial prediction of the tangled mucin network hypothesis was that the main parameter controlling mucus rheology should be hydration. According to Katchalski's observations in ion-exchange resins [20], the expectation was that, due to the polyionic properties of the mucins, the swelling of mucus should be governed by Donnan equilibrium [21]. Experimental tests confirmed this idea. Results of experiments conducted in mucus of the cow uterine cervix indicated that during swelling, the apparent diffusivity of the gel's network vary with the pH and decrease with the concentration of salt in the swelling medium [21].

2 Mucin Condensation and Decondensation

The finding of a tangle topology for the mucin network and the Donnan properties of the mucus gel brought us to the formulation of a new hypothesis to explain mucin secretion, mucus gelation and the regulation of the rheological properties of mucus. We proposed that mucins must be stored in a highly condensed form, and must swell upon release from the cell. Mucus would not be secreted as such, as it was believed at the time, but would be formed outside the cell by hydration of a condensed mucin network. Swelling of the mucin matrix – and thereby the rheological properties of the mucus – would be modulated by the pH and the ionic and polyionic composition of the liquid on the surface of the mucosa. The hydroelectrolitic composition of the liquid on the surface of the mucosa would be, in this case, the parameter under physiologic control [22, 23]. This hypothesis provided the first accurate explanation for the paramount defect in cystic fibrosis, a lethal genetic disease characterized by the presence of thick viscous mucus. We suggested that defective mucus rheology might not result from an imperfect mucin structure, as many thought then, but from a deficiency in the hydration of the mucin network. Based on the Donnan properties of mucus, we further inferred that incomplete mucus hydration could result from a faulty electrolytic composition of the water on the mucosal surface, and that the basic deficiency in this disease should be in the transmucosal movement of water and electrolytes, a finding that has now been confirmed [24].

Microscopic observations conducted in goblet cells grown in tissue culture provided unequivocal verification that mucins do indeed swell upon release (see Fig. 2). Measurements performed by digitizing video microscopic records of secreting goblet cells demonstrated that, during product release, the radius of the mucin matrix increases following a characteristic first-order kinetics, as predicted by Tanaka's theory of swelling of polymer gels [25]. These experiments also indicated that the diffusivity of the exocytosed mucin network, as estimated from the characteristic time τ of the first-order kinetics (see Fig. 2) and the final radius of the swelling mucin network, vary with the concentration of salt in the medium and exhibit the characteristic features of a Donnan equilibrium process [23, 26, 27].

2.1 Mucin Condensation and Storage

The finding that the granular content undergoes extensive swelling upon release provided objective evidence that the mucin network must be condensed inside the granule. Two important issues remained unresolved, however, one regarding the topology of packing inside the granule, the other regarding the mechanism of condensation inside the granule.

In the case of the giant mucin granule of the terrestrial slug *Ariolimax columbianus*, the swelling of the mucin matrix upon release is remarkably fast,

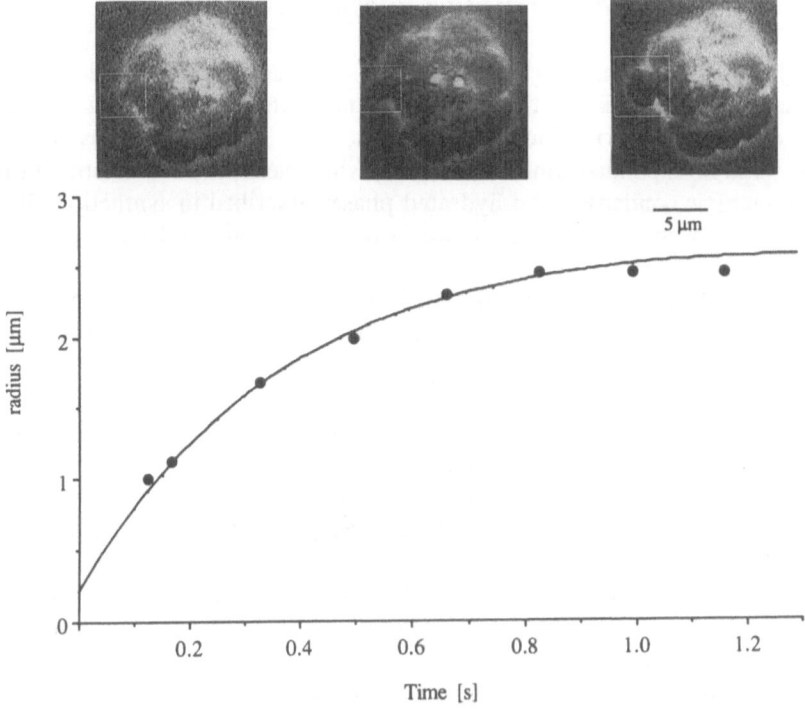

Fig. 2. Video micrograph of exocytosis in a goblet cell grown in tissue culture. Notice that the swelling of a secretory granule has been captured at 3 consecutive times (inset). The radial expansion of the exocytosed granules follows a typical first-order kinetics. The continuous line is a non-linear least square fitting to the data points to $r(t) = r_f - (r_f - r_i) \cdot e^{-t/\tau}$, where r_i and r_f are the initial and final radius of the granule, and τ is the characteristic relaxation time of swelling

driving a volume expansion of about 1000 fold in approximately 16 ms. These explosive rates of swelling suggest that condensed mucins must have some folding order which can explain the swift unhindered unfolding of the mucin network during product release. The observation that mucins in a constrained water space can spontaneously adopt a liquid crystalline order is consistent with the requirement of a non-random packing arrangement inside the mucin secretory granule [3].

A necessary condition for condensation of the mucin polymer matrix is the screening of the strong polyanionic charges of the mucin chains. The application of electron probe microanalysis to measure the elemental composition of secretory granules revealed that granules contain Ca in amounts more than sufficient to shield the polyanionic charges of the mucins. Calcium can reach up to 1 M (M/kg of dry material) in the giant mucin granule of the terrestrial slug *Ariolimax columbianus*, while in the mucin granule of goblet cells Ca was found to reach up to 120 Mm (Mm/kg dry weight) [28–31].

2.2 Polymer Gel Phase Transition in Mucin Release

Although charge screening is necessary, it is not sufficient to explain the condensation of mucins inside the secretory granule. An intriguing alternative is that the collapsed conformation of the mucin network – while stored inside the granule – and its expanded conformation following release could correspond to the characteristic condensed and hydrated phases described in synthetic polymer gel [32]. During exocytosis, the mucin network would undergo a typical polymer gel phase transition.

A direct implication of this hypothesis is that phase transition of the mucus gel should be reversible, and it should exhibit the characteristic features of a critical phenomenon. Studies conducted in isolated giant mucin secretory granules of the terrestrial slug revealed that hydrated mucin gels, released from individual secretory granules, can indeed be recondensed. Recondensation/decondensation is reversible and exhibits the typical features of a polymer gel phase transition. Namely, it is discontinuous, and is affected by pH, temperature and Ca^{2+} concentration in a fashion that mimics phase transition in synthetic polymer hydrogels [32, 33] (see Fig. 3).

The small mucin gels released from goblet cell granules can also be reversibly recondensed. However, the most striking finding in this case is that recondensation can be readily induced by equilibrating these small mucin gels in a medium that mimics the intragranular environment, namely, low pH (3.5), and high concentration of Ca^{2+} (150 mM) [34].

The verification that the mucin polymer matrix can undergo phase transition prompted the idea that the condensed and hydrated conformations of the

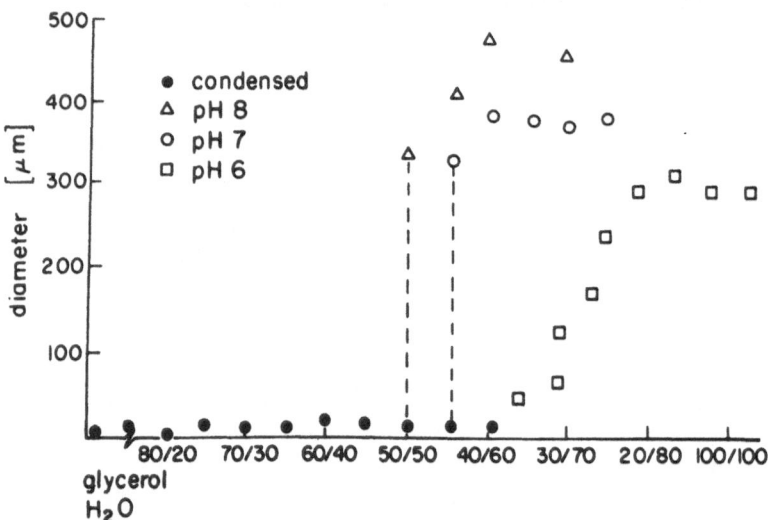

Fig. 3. Polymer gel phase transition of the mucin polymer matrix of a giant secretory granule of the terrestrial slug *Ariolimax columbianus*. Effect of pH

mucin network reflect the corresponding condensed and hydrated phases of a polymer gel. Thus, product release in exocytosis must result from a polymer gel phase transition [34–36]. Exocytosis starts with the formation of a secretory pore [37]. Once a water bridge between the intragranular and extracellular space is formed and becomes conductive, phase transition of the intragranular polymer network takes place. Driven by a Donnan potential, the intragranular polymer network undergoes dramatic swelling, which leads to the final release of the secretory material into the extracellular space. Phase transition is probably triggered by the exchange of intragranular divalent shielding cations by monovalent Na^+ from the extracellular space. In the goblet cell, this exchange is promoted by the Na^+/Ca^{2+} exchange properties of the mucin matrix [38, 39] (see Fig. 4).

Although the effect of Na^+/Ca^{2+} exchange in phase transition of mucin networks has not been systematically investigated, the role of ion exchange in triggering phase transition in exocytosis is best illustrated in experiments conducted in mast cells. In the mast cell granule, the polymer matrix is made out of heparin, a highly polyanionic proteoglycan. The corresponding shielding cation is histamine, which, at the low pH prevailing inside the granule, behaves

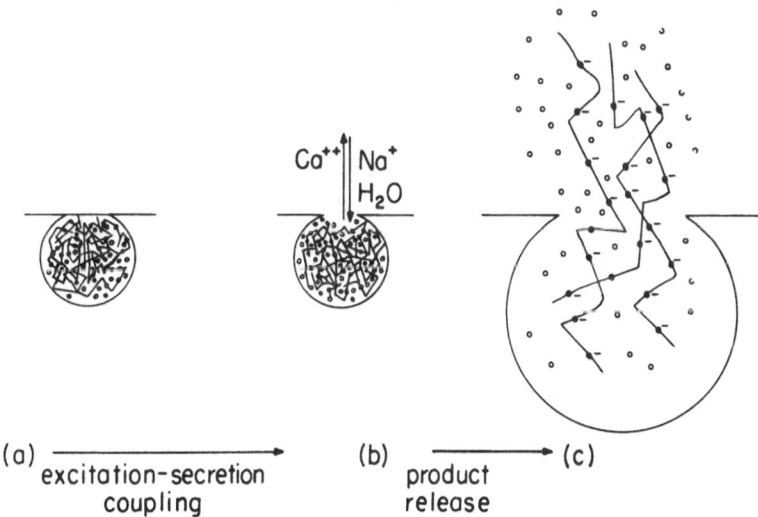

(a) ──────────────────▶ (b) ──────────▶ (c)
  excitation–secretion   product
      coupling   release

Fig. 4. This schematic representation of product release in secretion separate the stimulus coupling step from the actual mechanical phenomena that drive product release. The stimulus-coupling step encompass the events that start with receptor activation that leads to the docking of the secretory granule to the plasma membrane, the formation of a secretory pore, and the switching of the pore to high ionic conductance. Once a high conductance channel is established between intragranular and extracellular compartments, Na^+/Ca^{2+} ion exchange follows, triggering a phase transition that turns the mucin polymer from a condensed to a hydrated phase. At this stage, electrochemical energy stored in the mucin polymer network is transformed into mechanical energy, via electrostatic interactions, driving the final release of secretory products

as a divalent cation. Heparin, like mucins, also has divalent/monovalent (Na^+/histamine^{2+}) ion exchange properties [40]. As in mucin granules, the exocytosed heparin matrix can be similarly recondensed under conditions that mimic the intragranular milieu, in this case, 150 mM histamine at pH 5.5 or below. Condensation/decondensation in this instance is also reversible, and resembles a polymer gel phase transition. The critical role of Na^+/Ca^{2+} exchange in this phase transition system is illustrated by the observation that heparin matrices can remain in condensed phase even when equilibrated in distilled water. However, trace amounts of Na^+ in the medium can quickly trigger decondensation [2].

Recent findings indicate that a polymer gel phase transition of the heparin matrix of mast cell granules can also be induced electrically. This finding is not surprising, since it had been described in synthetic hydrogels. However, its significance is intriguing since it opens the possibility that the normal intra-granular/extracellular gradient of electric potential found in secretory cells might play a role in triggering the phase transition and product release [41].

One of the most attractive features of a polymer gel phase transition mechanism is that it can explain condensation and product release not only in goblet and mast cells, but also other secretory cells. Support for this argument comes from the finding that condensed, charge-shielded polyionic polymer networks and their corresponding shielding counterion are consistently found inside a broad range of secretory granules. In goblet cells, mucins are the polyanionic species and Ca^{2+} the counterion [28–31]; in mast cells, the polyanionic polymer is heparin and the shielding cation is divalent histamine [39]; in serous cells the polyanionic network is made of a proteoglycan and lyzosyme is the polycation [42]. The corresponding pair in parathyroid granules is the acidic protein SP-1 and Ca^{2+} [43]. In chromaffin granules, the condensed polyanion is chromogranin and catecolamines the shielding species [44]. The presence of polyanion-cation pairs have only been studied in a handful of secretory granules. However, acidic proteins similar to SP-1 have also been found in pancreatic islets, the anterior pituitary, celiac and mesenteric ganglia, in the gastric antrum, and the thyroid gland [43]. These countercharged pairs had long been thought to promote the formation of aggregates and reduce osmotic activity of the granular content [45]. However, the physicochemical mechanisms of condensation and decondensation of secretory products had not been explained.

The theory of polymer gel phase transition provides a novel paradigm of remarkable predictive power to explain both condensation and decondensation in secretion. It assigns specific functional roles to the polyanionic-network/cation pairs ubiquitously found in secretory granules, and provides a working framework to formulate new, testable questions to investigate further the mechanisms of storage and release in secretion.

The finding that intragranular polymer networks can undergo phase transition, and that product release in secretion might operate in a way similar to the newly developed drug release systems based on "intelligent" polymers, opens a promising horizon to learn from cells the implementation of drug release strategies that have been optimized by nature through the rigorous test of many thousands of years of evolution.

This work was supported by NASA Space Biology Program (Grant No NAG-9-604).

3 References

1. Flory PJ (1964) Principles of polymer chemistry. Cornell University Press, Ithaca
2. Fernandez JM, Villalon M, Verdugo P (1991) Biophys J 59: 1022
3. Viney C, Huber AE, Verdugo P (1993) Macromolecules 26: 852
4. Boat TF, Cheng PW (1980) Fed Proc 39: 3067
5. Boat TF, Cheng PW, Iyer RN, Carlson DM, Polony I (1976) Arch Biochem Biophys 177: 95
6. Carlstedt I, Sheehan JK (1984) In: Nuget J, O'Connor M (eds) Mucus and mucosa (Ciba Foundation Symposium 109). Pitman, London, p 157
7. Kaliner MA, Borson DB, Nadel JA, Shelhamer JH, Patow CA, Marom Z (1988) In: Kaliner MA, Barnes PJ (eds) The airways. Neural control in health and disease. Marcel Dekker, New York, p 575
8. Kent PW (1978) In: Nuget J, O'Connor M (eds) Symposium on respiratory tract mucus (Ciba Foundation Symposium new series). Pitman, London, p 155
9. Slayter HS, Lamblin G, Le Treut A, Galabert C, Houdret N, et al. (1984) Eur J Biochem 142: 209
10. Havez R, Roussel P (1976) In: Weiss EB, Sega MS (eds) Bronchial asthma: Mechanisms and therapeutics. Little, Brown, Boston, p 409
11. Kaliner M, Marom Z, Patow C, Shelhamer J (1984) J Allergy Clin Immunol 73: 318
12. Shelhamer JH, Marom Z, Logun C, Kaliner M (1984) Exp Lung Res 7: 149
13. Gibbons RA, Mattner, PE (1966) Int J Fertil 11: 366
14. Roberts GP (1976) Arch Biochem Biophys 173: 528
15. Lee WI, Verdugo P, Blandau RJ (1977) Gynec Invest 8: 254
16. Verdugo P, Tam PY, Butler JC (1983) Biorheology 20: 223
17. Harding SE, Rowe AJ, Creeth JM (1983) Biochem J 209: 893
18. Sheehan JK, Oates K, Carlstedt I (1986) Biochem J 239: 147
19. Silberberg A (1987) Biorheology 24: 605
20. Katchalsky A, Lifson S, Eisenberg HJ (1951) J Polymer Sci 7: 571
21. Donnan FG (1924) Chem Rev 1: 73
22. Tam PY, Verdugo P (1981) Nature 292: 340
23. Verdugo P (1984) In: Nuget J, O'Connor M (eds) Mucus and mucosa (Ciba Foundation Symposium 109). Pitman, London, p 212
24. Frizzell, RA (1988) Am Rev Respir Dis 138: S3
25. Tanaka T, Fillmore DJ (1979) J Chem Phys 70: 1214
26. Verdugo P, Aitken ML, Langley L, Villalon M (1987) Biorheology 24: 625
27. Aitken ML, Verdugo P (1989) In: Chantler EN (ed) Mucus and related topics. Plenum, New York, p 73
28. Verdugo P, Deyrup-Olsen L, Aitken ML, Villalon M, Johnson D (1987) J Dent Res 66: 506
29. Izutsu K, Johnson D, Schubert M, Wang E, Ramsey B. et al. (1985) J Clin Invest 75: 1951
30. Roomans GM, von Euler AM, Muller RM, Gilljman H (1986) J Submicroc Cytol 18: 613
31. Sasaki S, Nakagaki I, Mori H, Imai Y (1983) Jap J Physiol 33: 69
32. Tanaka T (1981) Sci Am 244: 124
33. Verdugo P, Deyrup-Olsen I, Martin AW, Luchtel DL (1992) In: Karalis TK (ed) Mechanics of swelling. Springer, Berlin Heidelberg New York, p 67
34. Verdugo P (1991) Am Rev Respir Dis 144: S33
35. Verdugo P (1986) Biophys J 49: 231a
36. Verdugo P (1990) Annu Rev Physiol 52: 157
37. Monck JR, Fernandez JM (1992) J Cell Biol 119: 1395

38. Forstner JF, Forstner GG (1975) Biochim Biophys Acta 386: 283
39. Crowther RS, Marriot C (1983) J Pharm Phamacol 36: 21
40. Uvnas B, Aborg CH (1977) Acta Physiol Scand 100: 309
41. Nanavati C, Fernandez FM (1993) Science 259: 963
42. Basbaum CB, Berthold J, Finkbeiner WE (1990) Annu Rev Physiol 52: 97
43. Cohen DV, Morrissey JJ, Hamilton JW, Shofstall RE, et al. (1981) Biochemistry 20: 4135
44. Winkler H, Westhead E (1980) Neuroscience 5: 1803
45. Palade H (1975) Science 189: 347

Received March 8, 1993

Novel Applications for Stimulus-Sensitive Polymer Gels in the Preparation of Functional Immobilized Biocatalysts

Etsuo Kokufuta
Institute of Applied Biochemistry, University of Tsukuba, Tsukuba,
Ibaraki 305, Japan

Functional immobilized biocatalysts may be taken to be immobilized biocatalyst systems with some beneficial functional capability other than the usually credited advantages obtained upon the immobilization of enzymes or microorganisms within appropriate polymers. This review aims to examine the preparation of functional immobilized biocatalysts using stimulus-sensitive polymer gels (SSPGs) that undergo continuous or discontinuous volume collapse in response to alterations in external stimuli. The use of SSPGs offers the following functional capabilities in immobilized preparations: (1) the enhancement of immobilized enzyme activity via the cyclic absorption of substrate solutions; (2) the control of immobilized enzyme reactions, especially their initiation and termination, using changes in temperature; and (3) the conversion of the energy of an immobilized enzyme reaction into mechanical work through discontinuous changes in gel volume. It has become apparent that all of these functions demonstrated by the immobilized preparations obtained rely on the characteristics of specially designed SSPGs: a lightly cross-linked copolymer gel consisting of N-isopropylacrylamide and acrylamide, the thermally controlled swelling/deswelling of which absorbs the substrate to facilitate its diffusion through the gel porosity; a thermosensitive gel consisting of poly(vinyl methyl ether) which is capable of dramatically reducing substrate diffusion above the phase transition temperature; and a polyelectrolyte gel which undergoes volume changes enzymatically at a certain temperature. In conclusion, the ideas or concepts reviewed here could play a part in the design and construction of functional immobilized biocatalysts.

Advances in Polymer Science, Vol. 110
© Springer-Verlag Berlin Heidelberg 1993

Abbreviations

AA	acrylic acid
AAm	acrylamide
Con A	concanavalin A
DSS	dextran sulfate sodium
DSS-gel	gel containing Con A/DSS complex
LCST	lower critical solution temperature
MAPTAC	[(methacrylamide)propyl]trimethylammonium chloride
MBA	N,N'-methylenebis(acrylamide)
MP	α-methyl-D-mannopyranoside
MP-gel	gel containing Con A/MP complex
NIPA	N-isopropylacrylamide
ONPG	O-nitrophenyl-β-D-galactopyranoside
PVMA	poly(vinyl methyl ether)
SSPG	stimulus-sensitive polymer gel
T_c	transition temperature
TMED	N,N,N',N'-tetramethylethylenediamine

1 Introduction

Immobilized enzymes and cells, which are now known as "immobilized biocatalysts," both provide us in general with the following advantages: (1) continuous operation becomes practical; (2) biocatalysts can be recovered and reused after reactions; (3) biocatalysts can be formed into the shapes, such as membranes or beads, required for specific reaction processes; (4) in some cases, the biocatalysts become stable with regard to changes in temperature, pH and inhibitor concentration. Immobilization methods can be classified according to the following scheme,

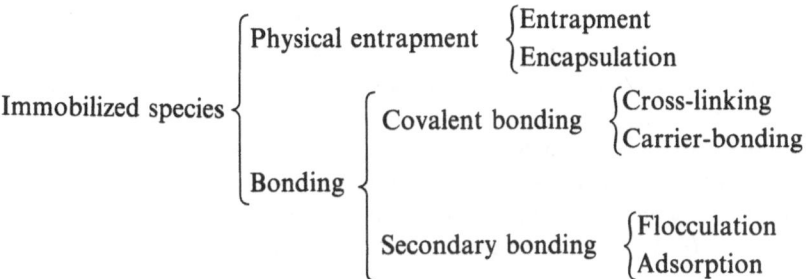

in which the type of enzyme or cell bonding in the immobilized biocatalyst is the dominant parameter. This scheme, however, is oriented more towards the final state of the immobilized biocatalyst and less towards the route along (or process by) which the catalyst was prepared. Both cross-linking and carrier-bonding methods cannot, in principle, be utilized in whole cell immobilization.

At a glance, the rapprochement between biochemistry and polymer chemistry seems to have played an important role in the methodological development of preparations for immobilized biocatalysts. A number of articles on the preparation and characterization of immobilized biocatalysts, together with their applications in a variety of fields besides synthetic chemical reactions – chemical and clinical analysis, medicine, and food processing, for example – have already been published. These results have been reviewed by many of the pioneers in this and related fields [1–20]. The technology for immobilizing enzymes and cells is believed to be relatively mature at this point. In addition, the nature of immobilized biocatalysts has become somewhat more transparent to us. The key now is to come up with new uses and new systems which can fulfill specific needs [21].

A renewed interest in this research field may lead to the construction of "functional" immobilized biocatalysts that *surpass the conventional definition, or usually credited advantages, of immobilized biocatalysts with regard to their capabilities as catalysts* [22–24], i.e. immobilized enzyme systems in which, for example, an enzymatic process can be controlled by externally applied stimuli such as light, electric fields, pH, temperature, and mechanical force. In such cases, what is crucial in system construction is not to rely on a possible alteration in the property of the biocatalyst (e.g. enhanced thermal stability) which has frequently been expected as a result of immobilization, but *to impose a new capability on a biocatalyst system as the result of a process of rational design.* As the principal means of reaching this goal, polymer chemists and scientists may consider the development of supporting matrices; such functional polymers as pH-sensitive microcapsules and reversibly soluble polyelectrolyte complexes have been employed for this purpose.

Since the original use by Bernfeld and Wan [25] of a gel – a cross-linked polymer of acrylamide (AAm) – for the immobilization of enzymes through physical entrapment, some polymer gels have frequently been employed in the preparation of immobilized biocatalysts. The best known example of such a case may be calcium alginate gel (see, for example, Refs. 26–37), intact [26–30] or chemically modified [31–37] forms of which have been applied in the entrapment of cells and enzymes. In previous gel-entrapped biocatalysts, however, the gels have for the most part played a role only as the supports 'maintaining' or 'holding' the biocatalysts, little attention having been paid to the application of the functional capabilities of gels in the preparation of immobilized biocatalysts.

Recent years have seen dramatic developments in the field of polymer gels, the majority of which have been successful syntheses of "stimulus-sensitive polymer gels" (SSPGs) that undergo continuous or discontinuous volume collapse in response to a change in externally applied stimuli. This marvelous property of polymer gels may be one of the most interesting and important subjects in the development of functional immobilized biocatalysts. This review aims to provide an up-to-date overview of novel applications of SSPGs in the preparation of functional immobilized biocatalysts and to discuss the utilization of the developing tools of the volume-phase transitions of polymer gels in the construction of such systems.

2 Outline of Phase Transitions of Polymer Gels

The swelling of gels, such as that observed when rubber is placed in benzene, has long been a well-known phenomenon. In 1978, Tanaka [38] discovered the discontinuous volume change (i.e. volume collapse) of a covalently cross-linked AAm gel in an acetone/water mixture when varying the temperature or composition of the mixture. This phenomenon is now called the volume-phase transition of gels and is observed in many gels (i.e. SSPGs) made of synthetic and natural polymers. The phase transition is accompanied by reversible, discontinuous (or in some cases, continuous) volume changes, often as large as several hundred times, in response to small variations in the external conditions surrounding a gel [39]. Variables that trigger the transition include temperature [38, 40–43], solvent composition [38, 40, 42, 44], pH [40, 45, 46], ion concentration [46, 47], small electric fields [48–50], and light [51, 52].

Theoretical studies on phase transitions were attempted by Tanaka et al. [38, 40, 43, 53] using the Flory–Huggins theory [54]. It has been shown as a result that a gel undergoes either a continuous volume change or a first-order discontinuous phase transition depending on the proportion of ionizable groups incorporated in the polymer network and on the stiffness of the polymer chains which constitute the network. The counterions of the ionized groups and the stiffness of the polymer chains increase the osmotic pressure acting to expand the polymer network, resulting in a discontinuous volume change. This situation is similar to a gas-liquid phase transition which can be either continuous or discontinuous depending on the external pressure exerted on the system.

Several other studies have also been made in an attempt to account theoretically for the phase transition in terms different from those of the Flory–Huggins theory. Otake et al. [55] thus proposed a theoretical model that takes hydrophobic interaction into account in explaining the thermally induced discontinuous volume collapse of hydrogels. In addition, Prausnitz et al. [56] proposed a lattice model, an improvement of which was made to explain the swelling curves of gels consisting of N,N'-methylenebis(acrylamide) (MBA)-crosslinked copolymers of AAm with [(methacrylamide)propyl]trimethyl-ammonium chloride (MAPTAC) [57].

$$CH=C-CH_3$$
$$|$$
$$O=C-NH-(CH_2)_3-N^+(CH_3)_3Cl^-$$
(MAPTAC)

which were measured as a function of the degree of cross-linking and the concentrations of NaCl solution.

In order to apply the phase transitions of gels in technological fields, however, a more generalized explanation was required. The authors [58] thus tried to account for the phase transition by hypothesizing a balance between the repulsion and attraction of the cross-linked polymer chains in the networks which arise from a combination of four intermolecular forces: ionic, hydrophobic, van der Waals and hydrogen bonding. When a repulsive force, usually electrostatic in nature, overcomes an attractive force such as the hydrogen bonding or the hydrophobic interaction between network chains, gel volume should increase discontinuously in some cases or continuously in others. Conversely, a decrease in the volume may occur when the attractive force becomes dominant. The variables that trigger the transition influence these intermolecular forces and thereby the balanced state of the attractive and repulsive forces. This concept is extremely qualitative, but makes it possible for us to classify the swelling curves of gels observed under various conditions into four different types (see Fig. 1). This concept has also been applied in the preparation of a functional immobilized enzyme whose catalytic activity can be initiated and terminated by a change in temperature (see Sect. 4.1), as well as in the design of a gel system that is capable of converting immobilized enzyme reactions into mechanical work (see Sect. 5.1).

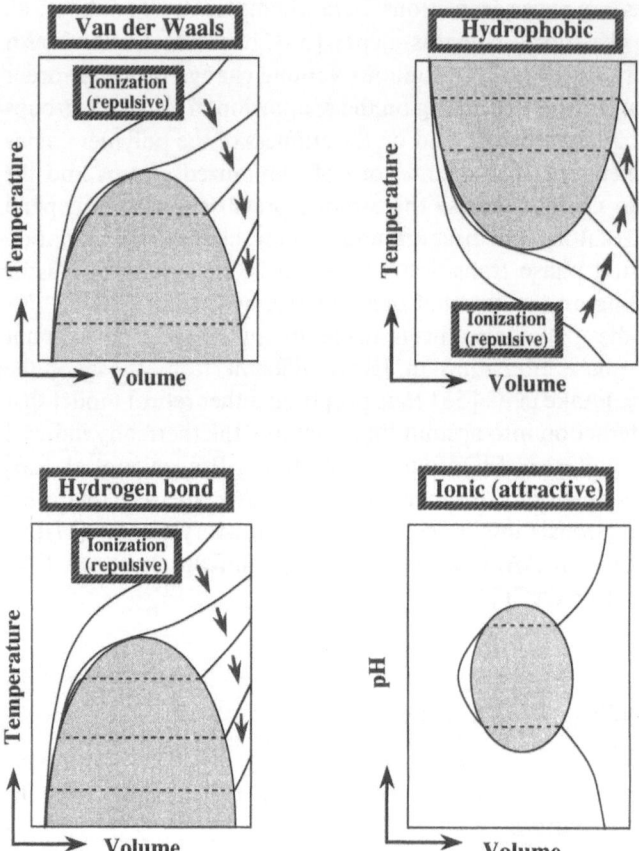

Fig. 1. Gel volume phase transitions induced by four types of intermolecular forces. The van der Waals interaction (attractive) causes phase transitions in gels in a mixed solvent system, in which a non-polar solvent influences the transition through a decrease in the dielectric constant. Gels such as N-isopropylacrylamide (NIPA) gel undergo phase transitions in pure water from a swollen state at low temperatures to a collapsed state at high temperatures due to hydrophobic interaction (attractive). Gels with cooperative hydrogen bonding as the attractive force, such as the inter-penetrating polymer network of acrylic acid (AA)/acrylamide (AAm) gel, undergo phase transitions in pure water (swollen state is the high-temperature state). The repulsive ionic interaction determines the transition temperature and the volume change at the transition, whereas the attractive ionic interaction is responsible for pH-driven phase transitions, as in gels consisting of AA/ [(methacrylamide)propyl]trimethylammonium chloride (MAPTAC). (From F. Ilmain, T. Tanaka and E. Kokufuta [58])

3 Gels Capable of Absorbing Substrate Solutions

When immobilizing biocatalysts within polymer gels using physical entrapment methods, we may take advantage of the great resistance to the diffusion of macromolecular substances due to the gel porosity. However, this limited diffusion within the gel phase also causes a reduced mass transfer rate for low molecular-weight compounds, which diminishes the chances for collision and reaction between the molecules lying within the gel and coming from the external bulk phase. For example, in immobilized enzyme reactions, the mass transfer resistances for substrate and product within the pores of the gel are generally responsible for a loss in activity upon immobilization [59]. One approach for enhancing the diffusion of substrates in immobilized reactions could involve the simulation [60, 61] of a biological transport system in which a variety of solutes penetrate membranes via a facilitated diffusion mechanism, which differs from more usual simple diffusion. It is well known that a mobile carrier, often a protein with a high affinity for the molecule to be transported, plays an important role in the facilitated diffusion process (see, for example, pp 266–275 in Ref. 62). On the basis of this idea, a model immobilized enzyme system has been designed and prepared [60, 61], which is capable of promoting the pore diffusion not only of a low molecular-weight substrate [61], but also that of a polymeric substrate [60]. In these examples of functional immobilized biocatalysts, however, polymer gels were still used as the support into which enzymes were coentrapped with carriers having a binding affinity towards the substrates.

3.1 NIPA/AAm Gel for Enzyme Immobilization

The first attempt at applying SSPGs in the field of functional immobilized biocatalysts was made by Park and Hoffman [63–65], who demonstrated the capability of thermosensitive gels for enhancing the diffusion of a substrate from the outside to the inside of a gel, which is based upon the possible absorption of the surrounding medium by the gel during a thermally controlled swelling process (see Fig. 2).

The gel was prepared in the form of beads with diameters ranging from 200 to 400 µm in the swollen state by an inverse suspension polymerization using paraffin oil as a continuous phase and Pluronic L-81 as a surfactant. A typical enzyme immobilization was carried out as follows [63]: a 0.1 M sodium phosphate buffer (pH 7.4, 40 ml) containing NIPA (7.18 g), AAm (0.5 g), MBA (cross-linker, 0.32 g), ammonium persulfate (initiator, 50 mg), and β-D-galacto sidase (1.8 mg) was immediately poured after preparation into 400 ml of paraffin oil containing the surfactant (0.1 ml); the polymerization was initiated by injecting N,N,N',N'-tetramethylethylenediamine (TMED, 0.5 ml) as an accelerator into the continuous organic phase where aqueous droplets of the solution with both the monomers and the enzyme had been formed by agitation (500 rpm). The reaction was performed in an ice-water bath, and nitrogen supplied continuously above the surface of the paraffin oil phase to avoid disturbing the beads with nitrogen bubbles.

The gel obtained shrank completely when warmed above a critical temperature (38.5 ± 1.5 °C), called the lower critical solution temperature (LCST), but reswelled when cooled below this point. This shrinking-swelling behavior was

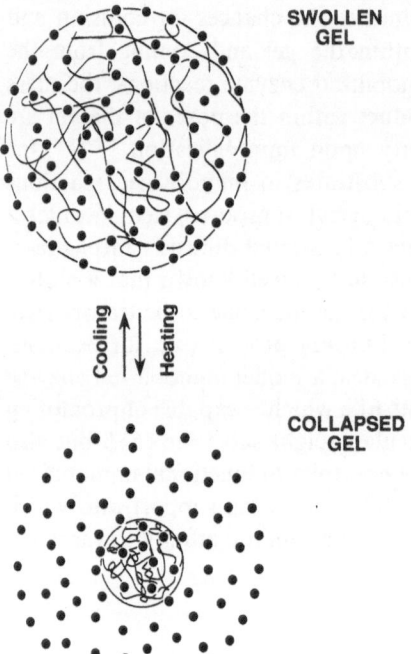

Fig. 2. Schematic illustration of cyclic pumping of surrounding medium (represented by ●) by thermosensitive gel during its thermally controlled swelling and collapse

reversible. The enzymatic hydrolysis of O-nitrophenyl-β-D-galactopyranoside (ONPG) by the gel beads with immobilized β-D-galactosidase was tested in a packed bed column reactor (continuous and single pass mode):

(ONPG)

The operation was either isothermal at 30 or 35 °C (both are lower than LCST) or cycled every 10 min at 1 °C/min between 30 and 35 °C. To assay the activity of the immobilized enzyme, the conversion of the substrate into the products was estimated on the basis of the molar concentration ratio of the outlet product to the inlet substrate (Fig. 3). The conversion cycled in accordance with the temperature cycling, and was at all times higher than that for isothermal operation at either 30 or 35 °C. The conversion also appeared to be at a maximum at 35 °C and at a minimum at 30 °C (each at 10 min intervals) during the thermal cycling operation. The shrinking of the gel upon heating squeezed

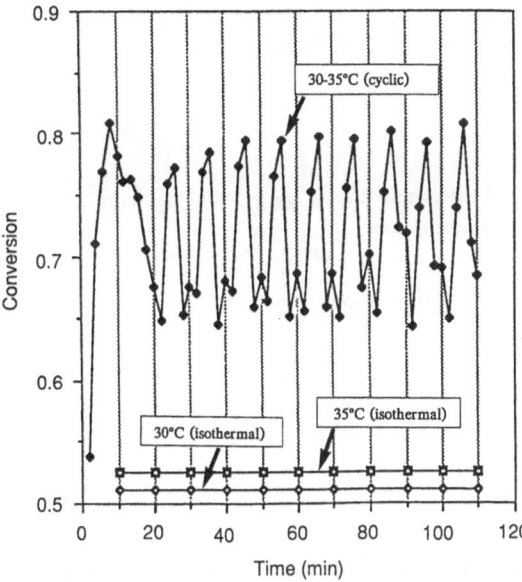

Fig. 3. Conversion of O-nitrophenyl-β-D-galactopyranoside (ONPG) into O-nitrophenol and D-galactose as a function of time in packed β-D-galactosidase-immobilized bed reactor operated isothermally at 30 °C and 35 °C or cycled between 30 and 35 °C. (From T.G. Park and A.S. Hoffman [63])

the product and any remaining substrate from inside the gel, leading to the observed increase in conversion, while the swelling of the gel upon cooling drew in the substrate along with any product in the surrounding fluid. Thus, mass transfer rates within the gel beads were greatly enhanced by the movement of water into and out of the gel. As a result, thermal cycling significantly elevated the overall reactor enzyme activity relative to the isothermal operation. On the other hand, a somewhat higher conversion at 35 °C relative to that at 30 °C was observed for the isothermal operation. At 30 °C, the gel was significantly more swollen than at 35 °C, and because of the greater volume and lower tortuosity of the pore structure, both the substrate and product diffusion rates into and out of the gel were expected to be higher at 30 °C than at 35 °C. In actual fact, however, the intrinsic diffusivities of both substrate and product, the intrinsic enzyme turnover rate, and the effective enzyme concentration within the gel phase were higher at 35 °C than at 30 °C. These various opposing factors tended almost to balance, with the increase in reaction kinetics during isothermal operation at 35 °C being more important than the decrease in diffusion rate, thus leading to the higher conversion at 35 °C.

3.2 NIPA/AAm Gel for Cell Immobilization

The present type of thermosensitive gel has been applied in the immobilization of *Arthrobacter simplex* cells, and the bioconversion of a steroid (hydrocortisone → prednisolone by 3-ketosteroid-Δ^1-dehydrogenase in *A. simplex*) was examined using the thermal cycling method [65]. In this case, the thermal cycling was also found to enhance the mass transfer of the substrate (hydrocortisone) into and the product (prednisolone) out of the gel, and thereby to increase the steroid conversion.

CH$_2$OH   CH$_2$OH

C=O   C=O

HO   ···OH   3-ketosteroid-Δ^1-   HO   ···OH

dehydrogenase

$\longrightarrow$

- 2H

O   O

hydrocortisone   prednisolone

3.3 Possibility of Other SSPGs

The cyclic pumping of the surrounding medium by a gel during its swelling-shrinking change may also be achieved using various stimulus-sensitive gels other than thermosensitive gels; for example, one can easily suggest that a photosensitive hydrogel [52] would permit the enhancement of immobilized enzyme activity through a possible photo-driven substrate-pumping mechanism after the successful entrapment of biocatalysts within the gel. In summary, SSPGs capable of cyclically pumping substrate solutions by means of their volume changes seem to be of considerable potential usefulness in the enhancement of the catalytic activities of a variety of immobilized biocatalyst systems.

4 Thermosensitive Gels Capable of Regulating Enzyme Reactions

The volume-phase transitions of polymer gels can be accompanied by variations in the mass-transfer rates of solutes from the outside to the inside of the gel, and vice versa. Gels that undergo a large volume change in response to external stimuli may therefore serve as supports for constructing various types of stimulus-sensitive immobilized biocatalysts. The gels studied thus far represent only two kinds of the thermosensitive type [66–68], both of which undergo sharp but not discontinuous volume changes with temperature. One is a lightly cross-linked copolymer gel consisting of NIPA and AAm which was used by Dong and Hoffman [67, 68] for the immobilization of asparaginase (L-asparagine amidohydrolase). From their detailed investigations on the temperature dependence of the activity of the immobilized enzyme, they proposed that the gel could be used to control the initiation and termination of immobilized enzyme reactions by temperature. However, this proposal was not confirmed by their experimental data.

4.1 PVME Gel for Enzyme Immobilization

The complete on/off control of enzyme activity using a thermosensitive gel has been performed by Kokufuta et al. [66], who prepared a gel with immobilized exo-1,4-α-D-glucosidase by cross-linking an aqueous poly(vinyl methyl ether) solution with γ-ray irradiation. The resultant gel exhibited a thermosensitive characteristic: it shrank above 38 °C and swelled below this transition temperature (T_c), as shown in Fig. 4(A). This behavior was reversible. The permeability of the gel for glucose thus changed dramatically above and below T_c: 2.0×10^{-6} $cm^2 \cdot s^{-1}$ at 32 °C and 7.3×10^{-11} $cm^2 \cdot s^{-1}$ at 42 °C. As a result, the immobilized preparation obtained displayed an excellent capacity for the on/off control of the enzymatic hydrolysis of maltose [see Fig. 4(B)]. When the immobilized enzyme was utilized, glucose formation from maltose halted at 42 °C but recommenced immediately when the temperature was lowered to 32 °C. Such initiation-termination control could be repeated reversibly throughout a single run of the measurements and reproduced without a serious loss in activity for at

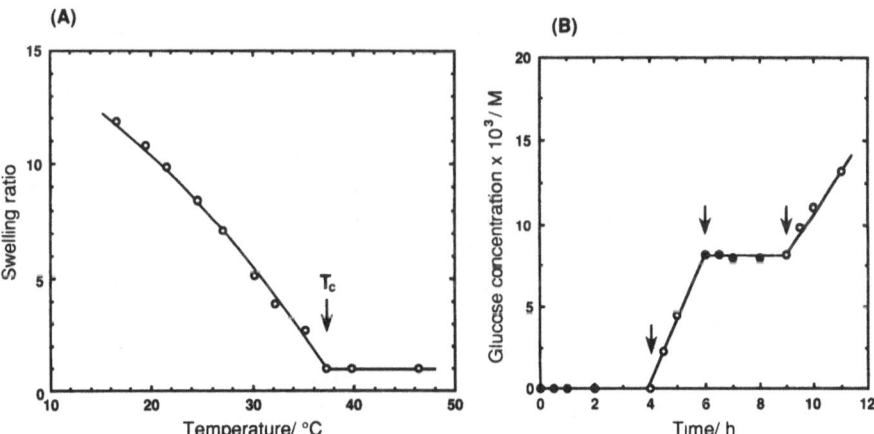

Fig. 4A, B. Application of thermosensitive poly(vinyl methyl ether) gel in the initiation-termination control of immobilized enzyme reaction: (**A**) Swelling curve of gel as a function of temperature, data normalized by dividing gel volume at an arbitrary temperature by volume of the completely collapsed gel at temperatures above 37 °C. (**B**) Temperature-responsive initiation-termination control of glucose formation by gel-entrapped exo-1,4-α-D-glucosidase, demonstrating that the enzymatic reaction was depressed at 42 °C (●) but commenced as the temperature jumped to 32 °C within 2 min (○). Activity of the immobilized enzyme can be switched on and off repeatedly by rapidly decreasing the temperature from 42 to 32 °C and rapidly increasing it from 32 to 42 °C at the times indicated by *arrows*. The average rate of glucose formation in the two "on" states during a single run at 32 °C changed from 58 ± 8 M·min^{-1} (initial run) to 45 ± 8 M·min^{-1} (final run) when the experiment was repeated in a series of 20 runs [results in (**B**) show initial run]. (From E. Kokufuta, O. Ogane, I. Ichijo, S. Watanabe and O. Hirasa [66])

least 20 runs carried out with a freshly prepared substrate solution. In conclusion, this result clearly demonstrates the potential utility of thermosensitive gels in the on/off regulation of immobilized enzyme reactions.

4.2 Possibility of Other Thermosensitive Gels

Very recently, Inomata et al. have reported that the T_c of thermosensitive gels consisting of the following N-substituted acrylamide derivatives varies depending on the species of the substituent groups [69]:

$$
\begin{array}{c}
-\text{(CH--CH)-} \\
| \\
\text{C}=\text{O} \\
| \\
\text{NH--R}
\end{array}
\qquad
R = \left\{
\begin{array}{ll}
-\text{CH}_2-\text{CH}_2-\text{CH}_3 & (T_c = 25\,^{\circ}\text{C, discontinuous}) \\[2mm]
-\text{CH} \big\langle {\text{CH}_3 \atop \text{CH}_3} & (T_c = 34\,^{\circ}\text{C, discontinuous}) \\[2mm]
-\text{CH} \big\langle {\text{CH}_2 \atop \text{CH}_2} & (T_c = 40\text{--}50\,^{\circ}\text{C continuous})
\end{array}
\right.
$$

This suggests the possibility of various types of thermosensitive gel for use in altering the temperature range in the on/off regulation of immobilized enzyme reactions. Thus, the use of thermosensitive gels in the field of stimulus-sensitive immobilized enzymes is expected to expand.

Prior to the above publications, Matsuoka et al. studied another type of thermosensitive immobilized enzyme using a liposomal membrane [70]. The membrane of a particular phospholipid such as dipalmitoyl phosphatidylcholine was known to undergo a phase transition at a specific temperature [71]; thus, the permeation of an aqueous solution containing substrates could be inhibited in the range below this phase transition temperature. Liposomes with immobilized urease were prepared on the basis of this idea using the following procedures: first, the enzyme was covalently bound to the purple membrane isolated from *Halobacterium halobium* with the aid of 1-cyclohexyl-3-(2-morpholinoethyl)-carbodiimide metho-*p*-toluene sulfonate; the purple membrane-bound enzyme was then incorporated into the liposomes of dipalmitoyl phosphatidylcholine by sonication. The preparation did not exhibit enzyme activity in the temperature range $< 42\,^{\circ}\text{C}$, which was close to its phase transition temperature; but when the preparation was heated to above $42\,^{\circ}\text{C}$, the amount of activity gradually increased. The on/off control of the enzymatic decomposition of urea could thus be performed by switching the temperature from 36.5 to $45.6\,^{\circ}\text{C}$, and again to $36.5\,^{\circ}\text{C}$. In this immobilized system, however, the repetition of this temperature cycle brought about a slight "leakage" of activity in the off condition [70]. Thermosensitive gels, with their greater resistance to the leakage of high molecular-weight protein molecules resulting from their cross-linked polymer networks, therefore seem to be superior to other thermosensitive types of supports in the on/off regulation of immobilized enzyme reactions.

5 Gel-entrapped Enzyme Systems with Biochemo-Mechanical Functions

The purpose of all the studies reviewed in the previous sections was to utilize immobilized biocatalysts in chemical conversion. On the other hand, new attempts have been made to directly convert the "energy" of immobilized enzyme reactions into mechanical work [24, 72, 73]. A system with such a biochemo-mechanical function can be prepared by entrapping enzymes into polymer gels that undergo volume-phase transitions in response to enzymatic changes. A general paradigm for the design and preparation of biochemo-mechanical systems using the entrapment of enzymes within SSPGs is discussed in this section.

Mechano-chemical or chemo-mechanical systems which exert mechanical energies in response to chemical changes have long attracted interest among scientists, medical researchers and engineers. Katchalsky and his coworkers [74–76] were pioneers in the development of such systems using polymer gels, and their work has been expanded further by various researchers (see Ref. 49).

Biochemo-mechanical systems, in which biochemical changes such as enzymic reactions are used in place of the usual chemical changes for creating mechanical energies, are taken to be one development of chemo-mechanical systems. Since biochemical reactions are generally believed to be uniquely specific, the creation of mechanical energies in a biochemo-mechanical system is expected to be responsive to a specific kind of molecule. This aspect is important not only technologically, but will also provide a basis for a better understanding of the marvelous energy-conversion mechanisms available only in biological systems.

5.1 NIPA/AA Gel with Immobilized Urease

One approach for constructing biochemo-mechanical systems may be the enzymatic control of the phase transition of gels. Among the many variables that trigger phase transitions, solvent composition and solution pH seem to be controlled enzymatically. It also seems, based on the concept described in the last paragraph of Sect. 2, that the phase transition occurring as the result of a change in the balance between the ionic (repulsive) and hydrophobic (attractive) forces can be regulated through an enzyme-induced change in pH or solvent composition. Two gel systems have been designed according to these considerations: a copolymer gel [73] of acrylic acid (AA) and NIPA into which urease has been entrapped; and an NIPA gel containing immobilized esterase [72].

The details of the experiments and the results for the copolymer gel with immobilized urease are as follows: The gel was obtained by gelling an aqueous solution (1 ml) containing NIPA (75.6 mg), AA (2.3 mg), MBA (cross-linker, 1.33 mg), and urease (20 mg). Ammonium persulfate (0.4 mg/ml) and TMED

(1.85 mg/ml) were used as the initiator and accelerator, respectively. The gelation was carried out at 0 °C for 1 hr in a test tube into which glass capillaries
with an inner diameter of 0.1 mm had previously been inserted. After the
gelation was completed, the gels were taken out of the capillaries and thoroughly washed with an NH_4Cl/HCl buffer solution (0.2 M; pH 4.0). All the gel
samples were cut into cylinders of ca. 2 mm length and stored at 3 °C before use.

Swelling curves were measured using the aforementioned buffer solutions,
both with and without urea as the substrate [Fig. 5A]. Gel diameters were

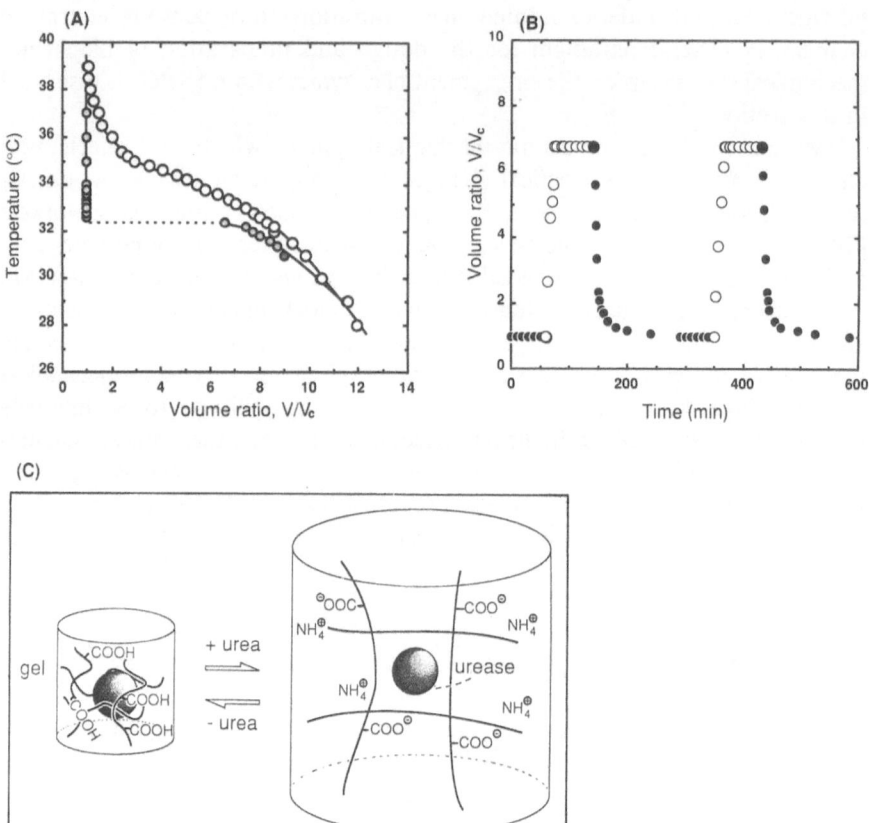

Fig. 5A–C. Biochemo-mechanical function of N-isopropylacrylamide (NPA)/acrylic acid (AA) gel
with immobilized urease: (A) Swelling curves of urease-loaded gel in 0.2 M ammonium buffer
(pH 4.0) containing (○) and not containing (●) 1 M urea as the substrate. The completely collapsed
volume (V_c) for each sample was determined at 50 °C. (B) Repeated swelling and shrinking control of
the urease-loaded gel at 33.4 °C by the alternate use of urea-free (○) and urea-laden 0.2 M
ammonium buffer solutions (●) at pH 4.0. The aqueous phase (80 μl) in the cell into which the gel
sample was immersed was quickly replaced by 4 ml of another aqueous solution within 2 minutes.
To avoid temperature change during replacement, both aqueous phases were kept at the same
temperature. (C) Schematic illustration of the present biochemo-mechanical system, in which
ammonia enzymatically produced from urea raises the pH of the gel phase and dissociates carboxyl
groups even when the ambient pH is kept at a low level (4.0) in the acidic range. This brings about
the swelling of gel, whereas the gel collapses in the absence of the substrate. (From E. Kokufuta,
Y.-Q. Zhang, F. Ilmain and T. Tanaka [73])

determined using a microscope with a calibrated scale, the temperature being controlled to within 0.01 °C between 25 °C and 40 °C. In the absence of urea, the gel underwent a discontinuous phase transition at 32.4 °C (T_c). In contrast, the presence of urea brought about a continuous transition, and the transition temperature was shifted into a higher range. Taking these results into account, the reversible regulation of the gel volume was attempted by means of the alternative immersion of the gel into buffer solutions with and without urea [Fig. 5B]. At a temperature of 33.4 °C, which was higher than the T_c, the gel shrank in the absence of urea. The gel began to swell as soon as the aqueous solution in which it was immersed was replaced by the urea solution, and the swelling became complete within 15 minutes. The swollen gel collapsed again when the ambient solution was replaced with the urea-free buffer. The time needed for a complete collapse of the gel-size employed here was approximately 90 minutes. The swelling and collapsing were repeated several times with satisfactory reproducibility.

The results obtained can be interpreted as shown in Fig. 5C. The immobilized enzyme catalyzed the hydrolysis of urea into ammonia and carbon dioxide:

$$(NH_2)_2C{=}O + 3H_2O \;\rightarrow\; CO_2 + 2NH_4OH$$

The carbon dioxide produced was soluble in water and partially turned into HCO_3^- and CO_3^{2-}. The gel phase was saturated with the following species: NH_4OH, NH_4^+, HCO_3^-, CO_3^{2-}, and Cl^-, of which the NH_4OH, NH_4^+, and Cl^- originated in part from the buffer solution. The reaction naturally increased the pH of the gel phase due to an increase in the ammonia concentration. The carboxyl groups of the gel were then dissociated, and the electrical force arising from the dissociation of the carboxyl groups overcame the hydrophobic interaction between the network chains. This, of course, brought about an increase in the transition temperature, as was previously expected. As a result, the gel was swollen when a certain amount of urea was present, but collapsed in its absence if the temperature was kept within a suitable range above T_c.

The enzymatically driven gel not only provides a simulation of biological energy-conversion systems, but is also available for technological applications. For example, gels undergoing reversible swelling and shrinking changes in response to enzyme reactions such as biochemical stimuli may serve as drug delivery devices. In fact, the release of a protein such as insulin from the present gel system occurred when the substrate was added into an aqueous bulk solution to initiate the enzyme reaction within the gel (see Fig. 6) [77].

5.2 NIPA Gel with Immobilized Concanavalin A

As described above, volume-phase transitions in gels with immobilized enzymes are available for the biochemical creation of mechanical energies when coupled with enzymatic changes within the gel phase. In the design of such immobilized enzyme systems, the concept of controlling the phase transition threshold by

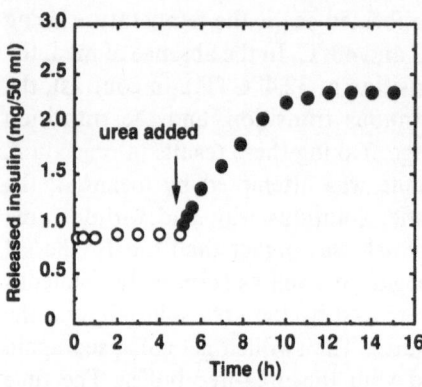

Fig. 6. Biochemo-mechanically controlled protein release using urease-loaded gel. The experiment was carried out in a 0.2 M ammonium buffer maintained at 35 °C using insulin (Mw = 5733) as the protein solute. (E. Kokufuta, S. Matsukawa, T. Ebihara, and K. Matsuda [77])

changing the balance of the repulsive and attractive forces between the polymer chain networks which constitute the gel is accepted as a general rule. This concept has also been applied in the construction of a gel system with immobilized concanavalin A (Con A; a sugar-binding protein or lectin, but not an enzyme) [78] in which it was possible to convert the difference in the biochemical affinities of this lectin towards two sugar molecules, α-methyl-D-mannopyranoside (MP; neutral) and dextran sulfate sodium (DSS; ionic), into mechanical energy using the phase transition (Fig. 7A).

MP

DSS
(R: H or -SO₃Na)

A gel containing a Con A/DSS complex (DSS-gel) was prepared by first making the complex and then immobilizing it in the NIPA gel. Using the complex rather than free Con A provided a marked increase in the swelling of the gel, presumably because of the protection against the chemical alteration of the active site of the Con A protein during gelation. Con A (0.4 mg) and DSS (40 mg) were dissolved in 1 ml of water. This was mixed with 1 ml of an aqueous

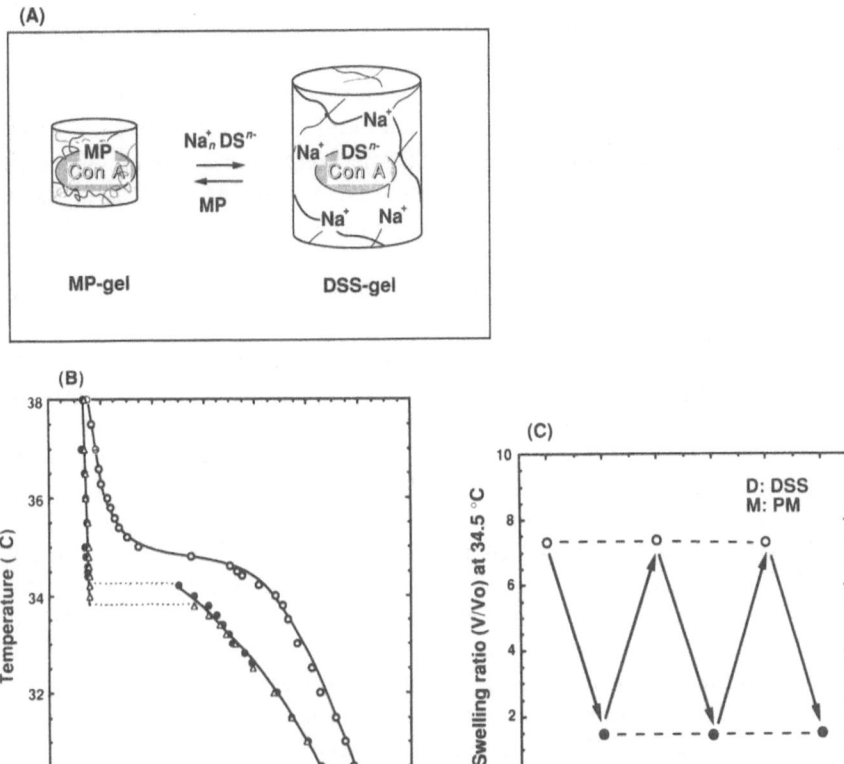

Fig. 7A–C. Biochemo-mechanical function of NIPA gel with immobilized concanavalin A: (A) Schematic illustration of saccharide-responsive, reversible swelling of a NIPA gel loaded with concanavalin A. $Na_n^+DS^{n-}$ is dextran sulfate sodium (DSS). (B) Temperature dependence for equilibrated volume of NIPA gel including the Con A-DSS complex (DSS-gel, ○), MP (MP-gel, ●), and free of both DSS and MP (△). The latter was prepared as a control sample as described in the text except for the use of an aqueous Con A solution instead of the Con A-DSS solution. Hysteresis was observed in the volume changes of the MP-gel and free-Con A gel on heating and cooling, indicating a discontinuous phase transition. The diameter of each gel in the collapsed state, determined at 50 °C, was $d_0 = 0.074$ mm; the volume of this gel is denoted V_0. The concentration of dry matter in the collapsed state was estimated from the preparation recipe to be 90 wt %. (C) Repeated swelling/shrinking control at 34.5 °C by alternate binding of DSS and MP to gel-entrapped Con A. (E. Kokufuta, Y.-Q. Zhang and T. Tanaka [78])

solution containing NIPA (156 mg), MBA (2.66 mg), and TEMED (4.8 µl). The solution was degassed, and 0.4 mg of ammonium persulfate was added to induce gelation. Gelation took place at 0 °C in glass capillaries with 0.1 mm inner diameters. After 1 h, gelation was completed and the gels removed from the capillaries and washed thoroughly with water. The samples were cut into 2 mm long cylinders and stored at 3 °C before use.

Changes in the volume of the DSS- and MP-gels in water were measured as a function of temperature. The MP-gel was obtained by soaking the DSS-gel in an aqueous MP solution (0.1 M) for 1 h, followed by washing with water. Gel diameters were determined using a microscope with a calibrated scale, the temperature being controlled to within 0.1 °C. The MP-gel underwent a discontinuous phase transition at 34.2 °C (Fig. 7B). Its swelling curve was similar to those for the pure NIPA gel without Con A and for the gel with free Con A. The transition temperatures of the pure NIPA gel and the gel with free Con A were the same: 33.8 °C. In contrast, when DSS ions were bound to Con A, the gel underwent a sharp but continuous volume change at 34.8 °C; that is, 0.6 °C higher than the transition temperature of the MP-gel. This difference is significant and reproducible.

Control of the swelling and shrinking of the lectin-loaded gel at 34.5 °C was checked by the repeated, alternate binding of DSS and MP to the gel-immobilized Con A (Fig. 7C). The MP-gel was converted into a DSS-gel by soaking it in a 2 wt % DSS solution for 1 h, then again converted into an MP-gel in the same manner as described above. The gel swelled to five times its original volume when DSS was bound to Con A, whereas the replacement of DSS by MP bought about the collapse of the gel. These swelling and shrinking cycles could be repeated with excellent reproducibility.

The swelling and shrinking changes of Con A-loaded NIPA gel are presumably determined by a balance between the pressures due to the repulsive electrostatic interactions between ionized dextran sulfate (DSS^{n-}) molecules, the osmotic pressure owing to counter ions (Na^+) from DSS, and the attractive hydrophobic interaction among NIPA residues. The first two factors act to raise the transition temperature. It is interesting to note that the phase transition, discontinuous for the MP-gel and the gel with free Con A, becomes continuous when DSS is bound to Con A. The introduction of ionizable groups into the gel network by copolymerization has been known to make the transition discontinuous and to shift it towards larger volume changes. In the present system, the DSS molecules have ionized sulfate groups which bind to the lectin entrapped in the network. This may cause a localized rise in ion concentration within the gel phase and thus lead to a range of 'local' transition temperatures, resulting in a continuous transition.

Certain other saccharides, to which ionizable groups such as $-NH_2$, COOH and $-OP(=O)(OH)_2$ have been bound chemically are also currently available. Various lectins have been isolated and characterized in terms of their interaction with saccharides. Therefore, different combinations of saccharide-lectin systems are available for designing saccharide-sensitive gels. More generally, to design a gel that undergoes a discontinuous volume transition in response to a molecular stimulus may be achieved by embedding into the gel an active element that interferes with gel equilibrium in response to the presence of 'stimulus' molecules.

In concluding this section, the author wishes to note that prior to the studies outlined above, two different research groups [79, 80] had attempted to control the volume of a gel or gel-like polymer membrane enzymatically. However, their purpose was not to convert the energy of enzyme reactions into mechanical work but to control the release of chemicals through the gel porosity. The phase transition was not taken into account in their design and preparation; thus, the swelling ratio of the matrices used in their release control experiments was only a few percent.

6 Conclusions and Topics for Future Research

The entrapment of biocatalysts within appropriate polymer gels is a very useful tool for the preparation of immobilized biocatalysts. Gel-entrapment techniques and the characteristics of entrapped preparations have been studied extensively, and the results of these studies partially utilized in industrial applications. A renewed interest in this research field may lead to the construction of "functional immobilized biocatalysts" that can be defined as immobilized biocatalyst systems with some beneficial functional capability other than the usually credited advantages obtained upon immobilization. The present review has dealt with several ideas for the preparation of functional immobilized bio-catalysts using stimulus-sensitive polymer gels (SSPGs), through which the following functional capabilities have been imposed on immobilized prepar-ations: (1) thermally enhanced immobilized enzyme activity via the cyclic absorption of substrate solutions; (2) thermal control of the initiation-termin-ation of an enzyme reaction via dramatically reduced substrate diffusion above the phase transition temperature; and (3) conversion of the energy of an immobilized enzyme reaction into mechanical work through a discontinuous change in gel volume.

All of the articles cited here have taken the developing tools for the design and syntheses of SSPGs into consideration in order to impose special functions on immobilized systems. Further developments in the field of SSPGs would make much more important contributions to the functionalization of immobil-ized biocatalysts. For example, the immobilization of an enzyme into an SSPG with a biochemo-mechanical function may enable the control of the initiation-termination of an enzyme reaction via the volume-phase transition in response to a biochemical stimulus. Since in living systems biochemical stimuli trigger the initiation and then termination of specific enzyme reactions, such SSPG-entrapped enzyme systems may be looked upon as biomimetic materials, and expected to be subject to application in clinical and medical fields; e.g., a drug delivery device which functions in response to some pathogenic stimulus. In conclusion, the applications of SSPGs in the immobilization of biocatalysts represent a rapprochement between biochemistry and polymer chemistry that will play a role in the design and construction of functional immobilized biocatalysts.

Acknowledgements: The author wishes to thank R. Craig for his critical reading of this manuscript. The author's studies cited here were supported in part by a grant to E.K. from the New Energy and Industrial Technology Development Organization (NEDO), Japan.

7 References

1. Stark GR (ed) (1971) Biochemical aspects of reactions on solid supports. Academic, New York
2. Wingard LB (ed) (1972) Enzyme engineering, vol 1. Wiley (Interscience), New York
3. Terui G (ed) (1972) Fermentation technology today, Soc Ferment Technol, Jpn, Tokyo
4. Zaborsky OR (1973) Immobilized enzymes. CRC Press, Cleveland, Ohio
5. Pey EK, Wingard LB (eds) (1974) Enzyme engineering, vol 2. Plenum, New York
6. Olson AC, Cooney CL (eds) (1974) Immobilized enzymes in food and microbial processes. Plenum, New York
7. Salmona M, Saronio C, Garattini S (eds) (1974) Insolubilized enzymes. Raven, New York
8. Messing RA (ed) (1975) Immobilized Enzymes for Industrial Reactors. Academic, New York
9. Weetall H, Suzuki S (eds) (1975) Immobilized enzyme technology. Plenum, New York
10. Weetall HH (ed) (1975) Immobilized enzymes, antigens, antibodies and peptides – preparation and characterization. Dekker, New York
11. Chang TMS (ed) (1976) Biomedical applications of immobilized enzymes and proteins. Plenum, New York
12. Mosbach K (ed) (1976) Immobilized enzymes: methods in enzymology, vol 44. Academic, New York
13. Wingard LB, Katchalski E, Goldstein L (eds) (1976) Immobilized enzyme principles: applied biochemistry and bioengineering, vol 1. Academic, New York
14. Wingard LB, Katchalski E, Goldstein L (eds) (1979) Enzyme technology: applied biochemistry and bioengineering, vol 2. Academic, New York
15. Johnson JC (1979) Immobilized enzymes – preparation and engineering. Noyes Data Co., Park Ridge, New Jersey
16. Wingard LB, Katchalski E, Goldstein L (eds) (1981) Analytical applications of immobilized enzymes and cells: applied biochemistry and bioengineering, vol 3. Academic, New York
17. Chibata I, Wingard LB (eds) (1983) Immobilized microbial cells: applied biochemistry and bioengineering, vol 4. Academic, New York
18. Lashkin AI (ed) (1985) Enzymes and immobilized cells in biotechnology. Benjamin/Cummings, California
19. Bailey JE, Ollis DF (1986) Biochemical engineering fundamentals. McGraw-Hill, New York
20. Rosevear A, Kennedy JF, Cabral JMS (1987) Immobilized enzymes and cells. Adam Hilger
21. Maugh TH (1984) Science 223: 474
22. Kokufuta E (1990) Kobunshi 39: 367 (in Japanese)
23. Kokufuta E (1991) Hyomen 29: 180 (in Japanese)
24. Kokufuta E (1992) Prog Polym Sci 17: 647
25. Bernfeld P, Wan J (1963) Science 142: 678
26. Kierstan M, Bucke C (1977) Biotechnol Bioeng 19: 387
27. Larsson PO, Mosbach K (1979) Biotechnol Lett 1: 501
28. Ohlson S, Larsson PO, Mosbach K (1979) Eur J Appl Microbiol Biotechnol 7: 103
29. Brodelius P, Nilsson K (1980) FEBS Lett 122: 312
30. Nilsson K, Mosbach K (1980) FEBS Lett 118: 145
31. Takata I, Tosa T, Chibata I (1975) J Solid-Phase Biochem. 2: 225
32. Lim F, Sun AM (1980) Science 210: 908
33. Veliky IA, Williams RE (1981) Biotechnol Lett 3: 275
34. Birnbaum S, Pendleton R, Larsson PO, Mosbach K (1981) Biotechnol Lett 3: 393
35. Tanaka H, Kurosawa H, Kokufuta E, Veliky I (1984) Biotechnol. Bioeng 26: 1393
36. Kokufuta E, Yukishige M, Nakamura I (1987) J Ferment Technol 65: 659
37. Kokufuta E, Shimizu N, Tanaka H, Nakamura I (1988) Biotechnol Bioeng 32: 756
38. Tanaka T (1978) Phys Rev Lett 40: 820
39. Tanaka T (1981) Sci Amer 244: 124
40. Tanaka T, Fillmore DJ, Sun ST, Nishio I, Swislow G, Shah A (1980) Phys Rev Lett 45: 1639
41. Ilavsky M, Hrouz J, Ulbrich K (1982) Polym Bull 7, 107
42. Hirokawa Y, Tanaka T (1984) J Chem Phys 81: 6379
43. Hirotsu S, Hirokawa Y, Tanaka T (1987) J Chem Phys 87: 1392
44. Katayama S, Hirokawa Y, Tanaka T, Macromolecules 17, 2641 (1984)
45. Siegel RA, Firestone BA (1990) J. Contr. Release 11: 181

46. Beltran S, Baker JP, Hooper HH, Blanch HW, Prausnitz JM (1991) Macromolecules 24: 549
47. Ohmine I, Tanaka T (1982) J Chem Phys 11: 5725
48. Tanaka T, Nishio I, Sun ST, Ueno-Nishio S (1982) Science 218: 467
49. Osada Y (1987) Adv Polym Sci 82: 1
50. Hirose Y (1989) International workshop on polymer gels, Soc Polym Sci, Jpn. p 5
51. Suzuki A, Tanaka T (1990) Nature 346: 345
52. Mamada A, Tanaka T, Kungwatchakun D, Irie M (1990) Macromolecules 23: 1517
53. Hirokawa Y, Tanaka T, Sato E (1985) Macromolecules 12: 2782
54. Flory PJ (1953) Principles of polymer chemistry. Cornell Univ Press, New York
55. Otake K, Inomata H, Konno M, Saito S (1989) J Chem Phys 91: 1345
56. Prange MM, Hooper HH, Prausnitz JM (1981) AIChE J 35: 803
57. Hooper HH, Baker JP, Blanch HB, Prausnitz JM (1990) Macromolecules 23: 1096
58. Ilmain F, Tanaka T, Kokufuta E (1991) Nature 349: 400
59. See, for example: Srere PA, Mattiasson B, Mosbach K (1973) Proc Nat Acad Sci USA, 70: 2534
60. Kokufuta E, Jinbo E (1992) Macromolecules 25: 3549
61. Kokufuta E, Yamaya Y, Shimada A, Nakamura I (1988) Biotechnol Lett 10: 301
62. Metzler DE (1977) Biochemistry. Academic Press, New York
63. Park TG, Hoffman AS (1988) Appl Biochem Biotechnol 19: 1
64. Park TG, Hoffman AS (1990) J Biomed Mater Res 24: 21
65. Park TG, Hoffman AS (1990) Biotechnol Bioeng 35: 152
66. Kokufuta E, Ogane O, Ichijo I, Watanabe S, Hirasa O (1992) J Chem Soc, Chem Commun 416
67. Dong LC, Hoffman AS (1987) In: Russo P (ed) ACS Symp Ser No 350, p 236
68. Dong LC, Hoffman AS (1986) J Contr Release 4: 223
69. Inomata H, Goto S, Saito S (1990) Macromolecules 23: 4887
70. Matsuoka H, Suzuki S, Aizawa M, Kimura Y, Ikegami A (1981) J Appl Biochem 3: 437
71. Blok MC, Deenen LLM, Gier J (1976) Biochim Biophys Acta 433: 1
72. Kokufuta E, Tanaka T (1991) Macromolecules 24: 1605
73. Kokufuta E, Zhang YQ, Ilmain F, Tanaka T (1991) Polym Prepr Jpn 40: 4204
74. Kuhn W, Hargitay B, Katchalsky A, Eisenberg E (1950) Nature 165: 514
75. Steinberg IZ, Oplatka H, Katchalsky A (1966) Nature 210: 568
76. Sussman MV, Katchalsky A (1970) Science 167: 45
77. Kokufuta E, Matsukawa S, Ebihara T, Matsuda K (1993) ACS Porym Prepr 34: 980
78. Kokufuta E, Zhang YQ, Tanaka T (1991) Nature 351: 302
79. Ishihara K, Kobayashi M, Shinohara I (1983) Makromol Chem Rapid Commun 4: 327
80. Kost J, Horbett TA, Ratner BD, Singh M (1985) J Biomed Mater Res 19: 1117

Received January 18, 1993

Molecular Design of Temperature-Responsive Polymers as Intelligent Materials

Teruo Okano
Institute of Biomedical Engineering, Tokyo Women's Medical College,
Shinjuku-ku, Tokyo 162, Japan

In recent years, temporal control of drug delivery has been of interest in basic and applied fields as a new approach to achieve improved drug therapies. This new drug delivery may be achieved by utilization of "intelligent" polymeric materials having the functions of sensing, processing and acting. Stimuli-responsive polymers have been investigated as potential molecular devices to achieve intelligent drug delivery systems such as self-regulating and externally nodulated drug delivery systems.

In this chapter, temperature responsive swelling-shrinking changes in hydrogels were reviewed from the view point of polymer–water interaction and polymer–polymer interaction. In particular, characteristics of interpenetrating polymer networks (IPN) constructed of two distinct polymers which lead to novel temperature-responsive swelling changes were discussed.

Advances in Polymer Science, Vol. 110
© Springer-Verlag Berlin Heidelberg 1993

List of Abbreviations

PIPAAm   poly(N-isopropyl acrylamide)
RMA   alkyl methacrylate
PEO   poly(ethylene oxide)
APy   poly(acryloylpyrrolidine)
PMAA   poly(methacrylic acid)
PAAm   polyacrylamide
PAAc   poly(acrylic acid)
IPN-X   Interpenetrating polymer networks of poly(AAm-*co*-BMA)
            (BMA X wt %) and PAAc (AAm:AAc = 1:1)
NVP N-vinyl-2-pyrrolidone
PBA   *m*-aminophenyl-boric acid
PVA   poly(vinyl alcohol)

1 Introduction

Stimuli-responsive polymers which change their structure and physical properties in response to external signals are a new set of materials with interesting applications in biomaterials science and technology [1]. Particularly, intelligent drug delivery systems (DDS) such as self-regulating or auto feed-back drug release systems may be achieved by utilization of stimuli-responsive polymers. In the intelligent DDS, the system senses a signal caused by disease or malfunction (sensor function), judges the magnitude of the signal (processor function), and then acts to release the drug in direct response (actuator and effector function). When polymeric materials are designed from first principles at the molecular level to serve these three functions, they might be called "Intelligent Materials".

From this perspective, temperature-responsive hydrogels are reviewed as agents in the design of effective polymeric structures for thermo-responsiveness as "Intelligent Materials".

2 Poly(N-isopropyl acrylamide) Gels

Polymer hydration and dissolution in aqueous systems has been considered to depend on strength of interaction between polymer and water. This water-polymer interaction is a primary driving force causing solubilization-precipitation changes in linear polymers and swelling-shrinking changes in polymeric gels.

Poly(N-isopropyl acrylamide) (PIPAAm) demonstrates a lower critical solution temperature (LCST) in aqueous solution and its crosslinked networks show high swelling thermosensitivity in water. The thermosensitivity has been investigated [2, 3] by swelling measurements and characterization using differential scanning calorimetery (DSC). PIPAAm gels shrink in water at higher temperatures and demonstrate a sharp swelling transition in the vicinity of 32 °C. Below this temperature, the gels swell with decreasing temperature. We have prepared a crosslinked monolithic gel device in which indomethacin was uniformly dispersed using a copolymer of IPAAm with alkyl methacrylate (RMA) and achieved complete "on-off" regulation of drug release [2–6] in response to stepwise temperature changes between 20 °C and 30 °C as shown in Fig. 1. The release patterns at low temperature (the "on" state) have been analyzed using a two-layer model we have recently developed [7]. The "on-off" switching mechanism has been clarified in terms of a dense skin layer formed on the surface of the gel with shrinking upon increasing temperature. This surface skin layer stops drug release from inside the polymer matrix. The surface skin layer has been observed by optical [5] and scanning electron microscopy [8].

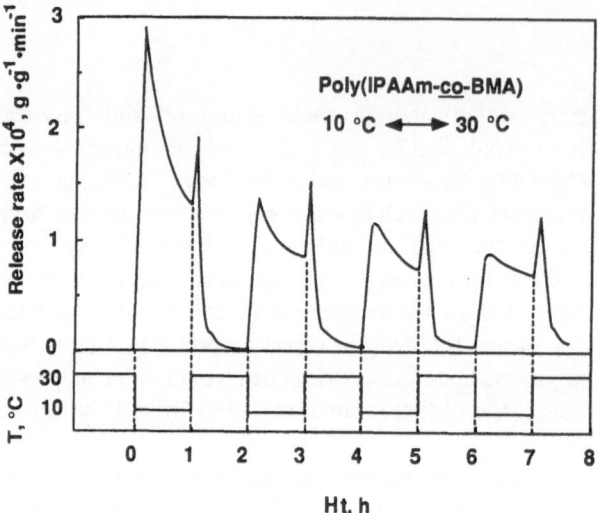

Fig. 1. Pulsatile release rate of indomethacin from poly(IPAAm-*co*-BMA) gel (BMA 5 wt %) in response to stepwise temperature changes between 10 °C and 30 °C

After polymer skin formation, gel bubbles form on the device surface because the skin layer prevents any further outflow of water and pressure to expel water is exerted on the surface from the inside after shrinking [5]. It is proposed that pressure is induced within the gel in the process of shrinking. This pressure may induce the outward convection of water. In the "on" state, drug is released from the surface by diffusion through the hydrated gel matrix. However, in the process from the "on" state to the "off" state upon increasing temperature, the drug must be released not only by diffusion but also by convective transport [9].

We have developed a drug release model for the shrinking process using the tortuous pore model [10] for solute transport across a membrane. Four decreasing drug release rate patterns were simulated from the model for different induction patterns of pressure. From measurement of swelling-deswelling kinetics and optical transmittance changes of gels in response to stepwise temperature changes, we have suggested that the surface skin layer formation process is controlled by changing the length of the alkyl side chain of alkyl methacrylate comonomer [8, 11]. Such changes in dynamic processes of skin formation may affect the induction pattern of pressure within the gel with shrinking. To realize experimentally the pattern of pressure induction assumed arbitrarily in our release simulations, we changed both the gel chemical structure and temperature differences. Release rates of indomethacin from the poly(IPAAm-*co*-RMA) gels after increasing temperature were continuously monitored by a flow cell system and compared with the simulation. Effects of the dynamic skin formation process on decreasing drug release rate patterns have been rationalized in terms of changing the copolymer alkyl side chain length and the degree of temperature difference between the gel "on" state and the "off" state [9].

3 Utilization of Polymer Complexes (or Complexation)

3.1 Interpolymer Interactions

When two different polymer solutions are mixed, they frequently undergo one or several distinct types of interaction, which in each case can lead to phase separation at polymer concentrations above a certain critical level [12]. In one case, two solution phases of approximately equal volume are formed, consisting of polymer A- and polymer B-rich solutions, respectively. This phase separation is called incompatibility, or simple coacervation. In the second case, two phases are formed but both polymers are concentrated in one of the phases (the "precipitate") while the other phase (the "supernatant") may be essentially polymer free. This separation is called complex coacervation. The two phase separation phenomena are shown in Fig. 2.

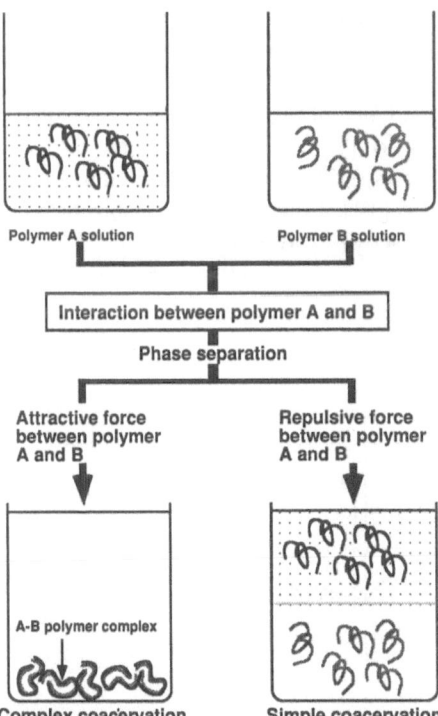

Fig. 2. Phase separations of interactive polymers

Interpolymer interactions are a fundamental driving force inducing both attractive and repulsive forces between polymers, often resulting in phase separation. By utilizing the change of interpolymer interactions induced by changing external signals, remarkable solubility changes in the polymers can be achieved. Further, interpenetrating polymer networks (IPNs) might be expected to exhibit significant shrinking-swelling changes due to large changes in interpolymer interactions. Namely, IPNs will exhibit shrunken states for cases of attractive forces between polymer A and polymer B, and fully swollen states for repulsive forces between the two polymers. Moreover, IPNs constructed from polymer A and polymer B would behave completely, differently then the corresponding random gels consisting of A and B.

3.2 Repulsive Interpolymer Interactions

We synthesized [13] IPNs composed of poly(ethylene oxide) (PEO) (polymer A) and poly(N-acryloylpyrrolidine) (PAPy) (polymer B). The IPN was synthesized by simultaneous crosslinked polymerization of APy and PEO. The overall reaction scheme for IPN synthesis by radical polymerization for APy (polymer A) and addition polymerization for PEO (polymer B) is shown in Fig. 3. This pair shows simple coacervation behavior in water. The IPN is constructed from PEO and PAPy networks as shown in Fig. 4. Chemically independent networks of polymer A and polymer B are interlocked and macroscopic phase separation in water swollen states is avoided.

Figure 5 shows the temperature dependence of the IPNs with various ratios of [EO]/[APy], kept in unit molar ratio [14]. The IPNs swell in lower temperature ranges and shrink in high temperature ranges. The curves show inflection points at around 35 °C. Figure 6 [14] shows the relationship between the swelling ratio at 10 °C and the [EO]/[APy] ratio within the IPNs. Swelling ratio increases as the [EO]/[APy] ratio nears 1 : 1. The IPN with an equimolar

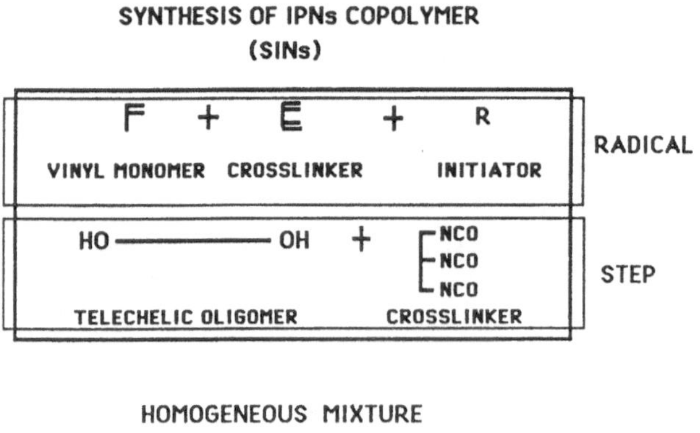

Fig. 3. Schematic illustration of SIN synthesis

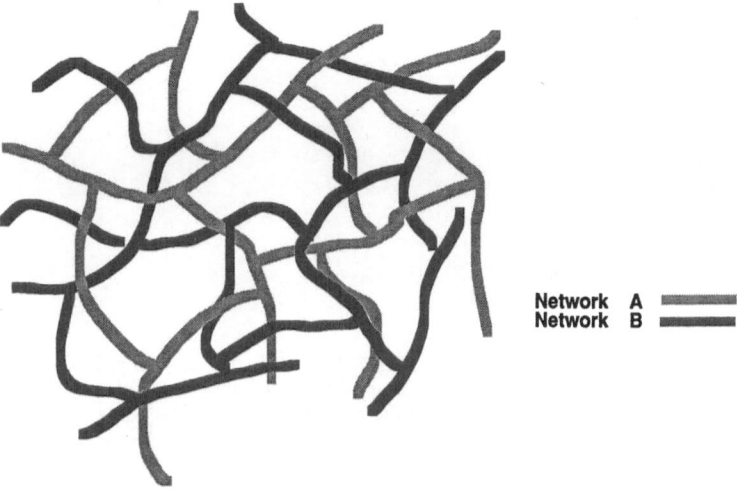

Fig. 4. Ideal formation of a full IPN

Network A ═══
Network B ═══

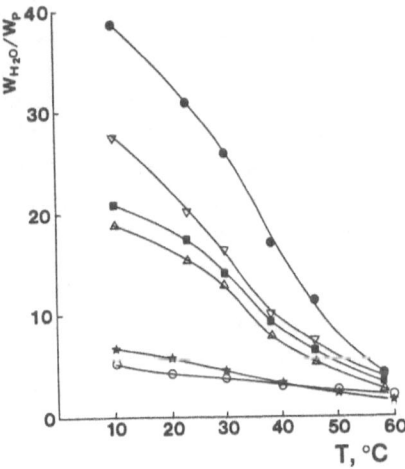

Fig. 5. Temperature dependence of equilibrium swelling ratio of poly(APy)/PEO IPNs in water; PEO repeating unit 20.6% mole fraction ($\triangle$); 36.3% mole fraction ($\triangledown$) 49.2% mole fraction ($\bullet$); 58.7% mole fraction ($\blacksquare$); crosslinked poly(APy) ($\star$); crosslinked PEO ($\bigcirc$)

ratio of [EO]/[APy] shows the largest swelling change with temperature change.

Homopolymer gels composed of PEO and PAPy show low swelling ratios, IPNs composed of both polymers, however, show higher swelling ratios which peak as the [APy]/[EO] ratio approaches 1:1. This result suggests that PEO chains and PAPy chains repel each other at lower temperatures when the two polymers are forced to occupy the same space by interpenetration. Since the IPN with an equimolar (PEO:APy) composition shows the highest swelling value, the two polymers apparently repel each other most significantly when the polymers are present in an equimolar ratio of the repeating units. The two polymer networks (two proton-acceptor type polymers) may be involved in a

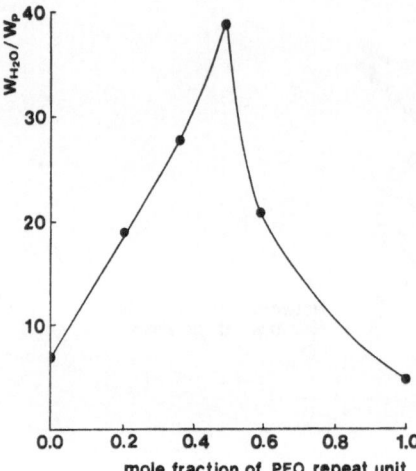

Fig. 6. Effect of the poly(APy)/PEO IPNs composition on the equilibrium swelling ratio at 10 °C

positive heat of mixing (endothermic mixing process). This indicates that the unfavorable enthalpy of mixing between the two polymers (coacervation) may cause increased chain repulsion in gels with a decrease in temperature, while the polymer–solvent interactions increase with decreasing temperature. Therefore, when the temperature of the systems is decreased, each network expands independently due to its increased affinity to water resulting in fully expanded structures both because of full hydration and repulsive interaction of the networks. This combined effect seems to alter swelling ratios and the thermosensitivity of swelling of the IPNs in comparison to that of the homopolymers.

In addition to the effects of chemical interpolymer interaction, structural effects from the physical entanglement of polymer A and polymer B in IPNs are another important factor causing such swelling behavior. Immiscible polymer chains maintain mutual segregation from each other, expanding their structure and absorbing water into the gel. Repulsive forces between the two polymeric chains effectively promote large gel swelling of IPN structures. Random copolymer gels composed of EO and APy show no effective interpolymer repulsion between chains expected as observed in the case of the IPN. Therefore, large swelling changes versus temperature can be also achieved in IPNs by utilizing structural features of entangled chains.

3.3 Attractive Interpolymer Interactions

Polymer complexation frequently leads to dehydration and precipitation of polymer [15–32] as shown in Fig. 2. To cause polymer complexation, attractive forces between polymers are needed, and the force must overwhelm the strength of interaction between polymer chains and hydrating water molecules. In this process, water molecules must be replaced by competing polymer contacts.

Possible sources for intermolecular interactions for interpolymer complexation are Coulombic forces, hydrogen bonding, van der Waals forces, charge transfer interactions, hydrophobic interaction and their combinations [33, 34]. Each interactive force has a distinct property and a dependence on external stimuli such as change of pH, salt concentration, temperature, and others [33, 34]. If combinations of various interactions are involved in complexation, it is important to consider how much each interaction influences the complexation and how the interactive force changes with stimuli. If some interactions offset each other enough to cancel out a net attractive force, complexation will remain improbable. When intermolecular interactions work cooperatively, complexation can be enhanced. Therefore, when we utilize these interactions as driving forces to promote complexation, we must ascertain what the main intermolecular interaction is, what unfavorable interactions are possible, and what methods are there for selectively eliminating unfavorable interactions. Determination of these parameters are very important in the design of materials with effective stimuli-responsiveness.

Osada et al. [16–21] reported the polymer complexation of poly(methacrylic acid) (PMAA) and PEO in gel systems. This pair of polymers is well known for forming polymer complexes by hydrogen bonding in solution [15–34]. Complexation and dissociation changes in the polymer complex are observed as a function of temperature [16–18, 20, 23]. In one of their reports, shrinking changes of PMAA gels immersed in PEO aqueous solution is discussed. Figure 7 [20] shows the results of the experiments; curve 1 refers to PMAA gel immersed in distilled water, and curve 2 refers to the gel immersed in PEO aqueous solution. In distilled water, the shrinking ratio simply increases with increasing temperature at first and then it simply decreases with decreasing temperature. In the PEO solution, however, the contraction increases with

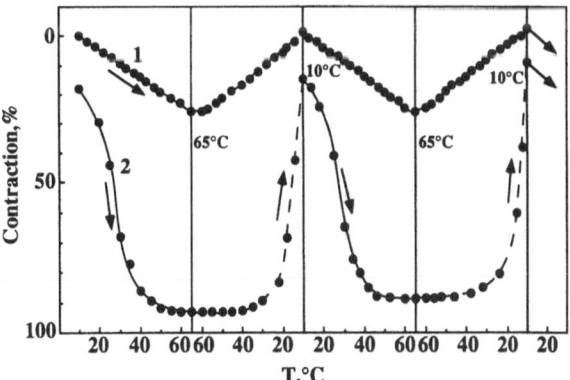

Fig. 7. Temperature dependences of contraction of PMAA membranes with various embedding fluids: (1) 70 ml of pure water, (2) 70 ml of 0.015 units mole/I PEG solution. Dry membrane: 10 mm wide, 23 mm long, 4.7 mg weight; load 490 mg; PEG molecular weight = 2000. Ordinate is expressed in % of the length of the dry membrane

temperature increase and shows a drastic change at around 30 °C. On cooling, the contraction curve shows a drastic change at around 20 °C. Curves 1 and 2 show the reversible contraction changes with temperature changes. A large difference in the response to temperature change between the two profiles exists.

The large contraction change shown by curve 2 is due to polymer dehydration or precipitation induced by complexation between PEO and PMAA. It is evident that the dehydration or precipitation is enhanced by complexation compared to curve 1. Furthermore, curve 2 indicates a reversible contraction/ recovery change, suggesting that complexation and dissociation occur reversibly with temperature changes within the PMAA gel. This drastic contraction change can be applied to temperature sensitive hydrogels as well.

4 Change of Association–Dissociation Interactions in Polymer Complexes

Osada et al. [20] performed the same experiments mentioned above using ethanol/water mixed solvents for the PEO solution. Incorporating ethanol into the system can inhibit hydrophobic interactions in polymer complexation. In a system having negligible hydrophobic interactions between polymer chains, the PMAA gel shrinks at lower temperatures and swells at higher temperatures; which is the reverse temperature tendency of curve 2 in Fig. 7. The temperature dependence in the ethanol/water mixture suggests that the main driving force in the system is hydrogen bonding. The result that the hydrogen bonding contributed to PMAA/PEO complexation as a driving force was supported by elemental analysis, pH change within the gel and IR measurements. The results suggest that PMAA and PEO units form a complex by hydrogen bonding at an equimolar ratio. In the complex, the carboxyl group of the PMAA and ether unit of PEO play the roles of hydrogen donor and acceptor, respectively. From these results it follows that PEO/PMAA complexation involves both hydrogen bonding and hydrophobic interactions, although the temperature dependences of each of the two driving forces is opposite. The temperature dependence of curve 2 indicates that the hydrophobic interactions act as a primary factor influencing the temperature dependence of the swelling changes in these gels.

There are some reports about PEO/PMAA IPNs being used as pH sensitive hydrogels [35–37]. Since PMAA has carboxyl groups, the solubility of PMAA strongly depends on pH of the external medium. Therefore, it is important to recognize that pH changes control the swelling state significantly. Kotaka et al. reported on IPNs composed of PEO and PMAA [35–37]. These IPNs showed deswollen states in the lower pH range and a swollen state in the higher pH range. The shrinking of the IPN is due to complexation of PMAA and PEO by hydrophobic and hydrogen bond interactions. The increase in swelling at higher pH is due to dissociation of the PEO/PMAA polymer complex. The polymer complex is forced to dissociate by increased hydration ionization, and solubilization of PMAA induced by external pH increases. In this case, the ionization/protonation changes of PMAA are the most important key to control the IPN swelling state. The swelling ratio of the IPN changes reversibly in response to pH changes in the external solution. In this case, the polymer complexation/dissociation changes are assisted by accompanying cooperative chain–chain interactions described below (zipper effect).

Polymer complexation efficiency depends on the molecular weight of the polymers [16–18, 20, 21, 25]. Polymers form tighter complexes with increasing molecular weight. When the molecular weights of the two polymers are sufficiently high, reversible complexation and dissociation occur in response to stimuli changes. However, if the relative molecular weights are too large, the resulting polymer complex does not dissociate because of irreversible association. For example, if the molecular weights are appropriate, complex formation/dissociation changes can occur as shown in Fig. 7. However, when the molecular weight of PEO is too large, polymer complexes do not dissociate, regardless of the temperature change. This point must be taken into account when designing materials using polymer complexation as an assembly tool.

Stable complex formation is desirable to allow a thorough homogeneous shrinking state when the polymers are incorporated into an IPN. Stable binding of the complex, however, may make it impossible to cause reversible complex formation/dissociation change. In such cases, the material composed of polymer complexes loses its value as stimuli-sensitive material. Control of effective complex dissociation and solubilization of the polymer, as well as stable complex formation make it possible to realize swelling changes by complex formation/dissociation change. Balance of complex stability before stimulation and solubility after stimulation are important factors for controlling material swelling.

5 Temperature-Responsive IPNs and Their Application to Switches for On-Off Drug Release

Hydrogen bonds, because of their relatively weak interaction, are stronger at lower temperatures and their strength decreases with increasing temperature. In the PEO/PMAA system, hydrogen bonding is referred to as one of the weaker driving forces among several kinds of forces possible for complexation. As mentioned previously, the temperature dependence of hydrogen bonding appears not to affect contraction behavior (curve 2 in Fig. 7). This is because the main driving force is hydrophobic interactions. To utilize hydrogen bonding for polymer complexation, it is important to inhibit other intermolecular interactions, especially hydrophobic interactions, because of possible reverse temperature dependencies. If the temperature dependence of the hydrogen bonding dominates the polymer complex formation/dissociation changes, the complex may form at lower temperatures and dissociate at higher temperatures. Furthermore, hydrogels composed of the polymer complexes may exhibit reverse swelling changes as shown in Fig. 7, showing drastic changes resulting from a zipper reaction along the associated polymer chains.

Polyacrylamide (PAAm) and poly(acrylic acid) (PAAc) are known as polymers which interact intra- or/and intermolecularly by hydrogen bonding [38, 40]. In aqueous solutions of a polymer binary system, polymer solubility was enhanced with increasing temperature because of breakage of intra- or/and intermolecular hydrogen bonding. On the other hand, strange solubility behavior versus temperature was observed in aqueous solutions of PAAm–PAAc and in related polymer solutions; which has not been observed in the polymer–water binary system [39]. We have studied the temperature dependence of polymer solubility in the PAAc–PAAm–water ternary system [40]. Figure 8 shows optical transmittance (T%) changes of the ternary aqueous system with temperature. The two polymers form polymer complexes by intermolecular hydrogen bonding at lower temperature as shown in Fig. 9, resulting in dehydration and precipitation (complex coacervation [39]). This precipitation (complex coacervation) leads to a turbid state and to a decrease of optical transmittance. The complex dissociates at a higher temperature due to breakage of the hydrogen bonds, followed by solubilization of the polymers. Furthermore, the transition between the complex formation state and dissociation state occurs at a critical temperature.

Hydrophobic interactions seem to have negligible effects on polymer complexation, since, in this case, this temperature-dependent solubility change shows a positive temperature dependence in optical T% changes. Therefore, hydrogen bonding forces are probably the primary intermolecular interactions.

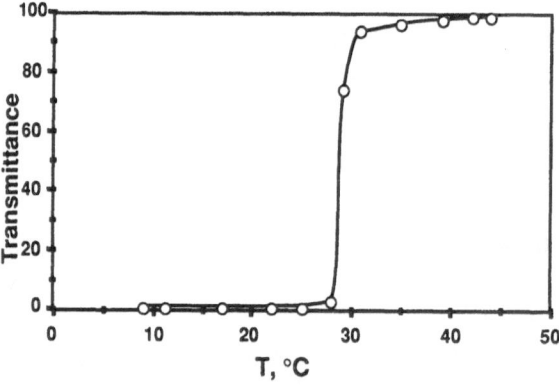

Fig. 8. Temperature dependence of transmittance for the system of PAAc-PAAm-water solution (at pH 3.17 adjusted by HCL). Polymer concentration (mass fraction in %); PAAc, 0.5; PAAm, 0.5

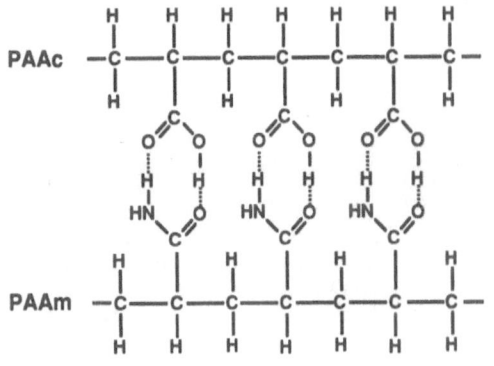

Fig. 9. Polymer complex formation between PAAc and PAAm by hydrogen bonding at low temperature

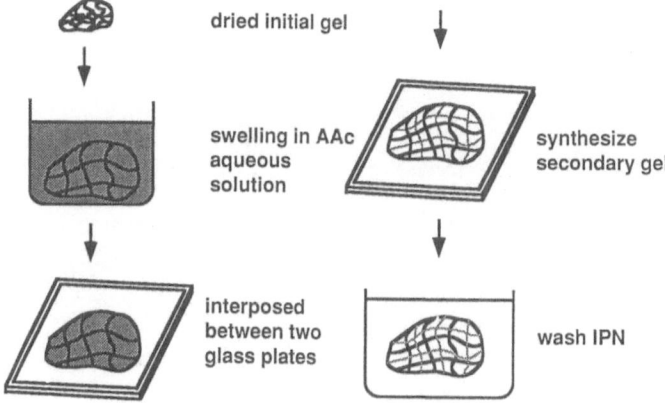

Fig. 10. Procedure for preparation of PAAm/PAAc based IPNs by means of SIPN method

Furthermore, this complexation/dissociation change may occur with an accompanying cooperative chain–chain zipper effect; the drastic change or transition of solubility implies extensive chain–chain interactions.

IPNs composed of PAAc and PAAm may shrink at a low temperature because of interpolymer complexes formed by hydrogen bonding as shown in Fig. 9. The complexes dissociate at higher temperatures due to breaking of hydrogen bonds, and IPNs swell rapidly above a critical transition temperature [40].

In poly(AAm-*co*-BMA)/PAAc IPNs (BMA is butyl methacrylate), the synthetic procedure applied was the sequential polymerization technique shown in Fig. 10. A monomer or set of comonomers is polymerized into a polymer gel. A second monomer is introduced and subsequently polymerized and crosslinked within the initial gel (AAm-*co*-BMA) matrix to form the IPN.

Temperature dependent swelling behavior of the poly(AAm-BMA) gel consisting 10% BMA (Initial gel), PAAc gel (secondary gel) and IPNs composed of poly(AAm-*co*-BMA-10) with PAAc are shown in Fig. 11 and compared with a corresponding random copolymer gel. The IPNs and random gels show

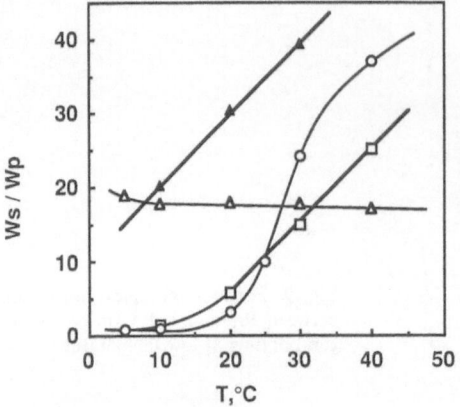

Fig. 11. Temperature dependence of swelling ratios of IPN-10 (○), poly(AAc-*co*-AAm-*co*-BMA) (BMA: 10 wt % for AAm) (□), poly(AAm-*co*-BMA) (BMA 10 wt %) (△) and crosslinked PAAc (▲)

Complex dissociation

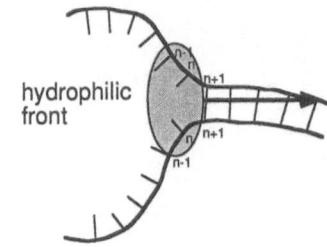

Complex formation

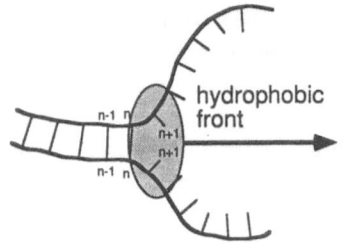

Fig. 12. Zipper effect in the process of formation and dissociation of interpolymer complex

distinctly different profiles of swelling temperature dependence, although both gel types show a positive temperature dependence. Only the IPNs show a sigmoidal alteration with a drastic transition zone within a certain characteristic temperature range. In the case of the random copolymer of PAAc and PAAm, swelling monotonously increases with temperature. Drastic alteration in the behavior of the swelling characteristics of IPNs is due to cooperative polymer–polymer interaction of PAAm and PAAc long chains causing drastic formation/dissociation change of the polymer complex. Figure 12 shows a model of polymer–polymer dissociation and complexation. In the case of dissociating polymer complexes, dissociation starts from an unstable binding site. The polymer–polymer binding sites are disrupted and the polymer complexes start to dissociate by substitution of complexed polymer with water molecules. The initial dissociation of the unit binding point makes it hydrophilic and this hydration initiates dissociation of the next binding point in rapid succession Dissociation of the nth bond facilitates dissociation of the (n + 1)th bond (zipper reaction) due to the hydration and electrostatic repulsive force between side chains of PAAc [40]. Phase transitions of swelling changes against temperature change of the IPNs (PIAAm-PAAc) were also observed by Tanaka and colleagues in the presense of urea [41]. The IPNs composed of poly(AAm-

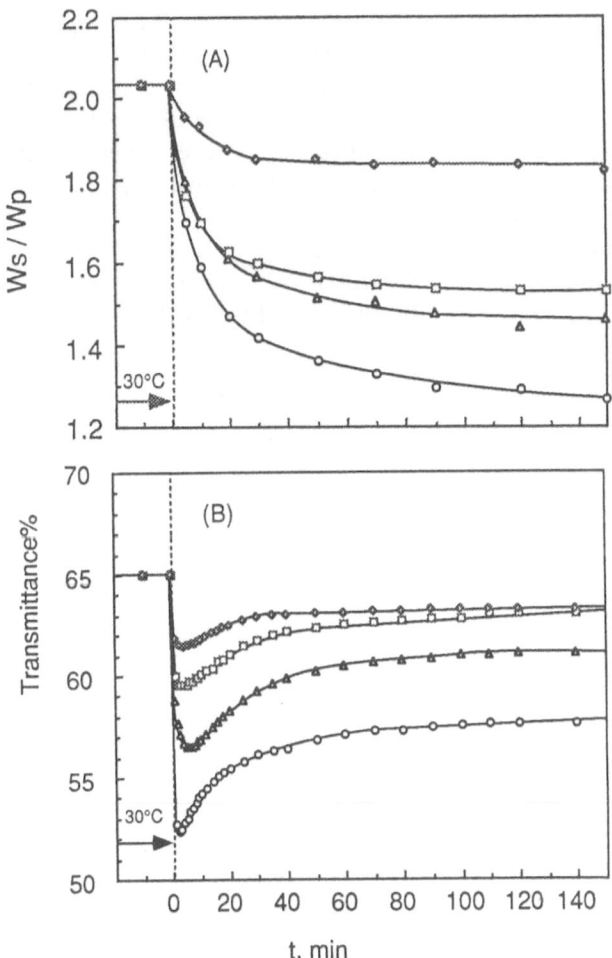

Fig. 13A, B. Time courses of swelling changes (A) and transmittance changes (A) of IPN-30 after temperature changes from 30 °C to 10 °C (O), 15 °C (△), 0 °C (□) and 25°C (◇)

co-BMA) and PAAc showed reversible swelling changes in response to temperature changes according to the alterations between polymer–polymer complexation and dissociation. Formation of the surface skin layer on the IPNs in
the shrinking process associated with a temperature decrease was suggested
from the results of water content (Ws/Wp) and transmittance (T%) changes as
shown in Fig. 13. The transparent gel at 30 °C became opaque immediately after
a temperature decrease from 30 °C to lower temperatures. This thermo responsiveness was evaluated quantitatively from the measurement of the T%
change at 500 nm with a spectrophotometer. Larger T% decreases were
observed with larger temperature decreases. The sharp decreases of T% immediately after cooling revealed the formation of a heterogeneous structure. IPN
shrinking started immediately from the surface upon decreasing of temperature
because the surface is the first part affected by temperature changes and has high
mobility to change polymer chain conformations. The degree of heterogeneity
caused by surface shrinking were largest at minimum T%, as shown in Fig. 14.
The T% increase after a minimum implies that the progression of homogeniz-
ation of the gel is due to a decrease of the degree of heterogeneous swelling inside
the gel. The strength and density of the shrunken surface layer depend on the
aggregation of dehydrated polymer chains induced by hydrogen bonding
complexations [42].

Figure 15 shows ketoprofen release from IPNs of poly(AAm-co-BMA) and
PAAc containing a 20% mass fraction of BMA (IPN-20) when the temperature
is varied between 10 °C and 30 °C. IPNs shrink gradually after the temperature

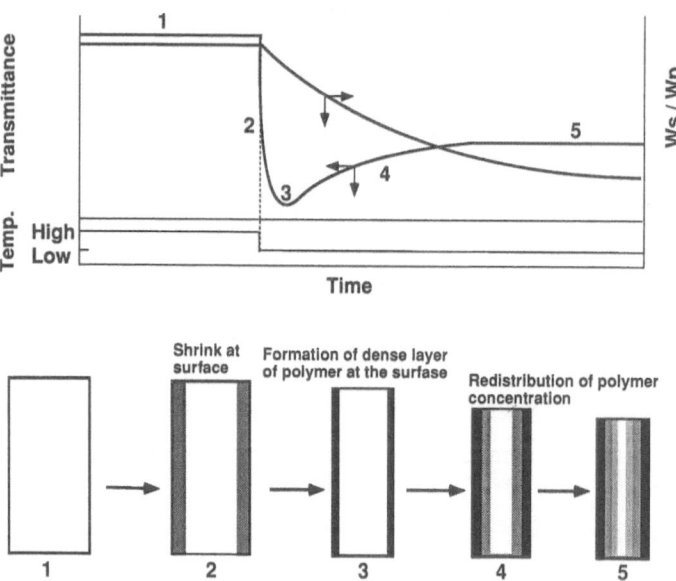

Fig. 14. Model of the shrinking mechanism of IPN

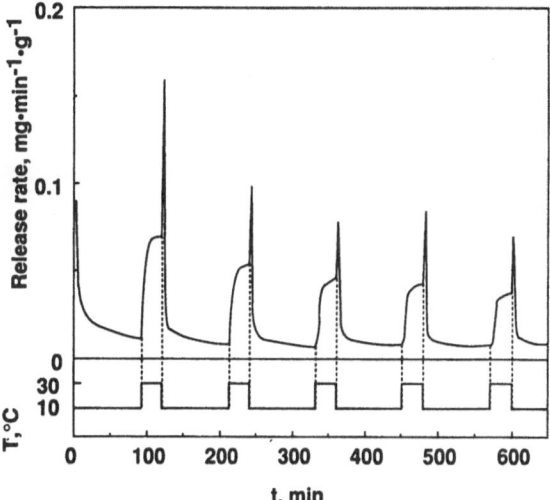

Fig. 15. Change of drug release rate from IPN-20 in response to step-wise temperature change between 10 °C and 30 °C in distilled water

is reduced from 30 °C to 10 °C, without any drastic changes. Release rates, however, demonstrate an immediate increase soon after temperature reduction to 10 °C, followed by a constant low release rate. The relationship between the shrinking profile and release profile suggests that modulation of on-off release does not necessarily require a large bulk swelling change. A phenomenon occuring primarily at the IPN surface seems to rather significantly relate to the "off" mechanism. Sharp rate increases probably result from the release of the drug being mechanically forced by IPN surface shrinkage and not by free diffusion. Larger release peaks are observed for larger temperature intervals in temperature variation [42].

6 Glucose-Responsive Polymer Complexes [43–45]

By utilizing the concept of polymer complexation and dissociation regulated by temperature as mentioned above, a glucose-responsive polymer complex has been studied. A novel polymer complex system sensitive to glucose was developed as a candidate material for formulating a chemically regulated insulin delivery system. Phenylboric acid is incorporated into water soluble polymeric chains as a sensor molecule for glucose; the molecule can form reversible covalent complexes with molecules which contain diol units. Borate, in the tetrahedral form, can exchange its hydroxyls with diol units in polyol substances to form a covalent five- or six-membered cyclic complex. The major structural

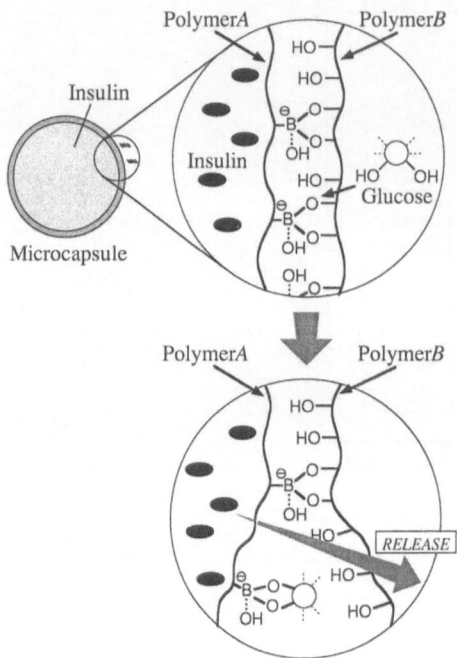

Fig. 16. Concept of glucose sensitive insulin release system using PVA/poly (NVP-10-PBA) complex system. (polymer capsule type)

requirement for complex formation is that the two hydroxyls are held in a coplanar configuration. As most of the closed-ring carbohydrates including glucose have a *cis*-diol moiety in their structure, they can form a relatively strong complex with borate. Reversible exchange of borate/diol complexes takes place by competing diol molecules which can form complexes more tightly. Based on this characteristic of borate/diol complexes, glucose-responsive polymer complexes have been studied. Borate copolymers (polymer A) (equimolar copolymer of N-vinyl-2-pyrrolidone (NVP) and m-aminophenyl-boric acid (PBA)), and poly(vinyl alcohol) (PVA) (polymer B) were used as components. A glucose-responsive complex was formed as shown in Fig. 16 [45]. Reversible viscosity changes due to a reversible complex formation have been observed in response to glucose concentration. These polymeric complexes have been applied as interpenetrating polymer networks to achieve pulsatile insulin release in response to glucose concentration.

7 References

1. Okano T, Bae YH, Kim SW (1990) In: Kost J (ed) Pulsed and self-regulated drug delivery. CRC Press, Boca Raton, FL, p 17
2. Bae YH, Okano T, Kim SW (1990) J Polym Sci B Polym Phys 28: 923

3. Bae YH, Okano T, Kim SW (1991) Pharm Res 8: 531
4. Bae YH, Okano T, Kim SW (1987) Macromol Chem. Rapid Commun 8: 481
5. Okano T, Bae YH, Jacobs H, Kim SW (1990) J Controlled Release 11: 255
6. Bae YH, Okano T, Sakurai Y (1991) Pharm Res 8: 624
7. Yoshida R, Sakai K, Okano T, Sakurai Y (1991) Polym J 23: 1111
8. Yoshida R, Sakai K, Okano T, Sakurai Y (1992) J Biomater Sci Polym Edn 3: 243
9. Yoshida R, Sakai K, Okano T, Sakurai Y (1992) Ind Eng Chem Res 31: 2339
10. Sakai K, Ozawa K, Mimura R, Ohashi H (1987) J Membrance Sci 32: 3
11. Yoshida R, Sakai K, Okano T, Sakurai Y (1991) J Biomater Sci Polym Edn 3: 155
12. Molyneux P (1984) Interactions between polymers. In: Water-soluble synthetic polymers: Properties and behavior, vol II, CRC Press, Boca Raton, FL, USA, p 159
13. Mukae K, Bae YH, Okano T, Kim SW (1990) Polym J 22: 206
14. Bae YH, Okano T, Kim SW (1988) Makromol Chem Rapid Commun 9: 185
15. Baranovsky VY, Litomanovich AA, Papisov IM, Kabanov VA (1981) Euro Polym J 17: 969
16. Osada Y (1979) J Polym Sci 17: 3485
17. Osada Y, Sato M (1976) J Polym Sci Polym Lett Ed 14: 129
18. Osada Y, Sato M (1980) Polym Lett Ed 14: 1057
19. Osada Y, Takeuchi Y (1983) Polym J 15: 279
20. Osada Y (1977) J Polym Sci Polym Chem Ed 15: 255
21. Osada Y (1980) J Polym Sci 18: 281
22. Bailay FE, Lundberg RD, Callard RW (1964) J Polym Sci Part A, 2: 845
23. Smith KL, Winslow AE, Petersen DE (1959) Ind Eng Chem 51: 1361
24. Litmanovich AA, Papisov IM, Kabanov VA (1981) Euro Polym J 17: 981
25. Iliopoulos I, Audebert R (1985) Polym Bull 13: 171
26. Bedner B, Li Z, Huang Y, Chang LCP, Morawetz H (1985) Macromolecules 18: 1829
27. Oyama HT, Tang WT, Frank CW (1987) Macromolecules 20: 474
28. Heyward JJ, Ghiggino KP (1989) Macromolecules 22: 1159
29. Oyama HT, Hemker DJ, Frank CW (1989) Macromolecules 22: 1255
30. Bednar B, Morawetz H, Shafer JA (1984) Macromolecules 17: 1634
31. Abe K, Koide M, Tsuchida E (1977) Macromolecules 10: 1259
32. Iliopoulos I, Audebert RA (1991) Macromolecules 24: 2566
33. Bekturov EA, Bimendina LA (1981) Adv Polym Sci 41: 99
34. Tsuchida E, Abe K (1982) Adv Polym Sci 45: 1
35. Adachi H, Nishi S, Kotaka T (1982) Polym J 14: 985
36. Nishi S, Kotaka T (1985) Macromolecules 18: 1519
37. Nishi S, Kotaka T (1986) Macromolecules 19: 978
38. Silberberg A, Eliassaf J, Katchalsky A (1957) J Polym Sci 23: 259
39. Klenina OV, Fain EG (1981) Polym Sci USSR 23: 1439
40. Katono H, Maruyama A, Sanui K, Ogata N, Okano T, Sakurai Y (1991) J Controlled Release, 16: 215
41. Ilmain F, Tanaka T, Kokufuta E (1991) Nature, 349: 400
42. Katono H, Sanui K, Ogata N, Okano T, Sakurai Y (1991) Polymer J 23: 1179
43. Kitano S, Kataoka K, Koyama Y, Okano T, Sakurai Y (1991) Makromol Chem Rapid Commun 12: 227
44. Kitano S, Hisamitsu I, Koyama Y, Kataoka K, Okano T, Sakurai Y (1991) Polym Adv Tech 2: 261
45. Kitano S, Koyama Y, Kataoka K, Okano T, Sakurai Y (1991) J Control Release 19: 162

Received 1 July 1992

Phase Transition in Gels of Sub-Millimeter Size Induced by Interaction with Stimuli

Atsushi Suzuki
Department of Materials Science, Faculty of Engineering, Yokohama National University, 156 Tokiwadai, Hodogaya-ku, Yokohama 240, Japan

Phase transition in response to infinitesimal change of external stimuli has been observed universally in various gels made of synthetic and natural polymers. The phase transition properties of gels observed in equilibrium and their dynamic and kinetic behavior are determined by the interaction of polymer networks and a liquid. They were extensively studied in the 1980s. Recently, simple technological improvement using gels of sub-millimeter size has made it possible to observe new phenomena, which opens a new scientific field. After a brief introduction to the recent development of investigations using such tiny gels, this paper will describe three principal ways in which gels exhibit phase transition by the interaction with external stimuli. (1) *Visible light-induced phase transition in gels.* Local heating of a thermoresponsive network by visible light can induce phase transition. (2) *Uniaxial stress-induced phase transition in gels.* Uniaxial stress by applying weight can change the transition temperature of thermoresponsive gels. (3) *Salt effects on the phase transition in non-ionic gels.* Adding salts can change the transition temperature and discontinuity. The experimental facts and their interpretation will be proposed on the basis of the mean field equation of state of gels. These principal experiments have been presented in order to demonstrate the recent development of investigations.

Advances in Polymer Science, Vol. 110
© Springer-Verlag Berlin Heidelberg 1993

1 Recent Understanding of the Fundamentals of Phase Transition in Gels Induced by Interaction with Stimuli

1.1 General Introduction

It is now well established that gels can reversibly swell and shrink by as much as several hundred times in response to infinitesimal changes in environmental conditions [1–12] such as temperature, solvent composition, pH, ionic strength, electric field, and light. The experimental study of volume phase transition (discontinuous volume change) in polymer network systems was started in 1973 when the fluctuation of polymer networks was observed using dynamic light scattering technique [13], which was extended by the detection of the critical behavior of gels [14]. All such parameters used so far to induce the phase transition of gels are physical, chemical, and biological stimuli, and their combinations. Since the discovery of discontinuous volume phase transition in 1978 [1] (which was theoretically predicted in 1968 [15]), the physico-chemical principles of gel phase transition have been established to some extent [16–18, and references therein], problems still remain to be solved to make the physical, chemical, and biological applications possible. Using the properties of large (discontinuous or continuous) volume change, the application of gels in devices such as actuators, artificial muscles, controlled delivery systems, sensors, selective pumps and controlled molecular separators have been suggested [19, 20]. The practical commercial uses of gels based on the volume phase transition, however, have yet to emerge. To make such systems technologically useful, it is quite important to establish the further understanding of fundamentals of the volume phase transition of gels.

Recently, it has been recognized that there are fundamentally four intermolecular forces [21]; van der Waals force, hydrogen bonding [22], hydrophobic interaction [23], and the electrostatic interaction which all contribute to the various types of the phase transition in polymer gels. The combination and proper balance of these four forces was shown to lead not only to a single volume phase transition between the swollen and collapsed phases [1, 15], but also to multiple phase transitions between various stable phases characterized by a district degree of swelling [24]. These new developments allow us to find the principle that governs the molecular interactions responsible for the specific molecular recognition only known to biological polymers. It is a most promising and extremely useful principle that can be applied directly to the understanding of life processes as well as the possible creation of new technologies [25]. It should be noted that some of these developments have been realized only by using gels of sub-millimeter size [12].

1.2 Recent Investigations of Gels of Sub-Millimeter Size

In 1989, Suzuki and Tanaka succeeded in extracting gels from glass capillaries of sub-millimeter size and in developing the specially designed measuring cell for the gels (mentioned in Sect. 2.2), which now allows us to measure a swelling phase diagram one hundred times faster than was possible previously. This is because the time needed for swelling and shrinking is proportional to the square of the gel diameter [26] and the diameter of the previously used gel was 1 mm. This means that a set of experiments, which used to take three months (or fifteen years after the discovery of the phase transition [1]), can now be done in only one day (or two months). This simple technological improvement is quietly revolutionizing the field which is related to the measurement of the gel volume. This technique was applied for the first time to the development of a photoresponsive gel [12], which reversibly changes its volume in response to visible light illumination. It is no exaggeration to say that this work succeeded only by using the technique. In so far, experiments using gels of sub-millimeter size have been conducted on the following subjects:

(1) *Light-induced phase transition in gels due to local heating of the polymer network* [12]. The combination of thermosensitive gel and light absorbable molecules makes it possible to induce phase transition of gels by visible light. When the gel absorbs light energy, the temperature of the gel will be partially raised by local heating and equilibrium state will be determined by the competition with thermal diffusion. If the temperature of the gel is higher than the transition temperature, the gel will exhibit phase transition and shrink. If the light intensity is decreased, the gel will swell again. The phase behavior, transmittance, and kinetic properties of such gels have been investigated.

(2) *Phase transition in a gel driven by hydrogen bonding as an attractive force responsible for the shrinking of the gel network* [21]. The volume phase transition behavior of gels under the effects of cooperative hydrogen bonding was observed, where discontinuous swelling transition occurred in slightly ionized gels as the temperature was increased. The gels are composed of interpenetrating polymer networks [22]. This new development has now provided us with the basic concept for understanding the various types of phase transition behavior.

(3) *Phase transition in gels in response to biochemical reactions* [27, 28]. Polymer gels were synthesized in which an enzyme (urease) or a biologically active protein (lectin) was immobilized. The volume phase transitions were observed in such gels when biochemical reactions took place. Such mechanobiochemical gels will be used in devices such as, sensors, selective absorbers, and biochemically controlled drug release.

(4) *Uniaxial stress-induced phase transition in gels* [29]. The changes in diameter and length of cylindrical N-isopropylacrylamide gels were measured under uniaxial stress along the axis. The transition temperature and discontinuity were increased with increasing stress, which indicates the possibility of the

uniaxial stress-induced phase transition of gels. With increasing stress, mechanical work at the phase transition temperature increased, and hereafter decreased under the larger stress. The peak of mechanical work at the transition point as a function of stress depends strongly on the degree of cross-linking.

(5) *Salt effects on the phase transition in non-ionic gels* [30]. Non-ionic *N*-isopropylacrylamide gel in water was found to undergo the reversible phase transition at room temperature (23 °C) by adding salt in the range of high concentrations. The swelling curves in many kinds of aqueous salt solutions were systematically determined using the tiny gels. Salt effects on the transition temperature and discontinuity are briefly discussed on the basis of the change of contact free energy for polymer chains; the structural changes of water molecule near polymer network were induced by ion hydrations which determine the ratio of the persistence length to the effective radius of the polymer chain.

(6) *Multiple phases and their phase transition in gels* [24]. It has been reported that more than two phases of swollen and collapsed states can be found in gels consisting of copolymers of randomly distributed positively and negatively charged groups. The combination of attractive and repulsive electrostatic interactions and hydrogen bonding is considered to result in the existence of such multiple phases.

The purpose of this paper is to demonstrate the recent experimental results on the volume phase transition in gels of sub-millimeter size which are mostly original ones studied by the author's group, and also to outline the recent progress in scientific understandings of stimuli-induced phase transition in polymer gels. The remaining sections will describe (1) photoresponsive gels, (4) deformed gels, and (5) salting-out of gels in that order. A summary of the results and suggestions for future investigations are given in the final section.

2 Phase Transition in Gels Induced by Illumination of Visible Light

2.1 Local Heating of a Polymer Network

Among the parameters which induce the phase transition in gels, light is a clean and easy parameter to control in contrast to other parameters such as, solvent composition, pH, ionic strength, and electric field. Thus, light sensitive gels will find a wide variety of applications; light can be introduced by an optical fiber and easily switched on and off with an extremely short switching time. So far, two methods of inducing volume phase transition using illumination have been reported:

(1) *Ionization by illumination of ultra-violet (UV) light* [11]; some molecules are ionized by illumination of UV light and gels show discontinuous phase

transition upon ionization of a polymer network. Making use of these phenomena, gels could exhibit phase transition induced with UV light. For example, the irradiation of UV light induces ionization of the leuco-derivative molecules incorporated in acrylamide gel [31], which in turn creates an internal ionic osmotic pressure within the gel network giving rise to a volume phase transition. Photosensitive gels of this particular type were synthesized by copolymerizing N-isopropylacrylamide and photosensitive leuco derivative molecule, {Bis(4-dimethylamino-phenyl)-4'-vinylphenylmethane leucocyanide}, and their volume phase transition brought about by UV irradiation and its switching off was subsequently reported [11].

(2) *Local heating by illumination of visible light* [12]; the combination of a thermosensitive gel and light absorbable molecules make it possible to induce phase transition of gels by visible light. When gel absorbs light energy, the temperature of gel will be partially raised by local heating of the network and reach the equilibrium state determined by the competition of the energy input with thermal diffusion. The swelling and shrinking process of the gel in response to light is governed only by the motion of the polymer network, if thermal diffusion is much faster than the collective diffusion of the gel network. When the light intensity is stronger than the critical value, the gel will undergo a phase transition and shrink. The gel will swell once again, when the light intensity is decreased. These phenomena were semi-qualitatively explained using the Flory–Huggins equation of state [32–34].

In this section, some properties of light induced phase transition of polymer gels in the latter systems are reported, i.e., the observation of the phase transition properties of light-sensitive gels in response to visible light illumination, the investigations of their light transmitting properties and their kinetics of the swelling as well as shrinking processes. They are described in detail as follows;

(1) Investigation of the volume phase transition of gels induced by temperature change without illumination of light; gels consisting of thermosensitive N-isopropylacrylamide polymer (main polymer constituent), N,N'-methylenebisacrylamide (cross-linker), and trisodium salt of copper chlorophyllin (chromophore) were designed in which the concentrations of chromophore and cross-linker were varied. The phase diagrams of such gels were determined by varying the temperature of the bulk medium.

(2) Determination of the light intensity dependence of the diameter of the thin cylindrical gels, the light transmitting properties of the thin film gels, and the kinetics of the swelling and shrinking process of the cylindrical gels upon illumination of light; the gel diameters were determined as a function of the light intensity at fixed temperatures. The relationship between the incident and the transmitted light intensity, and the time dependence of diameter on the swelling and shrinking process were also investigated.

(3) Establishment of the theory for such properties of light induced phase transition of N-isopropylacrylamide-chlorophyllin gels; the mechanism of the induction of phase transition is a result of local heating of the gel by light

illumination. The temperature and light intensity dependence of the gels were qualitatively described by the Flory–Huggins equation of state [32–34]. The increment of the temperature upon illumination, the transmitted light intensity and the kinetics were discussed on the basis of a simple phenomenological model.

2.2 Experimental Procedure

Copolymer gels consisting of *N*-isopropylacrylamide (main polymer constituent) and trisodium salt of copper chlorophyllin (chromophore) were prepared by a free radical copolymerization in distilled water under nitrogen atmosphere at 0 °C. The structure of the main polymer constituent and the chlorophyllin molecule are shown in Fig. 1. *N*-isopropylacrylamide was purified by recrystallization from the mixture of petroleum ether and toluene. *N*-isopropylacrylamide (NIPA, Kodak), *N,N'*-methylene-bis-acrylamide (BIS, cross-linker, Wako), trisodium salt of copper chlorophyllin (CH, Aldrich), and tetramethylethylenediamine (TEMED, accelerator, Wako) were dissolved in water. The gels were designated here as "xBISyCH", where x and y mean the respective amounts of reagents; NIPA was fixed at 7.8 g, BIS at 0.133x g, CH at 0.0722y g and TEMED was 240 µl, which were dissolved in 100 ml water. After the solution was fully saturated with nitrogen gas, ammonium persulfate (APS, initiator, Wako), 0.4 g, was added to the solution.

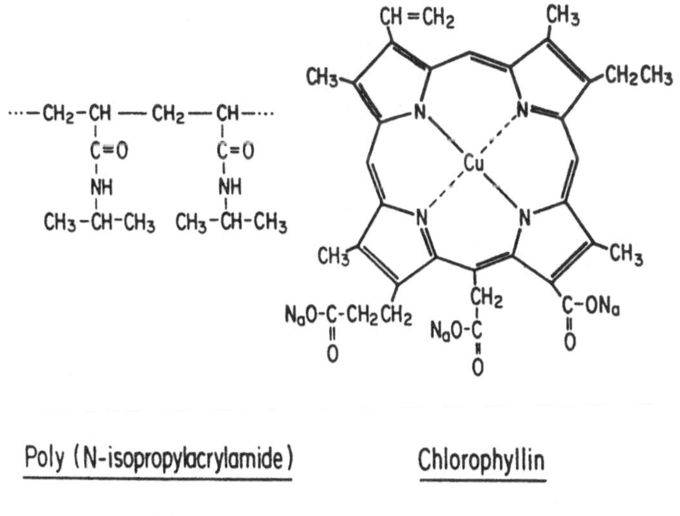

Poly (N-isopropylacrylamide)   Chlorophyllin

(a)   (b)

Fig. 1a, b. Chemical structure of poly *N*-isopropylacrylamide (**a**) and of trisodium salt of copper chlorophyllin molecule (**b**). The chlorophyllin molecule has a double bond which can be covalently connected to the polymer networks

Gels were synthesized in two shapes: thin cylindrical gels were prepared in glass microcapillary tubes with an inner diameter of 141.5 μm, and thin film gels were prepared in between a glass and a polymer-film plates with spacers of a 160 μm thickness. Thin capillaries and the sets of glass-(polymer-film) plates with spacers were inserted into the pregel solution. The solution was brought into the capillaries and into the spaces between glass and plate via capillary elevation. The gelation was carried out overnight at 0 °C. After gelation was complete, the capillaries were broken in order to remove the cylindrical gels, whereas the polymer-film plate was removed from the thin film gels which had adhered to the glass. The gels were then immersed in a large amount of deionized, distilled water to wash away residual chemicals and unreacted monomers from the polymer networks.

It is important to mention that the initiator, APS, was used in amounts ten times higher than was that for the standard concentration of the 1BIS0CH gel [8]. This was necessary to obtain gels with sufficiently high elasticity. The incorporation of chlorophyllin apparently made the gelation process less active than normal. Pure N-isopropylacrylamide, 1BIS0CH gels which were made with standard concentrations undergo slightly discontinuous phase transition. In contrast, all of the 1BIS systems except 1BIS0CH gels showed continuous volume changes, which indicate that a small amount of CH can make discontinuous volume transition to a continuous one. On the other hand, all of the gels with higher BIS concentration exhibit continuous volume changes, which may be reasonable, since the amount of the cross-linker is known to affect the sharpness of the volume phase transition [34].

The thin cylindrical gel was placed in a sealed glass microcapillary of inner diameter of 1.35 mm. As is illustrated in Fig. 2, the capillary was encapsulated in a transparent cell, within which water was circulated to control the temperature. The glass microcapillary with the thin film gel was directly encapsulated in the cell. The temperature was regulated within ± 0.05 °C. On the other hand, the film gel with glass plate was directly encapsulated in the transparent cell shown in Fig. 2. In both cases, the temperature of gels was not directly measured, but was obtained by measuring the difference between the circulated water temperature and the temperature of the places where gels were set in.

Light with a wavelength of 488 nm from an Argon ion laser was used as the light source. The incident beam with a Gaussian profile with a width of approximately 5 mm was focused on the cylindrical gel with a lens of focal length 240 mm, which produced a focused beam. The image of the gel was monitored using a video microscope system with which its diameter was determined. The incident and transmitted light intensities were measured using a light power meter. A few minutes of illumination was enough for both gels to reach their equilibrium states and this was confirmed by the kinetic experiments described in Sect. 2.6. In order to avoid unnecessary bleaching of the chromophore, illumination with a microscope incandescent lamp as well as with an Argon ion laser were used only when the measurements were carried out. The position where the laser light was focused was changed when the accumulated

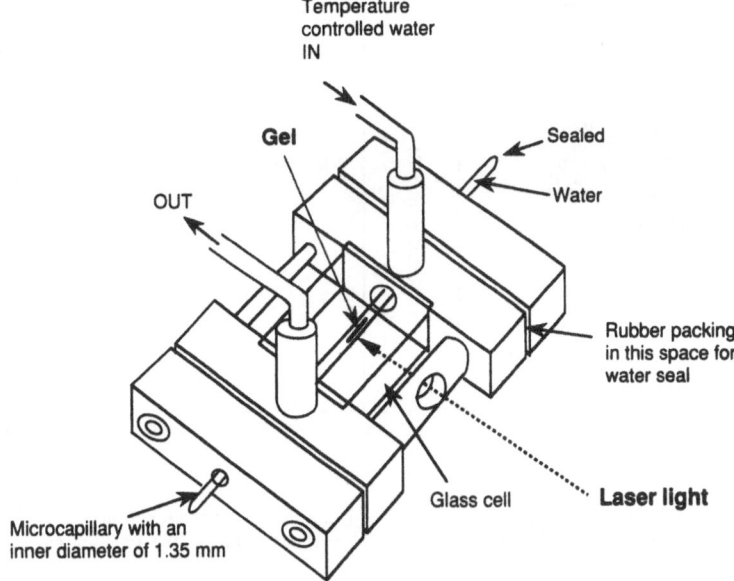

Fig. 2. Schematics of sample setup and the cell for measuring the diameter of the cylindrical gels

illumination time by laser light exceeded several minutes, in order to avoid bleaching effects.

2.3 Temperature Swelling Curves

Equilibrium diameters of the 1BIS10CH and 1BIS6CH gels in water are plotted as a function of temperature in Fig. 3(a) together with those of the pure 1BIS0CH gel. Similar plots of the 5BIS10CH, 5BIS6CH, and 5BIS0CH gels are shown in Fig. 3(b). As for the 1BIS system, the gel without a chromophore (1BIS0CH gel) underwent a discontinuous phase transition at around 33.6 °C, while the gel with a chromophore showed a continuous volume change; the gel underwent a sharp, but continuous volume change at approximately 34.5 °C (1BIS6CH) and 35.5 °C (1BIS10CH). The incorporation of chlorophyllin apparently enhanced the swelling curve to be more continuous, and made the swelling ratio larger and the transition temperature higher. The swelling curve became sharper at the transition temperature with increasing chromophore concentration. For the 5BIS system, on the other hand, the gels both with and without chromophore underwent a continuous phase change at around 34 °C; with increasing chromophore concentration the gel underwent a larger and sharper volume change, and at the same time the characteristic temperature at which the rate of volume change shows maximum (the point of inflexion in the volume-temperature curve) slightly increased. The incorporation of chlorophyllin ap-

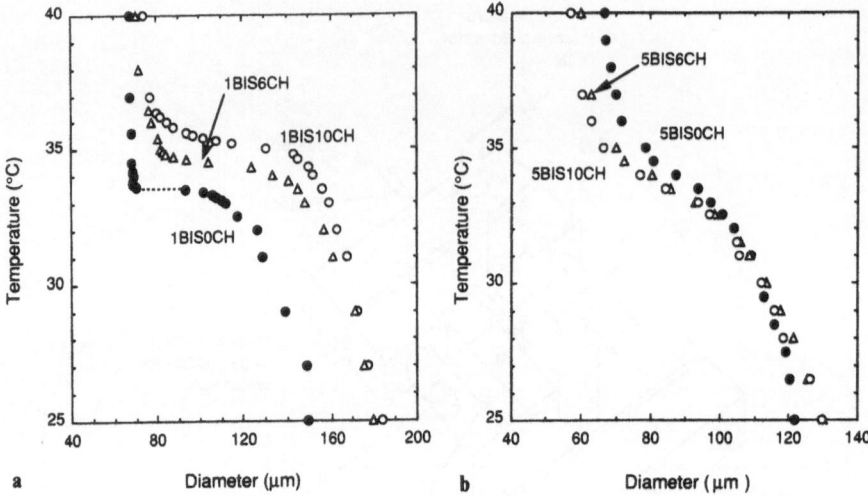

Fig. 3a, b. Diameter of N-isopropylacrylamide-chlorophyllin copolymer gel as a function of temperature. (a) 1BIS system, (b) 5BIS system. Effect of the CH concentration on the swelling curves depends strongly on the BIS concentration

parently enhanced the swelling curve to be less continuous and made the swelling ratio larger.

The equilibrium swelling curve of gels was successfully described on the basis of the Flory–Huggins mean field theory [32–34]; the swelling curve is given by equating the osmotic pressure of a gel to zero. According to the theory, the equation of state of gels is described as follows [2]:

$$
1 - \frac{\Delta F}{kT} = -\frac{v_1 v_e}{N\phi^2} \left\{ (2f+1)\left(\frac{\phi}{\phi_0}\right) - 2\left(\frac{\phi}{\phi_0}\right)^{1/3} \right\}
$$
$$
+ 1 + \frac{2}{\phi} + \frac{2\ln(1-\phi)}{\phi^2} \tag{1}
$$

where T is the absolute temperature, ΔF is the free-energy decrease associated with the formation of the contact, between polymer segments, k is the Boltzmann constant, N is the Avogadro number, v_1 is the molar volume of water, v_e is the total number of the elastically effective polymer chains in the gel, ϕ is the polymer network density, ϕ_0 is the density at Θ temperature where the gel network has an unperturbed configuration, and f is the number of counterions per chain. Now, if temperature dependence of ΔF is assumed to be $(\Delta H - T\Delta S)$ where ΔS and ΔH are the entropy and enthalpy of the polymer–polymer contact, the relation between T and ϕ is expressed as follows [10]:

$$
\frac{1}{T} = \frac{\Delta S}{\Delta H} + \frac{k}{\Delta H} \left[\frac{v_1 v_e}{N\phi^2} \left\{ (2f+1)\left(\frac{\phi}{\phi_0}\right) - 2\left(\frac{\phi}{\phi_0}\right)^{1/3} \right\} \right.
$$
$$
\left. - \frac{2}{\phi} - \frac{2\ln(1-\phi)}{\phi^2} \right]. \tag{2}
$$

For N-isopropylacrylamide gels the interaction parameters ΔS and ΔH are both negative, and the gels are swollen at low temperatures and collapsed at higher temperatures. For the sake of simplicity, let us rewrite Eq. (2) as

$$T = T_{gel}(\phi) . \tag{3}$$

This function increases monotonically with network density ϕ and has a large Maxwell's loop for the large ionization parameter f. The swelling curves of N-isopropylacrylamide gels at low ionization were essentially well described by this equation [10].

In the present systems, both N-isopropylacrylamide and bisacrylamide molecules are not ionizable in water and their contribution to osmotic pressure of gels are expected to be zero. As is shown in Fig. 1, a chlorophyllin molecule has three carboxyl groups, which are ionized in water. In order to examine the effect of ionization, swelling curves were determined for the gels in pure water (pH = 5.8) and in a NaOH solution (pH = 11.9). The gel placed in the higher pH showed a smaller volume change for the temperature range from 25 to 40 °C than the one at the lower pH. The overall behavior of phase transition, however, was not appreciably different between these gels at different pH's.

In order to examine the effect of the CH concentration on the swelling curve, the experimental results were fitted by Eq. (3). Among the parameters, v_1 is constant, f is proportional to the CH concentration, and $\Delta H/\Delta S$ is a measure of the transition temperature. As results of the best fits to the 5BIS system, both the total number of elastically active chains v_e and the reference network density ϕ_0 decreased with increasing CH concentration. These changes correspond to the fact that chlorophyllin makes the gelation process less active mentioned above. In Fig. 4, the theoretical curves were shown together with the experimental data. We are currently measuring systematically the swelling curves of

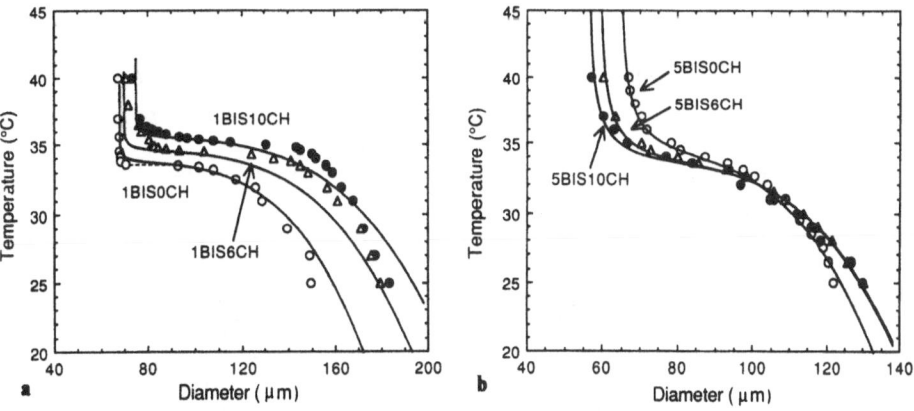

Fig. 4a, b. Theoretical fittings of temperature-swelling curves, diameter vs temperature of various gels synthesized by different schemes, constructed using the equations in the text. (a) 1BIS system, (b) 5BIS system

xBISyCH ($0.5 \leq x \leq 5$, $0 \leq y \leq 10$) to make clear the effects of BIS and CH concentrations on the temperature swelling curves, temperature vs diameter.

It is interesting to observe that in the 1BIS and 0.5BIS systems the gel exhibited a small hysteresis phenomena when the temperature was varied; hysteresis changes in swollen states were observed in temperature cycles between 25 and 40 °C, where the diameter became smaller than the initial value once after the gel collapsed. After one or two cycles, this hysteresis change could not be observed and hereafter we were able to observe a reproducible swelling curve. The appearance of the hysteresis was not significant enough to disturb the experiments. It should be noted that such a reproducible swollen diameter at a fixed temperature became larger and recovered its initial diameter when the gel was kept at the temperature for two months. It is desirable to investigate the physical basis of the hysteresis phenomena of gels by changing the parameters of surrounding condition at fixed temperatures.

2.4 Effects of Light Illumination

Upon varying the intensity of light at a fixed temperature of appropriate value, the gel underwent a discontinuous volume phase transition, typical examples of which are shown in Fig. 5. In both the 1BIS10CH and 5BIS10CH gels, the threshold light intensity decreased with increasing temperature, and at a critical temperature the transition became continuous. In the 1BIS10CH gel, the critical temperature is much higher, the critical light intensity is smaller, and the

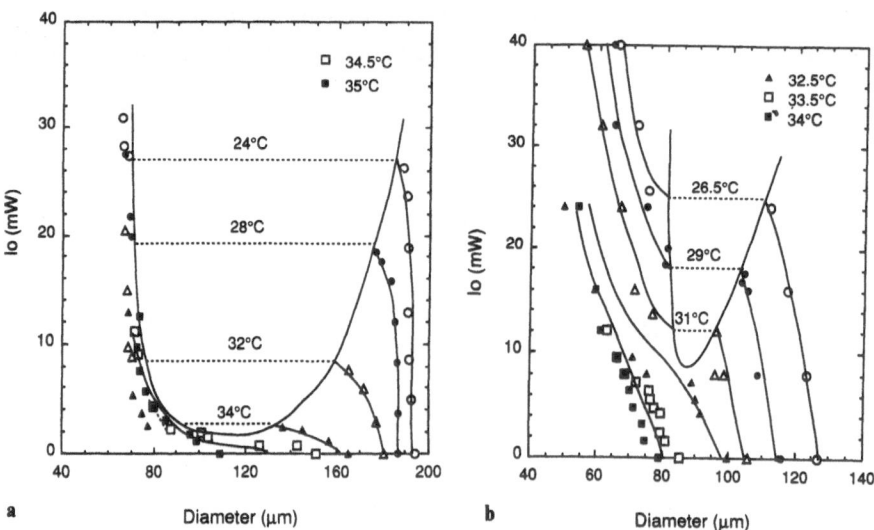

Fig. 5a, b. Diameter of N-isopropylacrylamide-chlorophyllin copolymer gel as a function of initial light intensity. (**a**) 1BIS10CH, (**b**) 5BIS10CH. The discontinuous change in the swelling curves of 1BIS10CH gels are much larger than those in 5BIS10CH

unstable region is much larger than those in the 5BIS10CH gel. In order to examine the relationship between the ambient temperature and the diameter with illumination, the temperature swelling curves of the 1BIS10CH and 5BIS10CH gels were plotted in Fig. 6 which was obtained from the data shown in Fig. 5. As can be seen in this figure, with illumination by light of small intensity the volume change was still continuous, but the swelling curve was shifted toward a lower temperature region. With a larger intensity than the critical value, the gel showed a discontinuous volume phase transition at a much lower temperature region. One can see that the discontinuous transition width of the 1BIS10CH gel is much larger with smaller critical intensity compared with those of the 5BIS10CH gel. The effect of light was largest near the transition, but was clearly observed in the entire experimental temperature range. Namely, in both the swollen and collapsed states there was always a shrinkage with illumination.

These features are qualitatively discussed as follows. When the gel is illuminated, the chromophore absorbs the light energy and the local temperature of the polymer network rises. The increment of the temperature should be proportional to the intensity of the incident light and also to the concentration of the chromophore, and therefore, to the polymer network density ϕ. It may thus be assumed that

$$T = T_0 + \alpha I_0 \phi , \qquad (4)$$

where T_0 is the ambient temperature, I_0 is the intensity of illuminated light, and α is a constant. Combining Eqs. (3) and (4), the relation between the controlled temperature T_0 and the gel density ϕ is obtained as follows;

$$T_0 = T_{gel}(\phi) - \alpha I_0 \phi . \qquad (5)$$

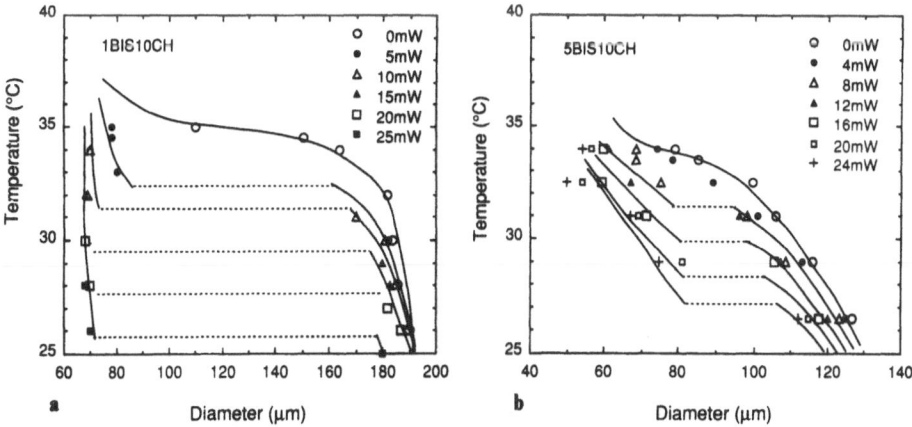

Fig. 6a, b. Diameter of N-isopropylacrylamide-chlorophyllin copolymer gel as a function of temperature upon illumination of light, which was obtained from Fig. 5. (a) 1BIS10CH, (b) 5BIS10CH

If T_0 is fixed in Eq. (5), the polymer network density ϕ can be calculated as a function of incident light intensity I_0. The rescaling of the swelling curves from polymer network density ϕ to gel diameter d which was measured in the present study is straightforward and is given by the following equation,

$$d = \left(\frac{\phi_0}{\phi}\right)^{1/3} d_0 , \qquad (6)$$

where d_0 is the gel diameter at Θ temperature. The light-swelling curves using diameter calculated by Eq. (6) are plotted in Fig. 7. The results of the 1BIS10CH gel coincide with the above experiments shown in Fig. 5(a), where the light intensity was varied at fixed temperatures. As for the results of the 5BIS10CH gel, the experimental result in Fig. 5(b) is qualitatively described by the calcu-

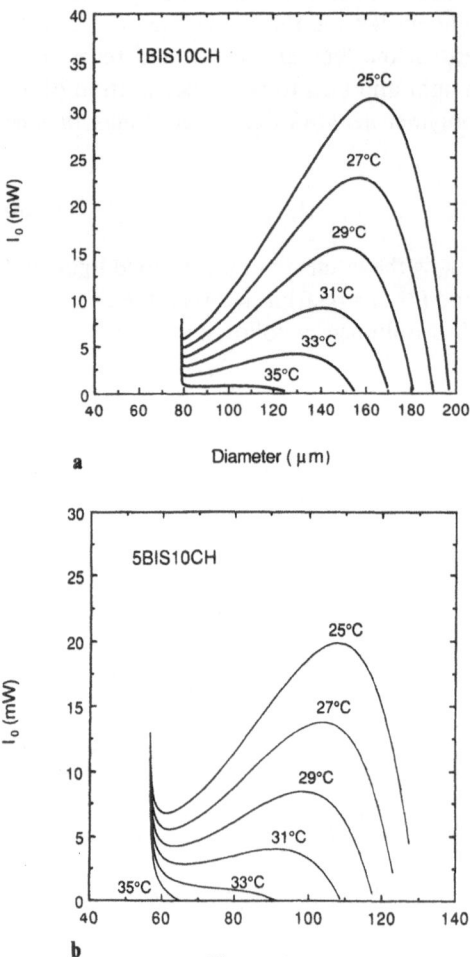

Fig. 7a, b. Theoretical light-swelling curves of a gel: Diameter vs initial light intensity constructed using the equations in the text. (a) 1BIS10CH, (b) 5BIS10CH

lation, while the unstable region in the experiment was much smaller than that of the prediction.

If I_0 is fixed in Eq. (5), the relation between T_0 and ϕ can be obtained, which explains two interesting features that should be noted in the observation of the light induced phase transition of the gel; first, the originally continuous transition without light illumination became discontinuous upon illumination with intensity larger than a threshold value, and second, the transition temperature was lowered upon light illumination. Equation (5) shows that the illumination enhances the Maxwell's loop, and therefore brings the gel state deeper into the unstable region inducing the discontinuous transition. At the same time, it lowers the transition temperature. The theoretical swelling curves using the above equations for various light intensities due to the contribution by light $(+ \alpha I \phi)$ to the swelling curves are shown in Fig. 8. The results of the 1BIS10CH

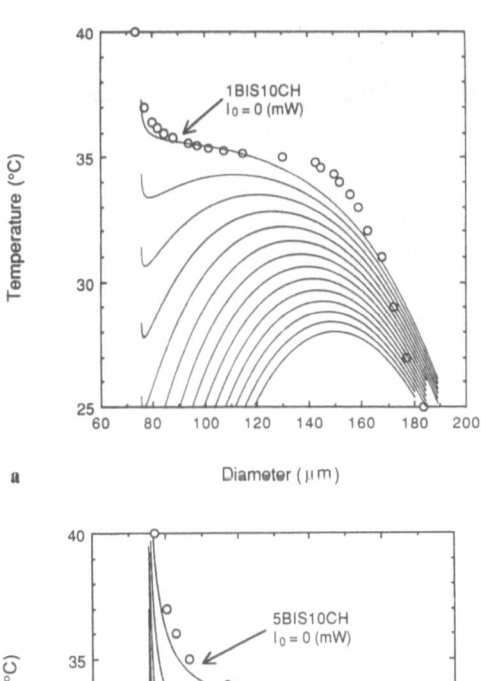

a

b

Fig. 8a, b. Theoretical temperature-swelling curves of a gel: Diameter vs temperature, for various light intensities, constructed using the equations in the text. The swelling curve on the top is without light which was determined by the experimental data (open circles), and the second is with light of a certain intensity and the third (or $(n + 1)$th) is with light twice (or n times) as much as the second one. The results predict excellently the experimental observations. (a) 1BIS10CH, (b) 5BIS10CH

agree qualitatively with the experiments as are shown in Fig. 6, nevertheless the predicted unstable region of the 5BIS10CH gel is much larger than that of the experiment.

2.5 Light Transmitting Properties

The intensity of the transmitted light, I, through the thin 1BIS film gels at a fixed temperature is plotted as a function of the incident light intensity I_0 in Fig. 9. In both systems with high and low chromophore concentrations, the transmitted light intensity showed a discontinuous decrease with increasing incident light intensity, and the threshold light intensity decreased with increasing fixed temperature. The discontinuous change corresponds to the volume phase transitions of gels. Increasing the chromophore concentration caused the transmitted light intensities to decrease both in the swollen and collapsed states and the threshold light intensity to decrease at the same fixed temperatures. It should be noted that absolute values of the threshold light intensity of the 1BIS10CH gel in Fig. 9(b) are different from those obtained for the cylindrical gel in Fig. 5(b). The discrepancy should be attributed to the differences of the sample shapes and sizes, i.e., cylindrical or film, 141.5 µm diameter or 160 µm thickness as prepared, and freely in water or adhered to glass.

In the following, the second term in Eq. (4) is discussed on the basis of a balance of absorption of light energy and heat dissipation. To determine the temperature increment, it is necessary to estimate absorption of light energy, heat dissipation, heat capacity of the gel, and thermal diffusion velocity. Accord-

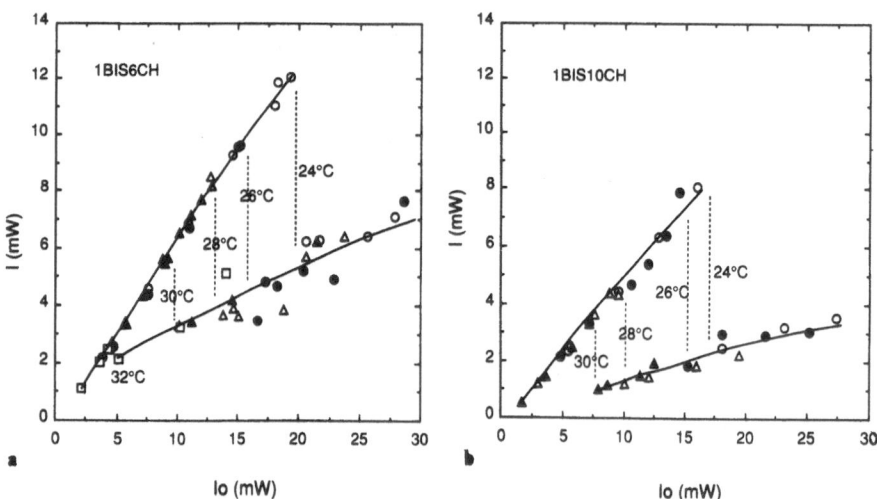

Fig. 9a, b. Transmitted light intensity through the thin film gels is plotted as a function of illuminated light intensity at fixed temperatures. (a) 1BIS6CH, (b) 1BIS10CH

ing to the Lambert–Beer equation, the transmitted light intensity, I, in the present system is expressed by

$$\log\left(\frac{I_0}{I}\right) = \alpha_0 \varepsilon \phi l \,, \tag{7}$$

where α_0 is a constant, ε is the molar absorption coefficient of the chromophore, and l is the thickness of the gel. In the present measurements, the beam diameter is constant and the cross section area in the gel cannot be changed, since one of the gel surfaces is adhered to the glass plate. Then the volume of the gel is proportional to its thickness, i.e., l is proportional to ϕ^{-1}. On the other hand, ε is almost constant, and therefore, the value of $(\alpha_0 \varepsilon \phi l)$ in Eq. (7) will not be strongly affected by the illumination of light. In order to examine this simple speculation, the internal transmittance, I/I_0 through the 1BIS6CH and 1BIS10CH thin film gels is plotted as a function of I_0 in Fig. 10. As is seen in this figure, the transmittance is almost constant with small I_0 and exhibits a discontinuous decrease with increasing I_0, which corresponds to a volume phase transition. With larger I_0 than the threshold intensity, it becomes almost constant. This observation cannot be explained by the above speculation. One of the possible reasons of the discrepancy is that the change of l is not exactly proportional to that of ϕ^{-1} especially at the transition point, since the gel is locally heated and only one of the gel surfaces is mechanically fixed, i.e., the number of polymer chains in light transmitted portion of the gel might change at the transition point. There is no significant difference in the width of the discontinuous jump at each temperature between both gels. The higher the concentration of chromophore, the smaller the absolute value of the transmittance and also the smaller the threshold light intensity at each temperature. If

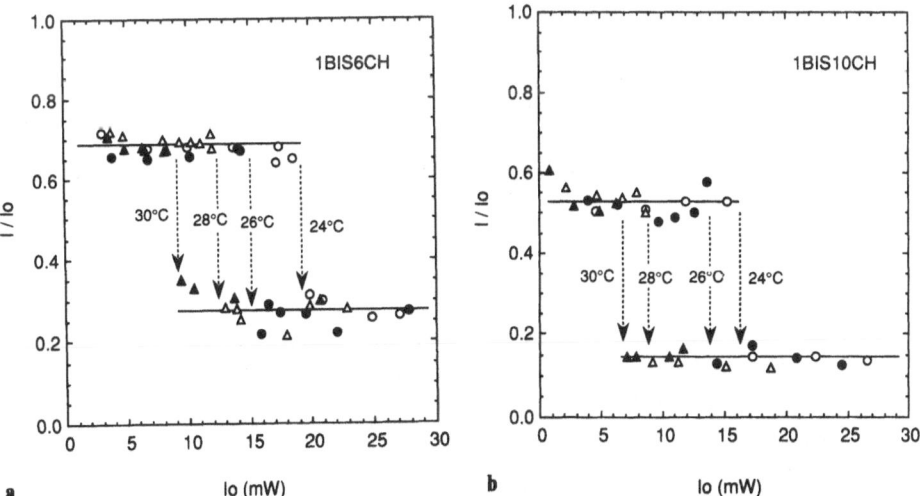

Fig. 10. Internal transmittance through the thin film gels is plotted as a function of illuminated light intensity at fixed temperatures. (**a**) 1BIS6CH and (**b**) 1BIS10CH

(εφl) is negligibly small, i.e., the penetrating depth of light in gel is small enough at a certain CH concentration, the loss of the light energy is proportional to the following:

$$I_0 - I \cong 2.303\alpha_0\varepsilon I_0\phi l \ . \tag{8}$$

From this equation, if the loss of the light energy is converted into the loss of thermal energy, the absorption of light energy per unit volume is proportional to ($\varepsilon I_0 \phi$). On the other hand, heat dissipation should be a function of absorption of light energy; the former increases as increasing the latter. The heat capacity of the gel and thermal diffusion velocity can be assumed to be constant, since the present gels consist mainly of water. Therefore, the expression of the temperature increment in Eq. (4) is valid in case that the heat dissipation is proportional to the absorption of light and that the heat capacity of the gel and thermal diffusion velocity are constant. These two assumptions are essentially justified in the present system.

It can be said that the above model agrees qualitatively with the experimental results, especially for the 1BIS10CH gel. As for the 5BIS10CH gel, the agreements between the experimental results and the theoretical model are poorer, as can be seen, for example, the discrepancies in the sizes of the unstable regions, between Fig. 5(b) and Fig. 7(b), or Fig. 6(b) and Fig. 8(b). It should be attributed to the fact that the CH concentration of the 5BIS10CH gel is much larger than that of the 1BIS10CH gel, e.g., at 25 °C it is twice as high, since the diameter of the 1BIS10CH gel is 1.3 times larger than that of the 5BIS10CH gel. The present model is valid when the CH concentration is small enough and/or the diameter of the gel is thin enough to satisfy the assumption that (εφl) is negligibly small.

2.6 Swelling and Shrinking Kinetics

In this section, the swelling and shrinking kinetics are described for the cylindrical 5BIS10CH gels. In Fig. 11(a), the typical time course of the change of diameter on the swelling process was plotted in a semi-logarithmic scale as a function of the displacement ratio of a fixed point on the network from its equilibrium location. Similar curves of the shrinking process are shown in Fig. 11(b). It is interesting to note that the swelling and shrinking processes follow clearly different equations for the time dependence; only on the swelling process, the volume change follow a single-exponential dependence of time.

If thermal diffusion is much faster than the collective diffusion of the gel network, the swelling and shrinking process of the gel in response to light should be solely governed by the motion of the polymer network. In accordance with the kinetic theory of collective diffusion in a gel [26, 36–38], there is a fast decay followed by a single exponential change with time which is valid for long cylindrical as well as spherical gels:

$$1 - \left| \frac{d(t) - d(0)}{\Delta d} \right| \sim \exp\left(-\frac{t}{\tau}\right) \tag{9}$$

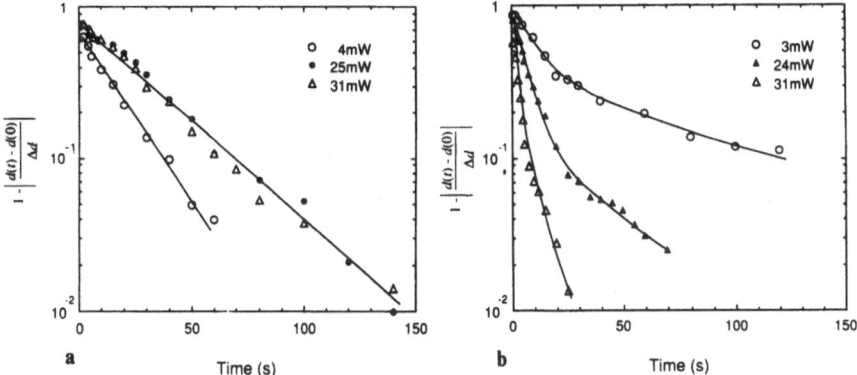

Fig. 11a, b. Time dependence of the swelling behavior of 5BIS10CH at 31 °C, (a) when shielded from light (b) when exposed to light, where d(t) and Δd denote the gel diameter at time t and the total displacement of diameter, respectively

where d(t) is the gel diameter at time t, d(0) is the initial diameter, Δd is the total displacement of diameter, and τ is the relaxation time.

As can be seen in Fig. 11(a), the present results of the swelling process studies confirm these assumptions, which indicates that the relaxation time of the thermal diffusion in the swelling process is negligibly small compared with that of network collective diffusion. On the other hand, the results of the studies of the shrinking process which is shown in Fig. 11(b) are not in agreement especially under light illumination with small intensity. The difference between the swelling and shrinking processes is due to the fact that in the shrinking process it takes a longer time to reach the thermal equilibrium state, and therefore the thermal diffusion time cannot be neglected. It may be reasonable, since in the shrinking process the gel temperature is equal to the ambient water temperature at t = 0, while only the gel temperature is raised in the swelling process. If the ambient temperature can be instantly lowered in the swelling process, one can expect the same time course as that in the shrinking process.

For the swelling process, one can calculate the relaxation time from the slopes in Fig. 11(a). Figure 12 shows the transition rate, τ^{-1}, as a function of the light intensity at several temperatures. At the lower temperatures, 22, 25, and 28 °C, gels exhibit phase transitions with increasing the illuminated light intensity, where τ^{-1} decreases rapidly as increasing the illuminated light intensity. Hereafter, it becomes constant when the light intensity exceeds each threshold value, which corresponds to the transition intensity in Fig. 5(b). On the other hand, at the higher temperatures, 31 and 33 °C, τ^{-1} is almost constant against the illuminated light intensity. These results indicate that the transition rate depends strongly on two factors; one is the kind of phase transition, continuous or discontinuous, and the other is the initial state of gels at t = 0, swollen and collapsed state.

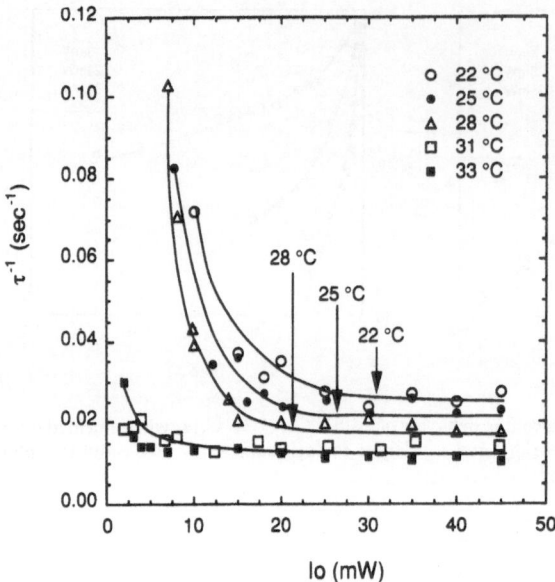

Fig. 12. The transition rate, τ^{-1} of the 5BIS10CH cylindrical gel as a function of the light intensity at several temperatures. The arrow indicates the threshold light intensity estimated from Fig. 5(b)

3 Phase Transition in Gels Induced by Uniaxial Stress

3.1 Mechanical Properties of Gels at the Transition Point

Since the discovery of the discontinuous volume phase transition in gels [1], practical commercial uses of gels in devices based on the volume phase transition have been suggested but have not emerged yet. To make such systems technologically useful, it is quite important to establish the fundamentals of the effect of pressure or extension on the phase transition of gels, since gels will be used under critical conditions where the stress should play an important role. A gel is composed of a three-dimensional networks of natural or synthetic polymers swollen in a liquid, therefore the volume phase transition in gels should not be strongly affected by applying hydrostatic pressure to the gel. However, if stress can be applied directly to polymer networks, the transition behavior can be expected to be changed. So far, the swelling of a neutral network under external load has been investigated on gels, and the mechanochemical efficiency was found to depend strongly on the cross-linking density, the solvent quality, and the external load [39]. On the other hand, the volume phase transition in cylindrical N-isopropylacrylamide gels under uniaxial stress was reported [40]; with increasing the stress along the axis, both the transition temperature and discontinuity observed in diameter increase, which is consistent with the phenomenological theory. In this section, the phase transition of the deformed gels of two types with high and low cross-linking densities are presented, which are under uniaxial stress of eight times larger than that used in the previously reported experiment [40]; not only the diameter but also the length of cylindrical N-isopropylacrylamide (NIPA) gels of sub-millimeter diameter were measured as functions of temperature under uniaxial stress, which demonstrates the possibility of the volume phase transition in gels induced by uniaxial stress. The results were discussed in terms of the phenomenological model [40] on the basis of the extended Flory-type free energy [41].

3.2 Experimental Procedure

Two types of NIPA gels with low (designated 1BIS gel) and high (designated 2BIS gel) density of cross-linking were prepared by a free radical copolymerization in water at $0\,°C$, using the established standard method [8]. Gels were synthesized in the following scheme; the number of monomers as the main polymer constituent in both gels was the same, and only the cross-linking density of the 2BIS gel was twice that of the 1BIS gel. By varying the temperature, the 1BIS gel exhibits a slightly discontinuous phase transition, while the 2BIS gel shows a continuous change. A glass microcapillary of inner diameter of $141.5\,\mu m$ was used to make a uniform cylindrical shape. The reagents and their amounts for preparation of the 1BIS/2BIS gels were as follows; purified NIPA, $690\,mM/672\,mM$ as the main polymer constituent, and N,N'-methylenebisacrylamide, $8.6\,mM/17.2\,mM$ as cross-linker, were dissolved in 100 ml water, to which 240 μl of N,N,N',N'-tetramethylethylenediamine as accelerator was added, and 40 mg of ammonium persulfate was used to initiate the reaction. The cylindrical gels were taken out, held in water for several days, and were dried in the air. As is illustrated in Fig. 13, one end of the gel was fixed to the weight using silicone glues and the other end was fixed to the capillary of inner diameter of $141.5\,\mu m$, which was set in the capillary of inner diameter of 2 mm with water. The capillary was encapsulated in a transparent cell, within which the temperature-controlled water was circulated. The temperature was regulated with an accuracy of $\pm 0.05\,°C$. After the gel reached equilibrium at each temperature, the diameter and length were measured by optical microscope. The swelling ratio, defined as (V/V_0), was calculated as $(d/d_0)^2 \cdot (l/l_0)$ where V, d, and l are the swollen volume, diameter, and length of the gel, respectively, and V_0, d_0, and l_0 are the respective initial values of the prepared samples. Note that for technical reason it was difficult to use the same gel to measure the weight dependence, then new samples were used for each determination of the weight. Nevertheless, one can believe that it does not distort the analysis, since all gels used in this experiment were synthesized from the same solution at the same time.

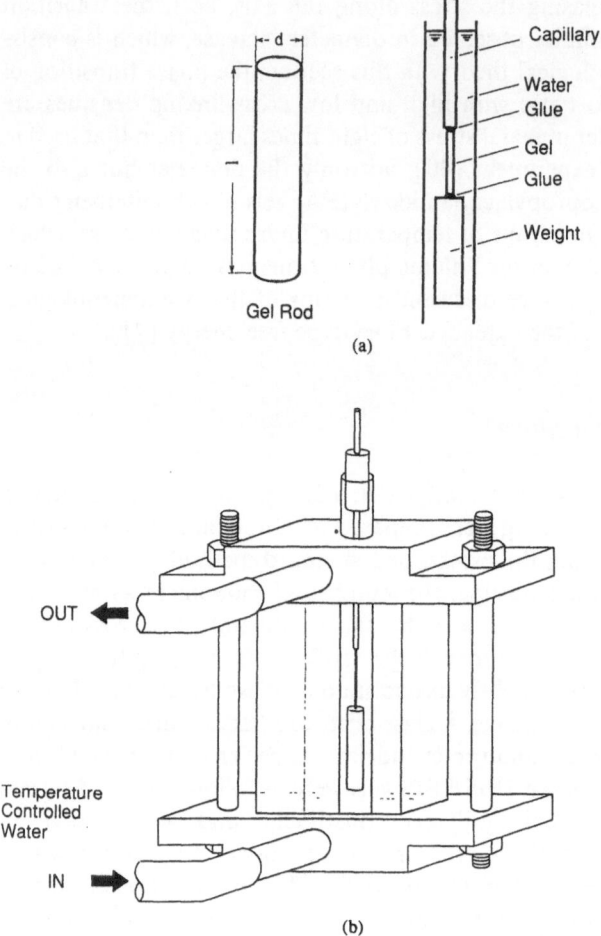

Fig. 13a, b. Schematics of sample setup (**a**) and the cell for measuring both the diameter and length of the cylindrical gels under uniaxial stress (**b**)

3.3 Effects of Uniaxial Stress on the Swelling Curve

First, the effects of uniaxial stress on the volume phase transition will be presented for two types of gels with different density of cross-linking by changing the temperature. Figures 14 and 15 are the temperature dependence of (d/d_0) and (l/l_0) under uniaxial stress on heating runs, respectively. In both gels, the total change in (l/l_0) at temperatures between 30 and 40 °C decreased with increasing the weight, while the change in (d/d_0) increased, and discontinuous changes in both directions occurred at exactly the same temperature within the present accuracy of temperature control. In these measurements, the discontinuous or continuous transition was determined by observing whether hysteresis

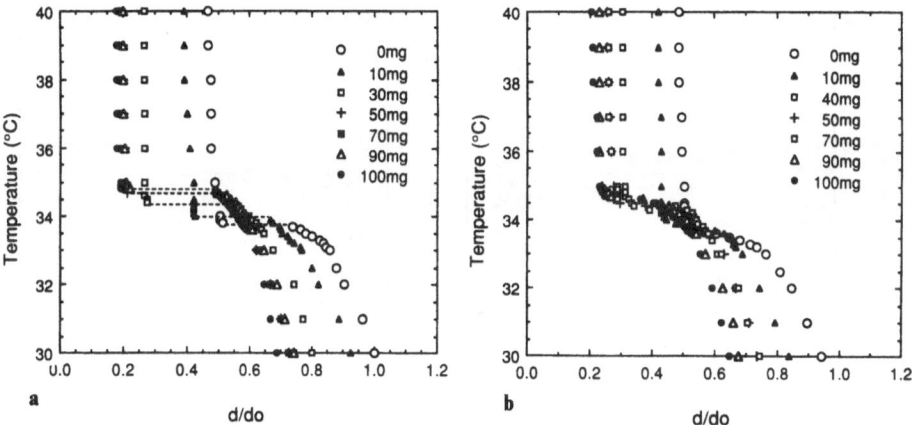

Fig. 14a, b. Temperature dependence of d/d_0 of (a) 1BIS, and (b) 2BIS gels under uniaxial stress on heating runs

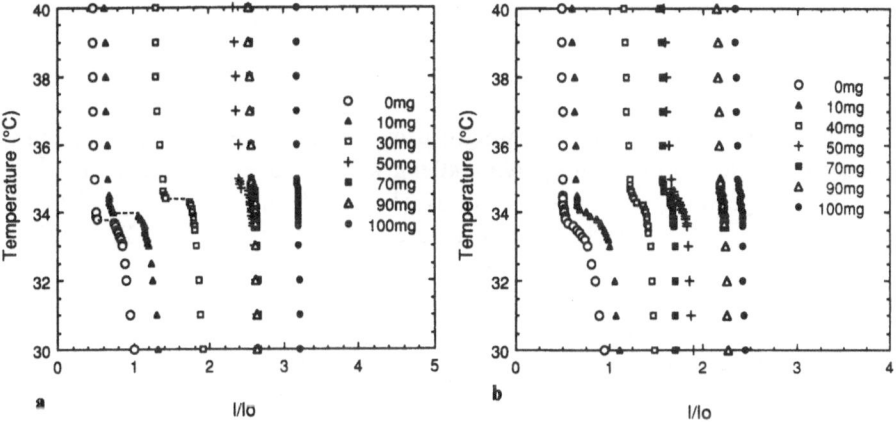

Fig. 15a, b. Temperature dependence of l/l_0 of (a) 1BIS, and (b) 2BIS gels under uniaxial stress on heating runs

occurs at the transition point or not [8]. Figure 16 shows the typical swelling curves on heating under uniaxial stress of the 1BIS and 2BIS gels with different cross-linking concentration, which was calculated by using the measured values of (d/d_0) and (l/l_0) shown in Figs. 14 and 15, respectively. Changes in the transition temperature and width for the 1BIS gels are plotted as functions of weight which is shown in Fig. 17. From this figure, one can see that the transition temperature of the 1BIS gel increases with increasing weight while the jump of the swelling ratio, $\Delta(V/V_0)$, at the transition point also increases in spite of the decrease in $\Delta(l/l_0)$. For the 2BIS gel, similar properties were observed, and as can be seen in Fig. 16(b) the originally continuous phase transition changed

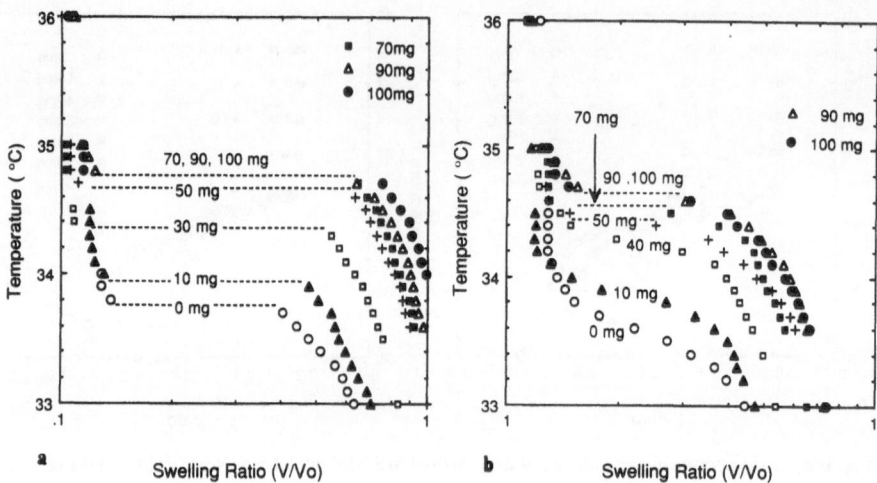

Fig. 16a, b. Temperature dependence of swelling curves of (a) 1BIS, and (b) 2BIS gels under uniaxial stress calculated from Figs. 14 and 15

to discontinuous one when the weight exceeded the threshold value; in this figure it is continuous when applying 40 mg weight that is $2.5 \times 10^4 \, \mathrm{N \cdot m^{-2}}$ at the time of gelation, while discontinuous when applying 50 mg weight. This is the first observation which agrees with the formerly reported prediction that a uniaxial stress can induce discontinuous phase transition in polymer networks [15]. It should be noted that $\Delta(V/V_0)$ and the transition temperature are approach constant values with larger weights.

Now, the swelling curves are analyzed by the Hirotsu–Onuki expression [40] on the basis of the extended Flory-type free energy [41]. The free energy of gels under uniaxial stress can be described as,

$$\frac{\Delta G}{kT} = N_s[\ln(1 - \phi) + \chi\phi] + \frac{N_c}{2}\left[\ln\left(\frac{\phi}{\phi_0}\right) - 3\right]$$

$$+ \frac{N_c}{2}(\alpha_\parallel^2 + 2\alpha_\perp^2) - \frac{F}{kT}(1 - l_0), \tag{10}$$

where N_s is the number of solvent molecules, N_c is the number of polymer chains, $\phi(\phi_0)$ is the volume fraction of polymer (the initial value as prepared), χ [42] is the polymer-solvent interaction parameter, F is the applied force, k is the Boltzmann constant, and $\alpha_\perp = d/d_0$, $\alpha_\parallel = l/l_0$. Analysis of the data was performed in a first approximation by assuming the following relations,

$$\frac{\phi_0}{\phi} = \frac{V}{V_0} = \alpha_\parallel \alpha_\perp^2, \quad N_s = \frac{(1 - \phi)V}{v_s} \tag{11}$$

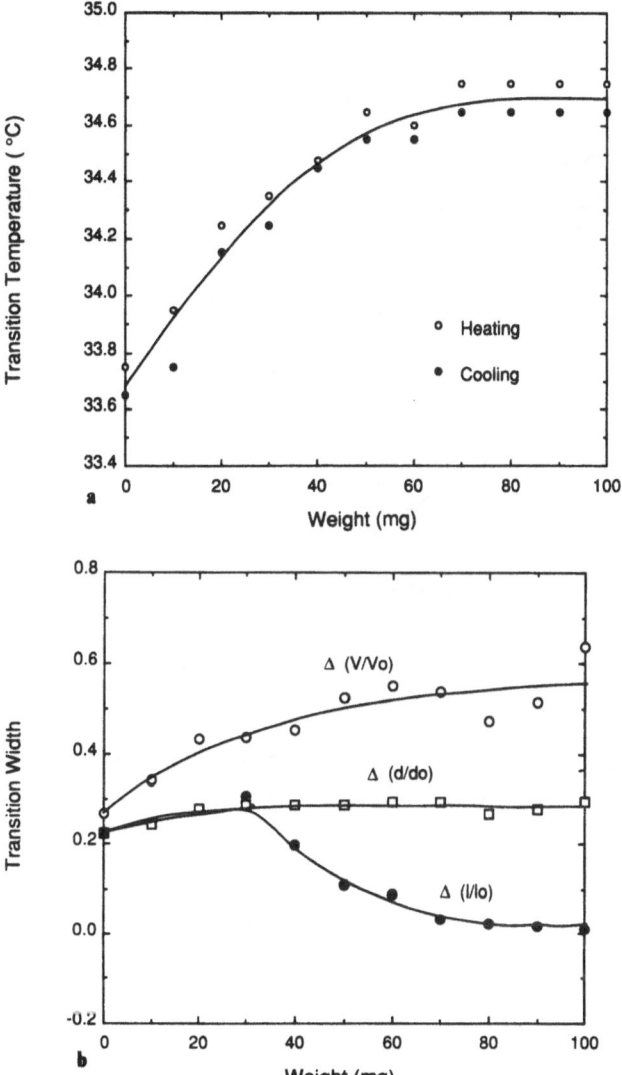

Fig. 17a, b. Uniaxial stress dependence of (**a**) the transition temperature, and (**b**) the transition width of d/d_0, l/l_0, and V/V_0 of the 1BIS gels

where υ_S is the volume occupied by one solvent molecule. The equilibrium conditions are expressed by

$$\frac{\partial \Delta G}{\partial \alpha_{\parallel}} = 0, \quad \frac{\partial \Delta G}{\partial \alpha_{\perp}} = 0 , \tag{12}$$

which give two equations both of which include three variables, ϕ, T, and $\alpha_{\perp}$ (or $\alpha_{\parallel}$). From these equations, the relation between ϕ and T, i.e., V/V_0 and T can

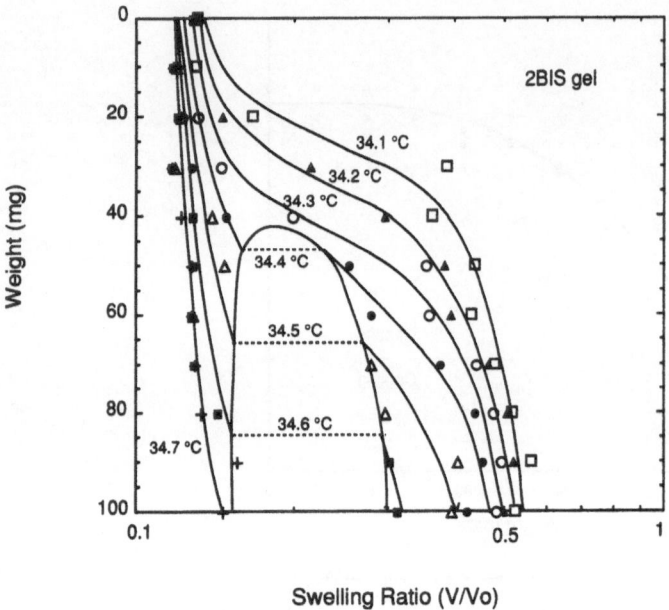

Fig. 18. Relation between the swelling ratio and applied weight of 2BIS gels obtained from Fig. 16(b)

be numerically calculated; first of all, in order to determine the materials constants in Eq. (10), the calculations on the results of the basic swelling curve in the absence of stress ($F = 0$) were carried out for various sets of the parameters, then, using the determined constants F was increased. The preliminary result of the 1BIS gel, the transition temperature of which is higher than the gel shown in Fig. 16(a), was obtained by calculating for a few values of F to fit the swelling curves. A fitting of the result is shown in Fig. 19 together with the experimental data. It should be noted that the phenomenological model [40] quite successfully describes the experimental data in a small range of weights. However, the phenomena observed in a large range of weights cannot be described by the present equations, i.e., the model cannot predict the saturation of both $\Delta(V/V_0)$ and the transition temperature with larger weights. It is desirable to analyze the resutls shown in Fig. 16 and also to investigate the effects of high uniaxial stress on the swelling curves.

3.4 Uniaxial Stress-Induced Phase Transition

Figure 18 shows the uniaxial stress-induced volume phase transition in the 2BIS gel obtained by replotting the measured values in the swelling curves in Fig. 16(b). As can be seen in this figure, at temperatures lower than 34.3 °C the swelling ratio, (V/V_0), increases monotonically with increasing weight. Discon-

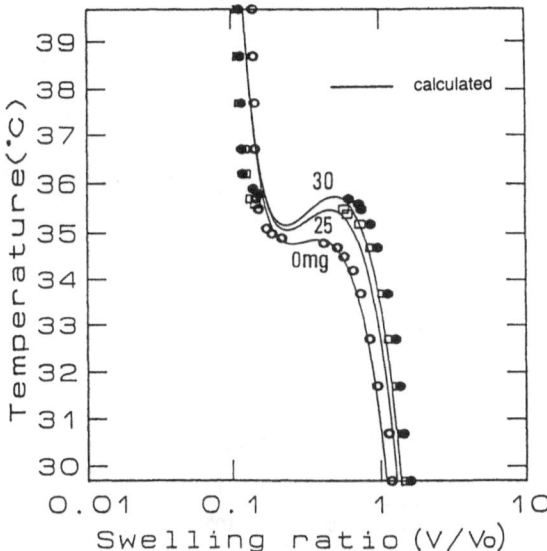

Fig. 19. Swelling curves of a 1BIS gel the transition temperature of which is higher than that shown in Fig. 17(a), and their calculated curves for low stress values

tinuous phase transition can be induced by applying uniaxial stress at higher temperatures above 34.4 °C. Above this temperature, the transition weight decreases and the discontinuity increases with increasing temperature. One can expect that there should be a critical point at the transition temperature between 34.3 and 34.4 °C. These features are very similar to those of gas-liquid phase transition induced by hydrostatic pressure, where one can consider that the liquid and gas correspond to the collapsed and swollen states of gels, respectively. The difference between these systems is in the temperature dependence, which can be explained by the opposite temperature dependence of attractive interactions between composite molecules, i.e., at higher temperatures the gas is stable, while the corresponding swollen states of the gel are stable at lower temperatures.

3.5 Mechanical Work at the Transition Point

The mechanical work at the transition point is expected to depend on the applied weight. As it was mentioned above, it was difficult to use the same gel in the course of the weight dependence experiment by using the present method. Therefore, it is not possible to discuss the absolute value of the work, since l_0 could not be fixed for technical reasons. It is, however, possible to calculate the

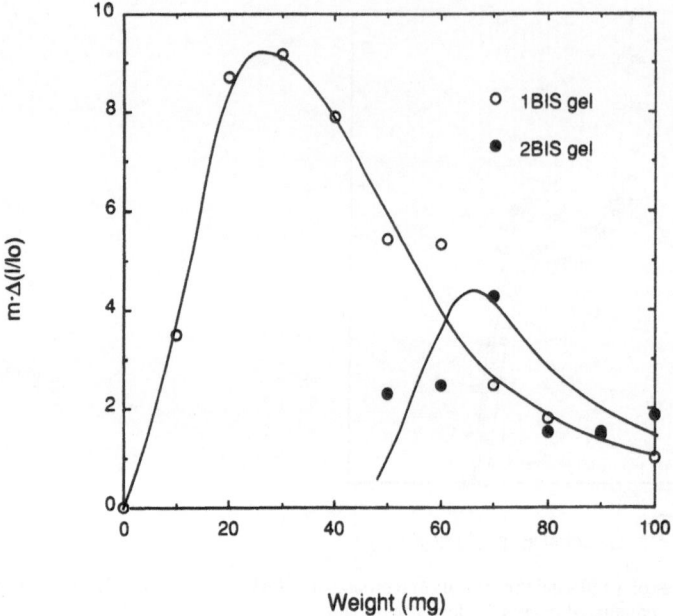

Fig. 20. Normalized mechanical works per unit length of gels at the transition temperature on heating as a function of applied weight

normalized work, $m \cdot \Delta(l/l_0)$, where m is weight and $\Delta(l/l_0)$ is the change in the reduced length at the transition point which is shown in Fig. 17(b). Figure 20 shows a plot of the normalized work, $m \cdot \Delta(l/l_0)$, as a function of weight. One can see that $m \cdot \Delta(l/l_0)$ has a maximum in each case of the 1BIS and 2BIS systems with increasing weight. It can be seen in this figure that the weight at the peak of $m \cdot \Delta(l/l_0)$ shifts to a larger value and that the peak height decreases with increasing the degree of cross-linking. These preliminary results suggest that the mechanochemical efficiency at the transition point is strongly affected by the cross-linking density of the networks as well as the applied stress.

4 Salt Effects on the Phase Transition in Non-Ionic Gels

4.1 Salting-out Effect in Gels

It has been well recognized that simple salts may drastically change solution properties of macromolecules such as solubility, precipitation temperature, viscosity, etc., and that their efficiency depends strongly on the nature of individual ions [43]. In general, the relative efficiency of different ions follow a well defined order of the lyotropic series, e.g., the effects of the alkali metals can be classified according to the Hoffmeister series [44]. Upon addition of a salt to an aqueous solution of a gel, a similar type of "salting-out" effect can be expected, the physical origin of which has not been clarified as yet. In fact, it was reported that addition of salt affects the phase transition in ionic gels [7]. Although ion-polymer interaction has been extensively studied for many kinds of non-electrolyte polymers, little attention has been paid to the influence of ions on conformational changes of non-ionic gels [45–47]. Among the external stimuli, it is important to understand the structural and dynamic changes of polymer gels induced by addition of salts in solvent with respect to many physical and chemical processes in biosystems.

In this section, the effects of salt on the swelling curve of a non-ionic gel are presented; the temperature dependence of the volume of N-isopropylacrylamide (NIPA) gels of sub-millimeter size are measured at several concentrations of aqueous salt solution. The changes are briefly discussed on the basis of changes of contact free energy for polymer chains due to changes of the molecular structure of water in the vicinity of polymer network chains induced by ion hydrations, using the Flory–Huggins equation of state of gels [32–34]. In the following, after the experimental procedure, the relation between diameter and salt concentration at room temperature is briefly discussed, which was measured on large cylindrical gels of 1.35 mm diameter.

4.2 Experimental Procedure

Gels were prepared by a free radical copolymerization in water at 0 °C. A capillary of inner diameter of 141.5 µm (1.35 mm only for the samples described in the next section) was used to make a uniform cylindrical shape. The reagents and their amounts were standard [8]: purified NIPA (Kodak), 690 mM as the main polymer constituent, N,N,N',N'-tetramethylethylenediamine (Bio-Rad or Wako), 240 µl/100 ml water as accelerator and ammonium persulfate (Mallincrodt or Wako), 40 mg/100 ml water as initiator. The cylindrical gels were taken out and kept in water for several days, hereafter, the gels were placed in water solutions of various salt concentrations. The diameter was measured after the gels reached equilibrium at each temperature. Temperature was controlled with an accuracy of ± 0.05 °C.

4.3 Phase Transition Induced by Adding Salts at Room Temperature

Figure 21 shows the relation between the normalized gel diameter, (d/d_0) and the salt molality at room temperature (23 °C) for typical type of inorganic salts which include selected combinations of cations with halogens or divalent

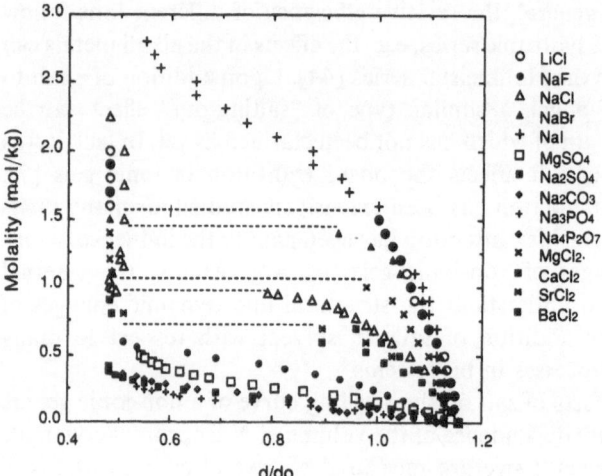

Fig. 21. Salt effects on the volume of N-isopropylacrylamide gels as functions of salt molalities at room temperature (23 °C)

anions (SO_4^{2-}, CO_3^{2-}), and sodium polyvalent anions (PO_4^{3-}, $P_2O_7^{4-}$). Notable features in this figure are that the gel diameters decrease with increasing salt molalities and approach to those of completely collapsed states, whereas the efficiency is strongly dependent on individual ions. Diameters of collapsed states remain nearly constant for all the salts at high molalities and were about 43% less than the initial swollen states in pure water. This ratio of the swollen to collapsed diameter (23 °C) is almost the same as that obtained in the temperature-swelling curve [8], i.e., the swollen diameter at 23 °C to collapsed one at 40 °C. Moreover, the salt effect of decreasing the volume is not simply proportional to salt molality, but has a similar role as temperature in the temperature-swelling curve [8]. For some limited salts with Cl^- and Br^-, gel diameters change discontinuously as functions of salt molalities. It is noted that the efficiency of salts does not follow the ionic strength as can be seen in Fig. 21, for example, in the almost same efficiency of Na_2SO_4 and $MgSO_4$.

Figure 22 shows the salt effects for monovalent and divalent cations (alkali and alkali earth metals) chlorides. For all these salts, the phase transitions took place reversibly and were discontinuous as functions of salt molality. The efficiency of salts in contracting gels was almost the same for alkali metals chlorides (monovalent cations) except Li^+, whereas it was slightly dependent on alkali earth metals (divalent cations) in inverse order of ionic radii, $Mg^{2+} < Ca^{2+} < Sr^{2+} < Ba^{2+}$.

Figure 23 shows the salt effects for sodium and potassium halogens. The efficiency of salts in contracting gels is strongly dependent on anions, and it increases with decreasing the ionic radius of the anion, $I^- < Br^- < Cl^- < F^-$. It is worth mentioning that gel diameters show continuous and discontinuous changes for F^-, I^- salts and Br^-, Cl^- salts, respectively.

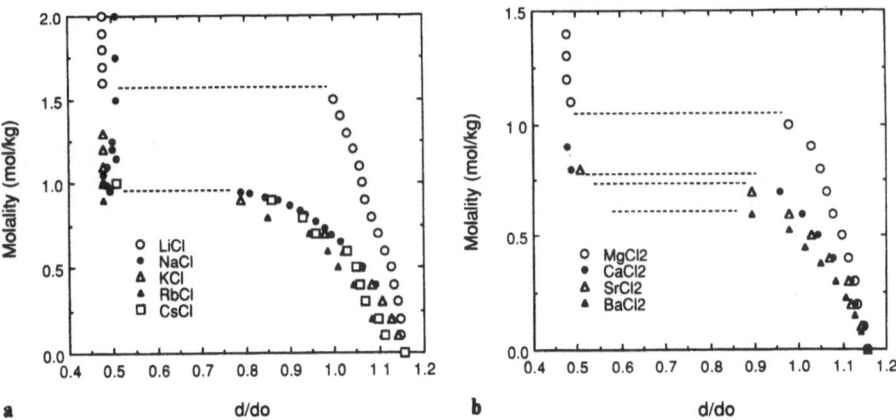

Fig. 22a, b. Effects of cations, alkali (**a**) and alkali earth (**b**) metals on the gel volumes. Effects of divalent anions seem to be in the reverse order to that expected on the basis of ionic radii

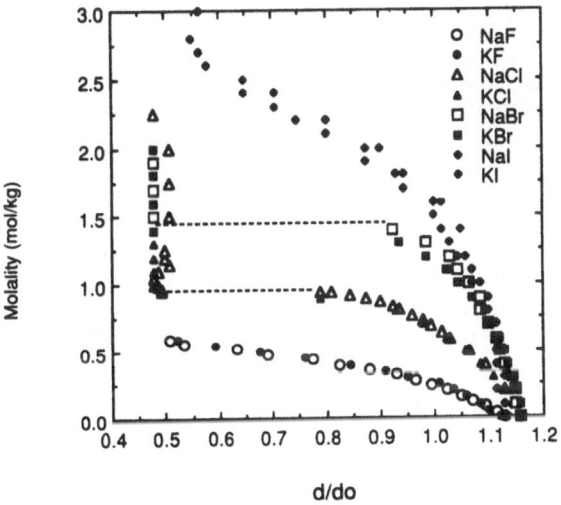

Fig. 23. Volume changes of gels upon addition of sodium and potassium halides. The efficiency of anions are in order of the ionic radii, while there are also markedly greater values for the divalent anions, sulfate and carbonate

It is well known that addition of neutral salts to polymer solutions reduces the overall dimensions of polymer chains (the salting-out effect) [43, 44]. In general, the reduction in chain dimensions is reflected in the polymer viscosity. An example of salt effects on water soluble polymer with non-ionic characters has been reported in the literature where the precipitation temperature and the viscosity of polyethylene oxide (PEO) were measured to interpret the unusual

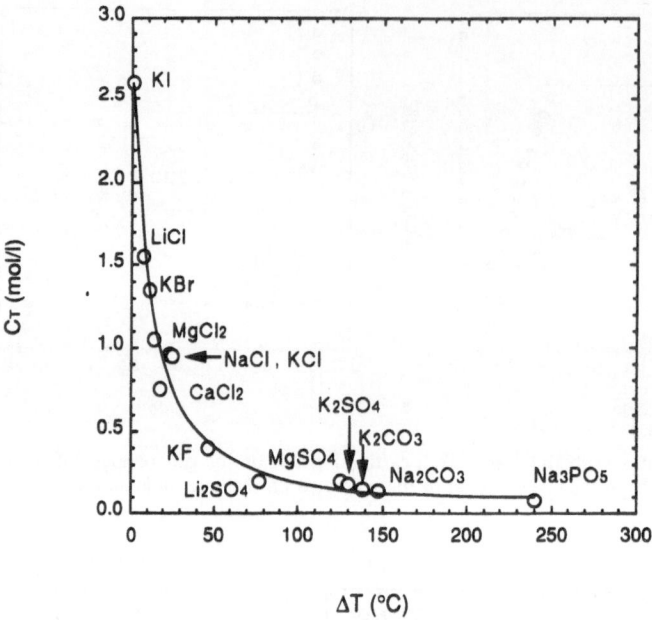

Fig. 24. Parallel effects of salts on the molar reduction in the precipitation temperature of polyethylene oxide in the salt concentration of $1 \, mol \cdot l^{-1}$, ΔT_m and the transition molality of N-isopropylacrylamide gel, C_T. In the case of continuous phase change, C_T was defined as the point of inflexion in the volume-molality curve

rheology of high molecular weight PEO in water solution [48]. In Fig. 24, the reported decrements of precipitation temperature of PEO by addition of $1 \, mol \cdot l^{-1}$ salts, ΔT, are plotted against the transition molalities of NIPA gels, C_T. The salt efficiency of reducing the precipitation temperature is found to correspond strongly to that in contracting gels, which indicates that a similar mechanism exists for water soluble non-ionic hydrophobic polymers including PEO and NIPA. It should be noted that its origin depends on water structure (the cause of order in water) on hydrophobic groups and hydration of ion [49]. In a simple model, the ratio of the number of bound water molecules to that of free water molecules decreases with increasing temperature, while it also changes by adding salt at a fixed temperature. Combining these phenomena, each salt effect should be characterized by a transition temperature which is related to the contact free energy. Moreover, the above results indicate that the discontinuity of the phase transition is strongly dependent on the individual anions, which suggests that addition of a salt also changes the persistence length and/or the effective radius of the polymer chains, which is characterized by the S value in the scaling equation [2, 16–18]. The results at room temperature suggest that two parameters, contact free energy and the S value in the scaling equation determine the transition molality of salt and the discontinuity of transition, respectively – these will be discussed later.

4.4 Classical Electrostatic Model

The theoretical picture is expressed by the Debye and McAulay equation [50], which relates the salt effects on nonelectrolytes to the dielectric constant of the solution, ε and ionic radius, r_j of individual ions. In this theory, the ions make the solution a poorer solvent for a nonelectrolyte, so that its activity coefficient, f_i, is increased as follows;

$$\ln(f_i) = \frac{\alpha}{2kT\varepsilon_0} \sum \frac{c_j z_j^2}{r_j} \tag{13}$$

where α is the dielectric decrement due to the nonelectrolyte, ε_0 is the dielectric constant of the solution, c_j is the concentration of an ion, and z_j is its charge number. According to this expression, it can be expected that the efficiency of polyvalent ions should be of much higher than that of monovalent ions and that the smaller the ion radius, the greater the efficiency. The present results are consistent with the former expectation and indicate that the effect of anions is more essential than that of cations. It should be, however, noted that the nature of salts is much more complicated than that expected from the Debye–McAulay theory [32], as can be seen in Fig. 22(a), for example, in the almost same efficiency of alkali metal chlorides with having different α, ε_0, and r_j in each system.

Salt effects on the gel volume in the present systems is expected to be concerned with polymer–polymer interaction itself and the combination of three competitive interactions, i.e., polymer–water, polymer–ion and water–ion. Direct interaction of polymer–water is most important with respect to the shrinking force which depend on the association and release of structured water from the hydrophobic groups. The effect of each inorganic salt on the volume phase transition of water soluble non-ionic gels is mainly due to the changes of structures of bound water, since disturbance of water structure [49] by adding ions in each system induces contact between polymer chains, causing entropy change. Indirect interactions of polymer–ion and water–ion are also important, since those are related with the structure change of bound water. In each system of salt solutions, the physico-chemical properties are quite different – dielectric relaxation time and static dielectric constant, relative affinity of water for hydrophobic groups, overall polarity of water, etc. Nevertheless, the volume change by adding ions in each system is very similar to that caused by changing temperature, though transition salt molality and transition discontinuity depend on the individual salt which cannot be explained by such well-known parameters. Among the properties, hydration and water structure [49] are essential to understand the volume phase transition.

4.5 Effect of Sodium Halides and Sodium Salts with Organic Molecules

Typical examples of the molality dependence of the swelling curves are shown in Fig. 25 where the swelling curves of NIPA gels immersed in four simple aqueous inorganic salt solutions are demonstrated for various molalities. In each case, with increasing salt molality, the transition temperature shifts to the lower

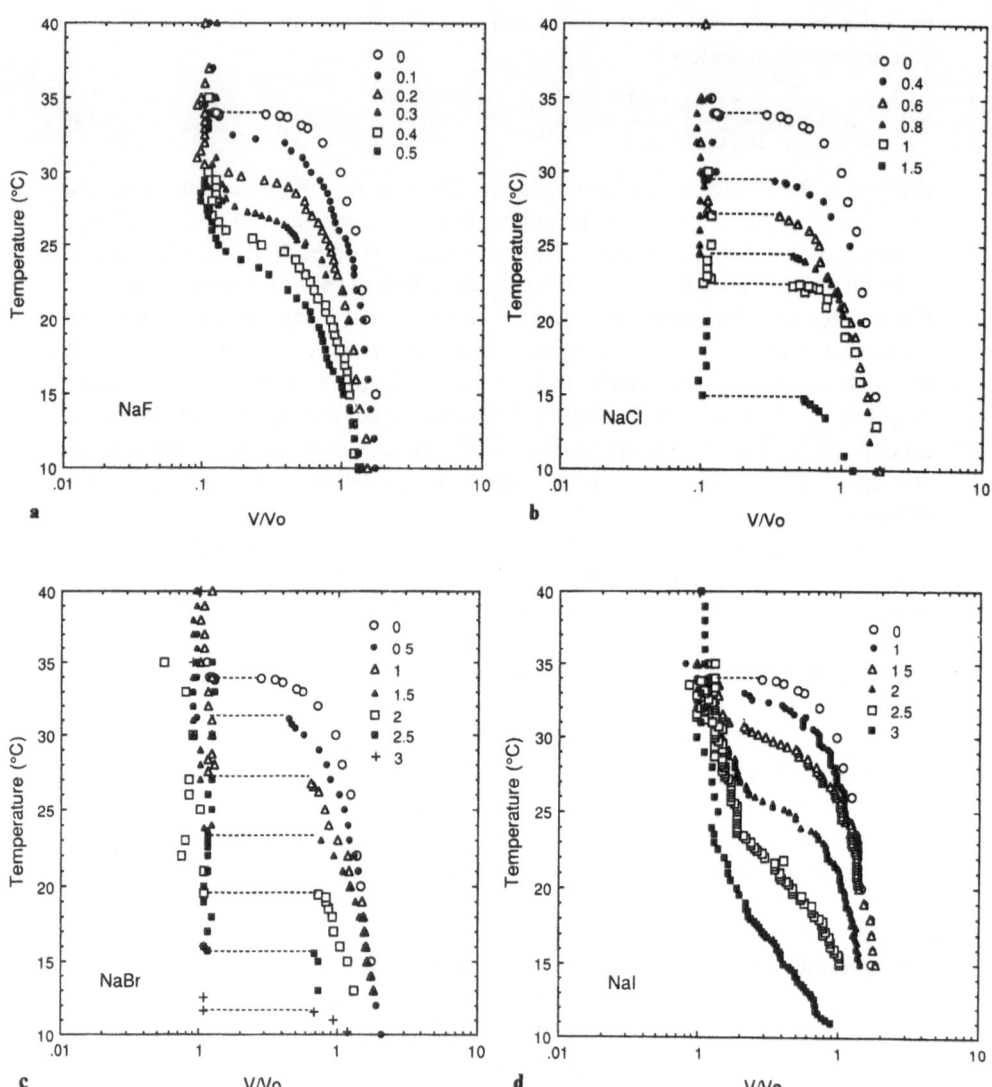

Fig. 25a–d. Swelling curves in aqueous solution with various molalities (mol/kg) of sodium halides, (a) NaF, (b) NaCl, (c) NaBr, and (d) NaI

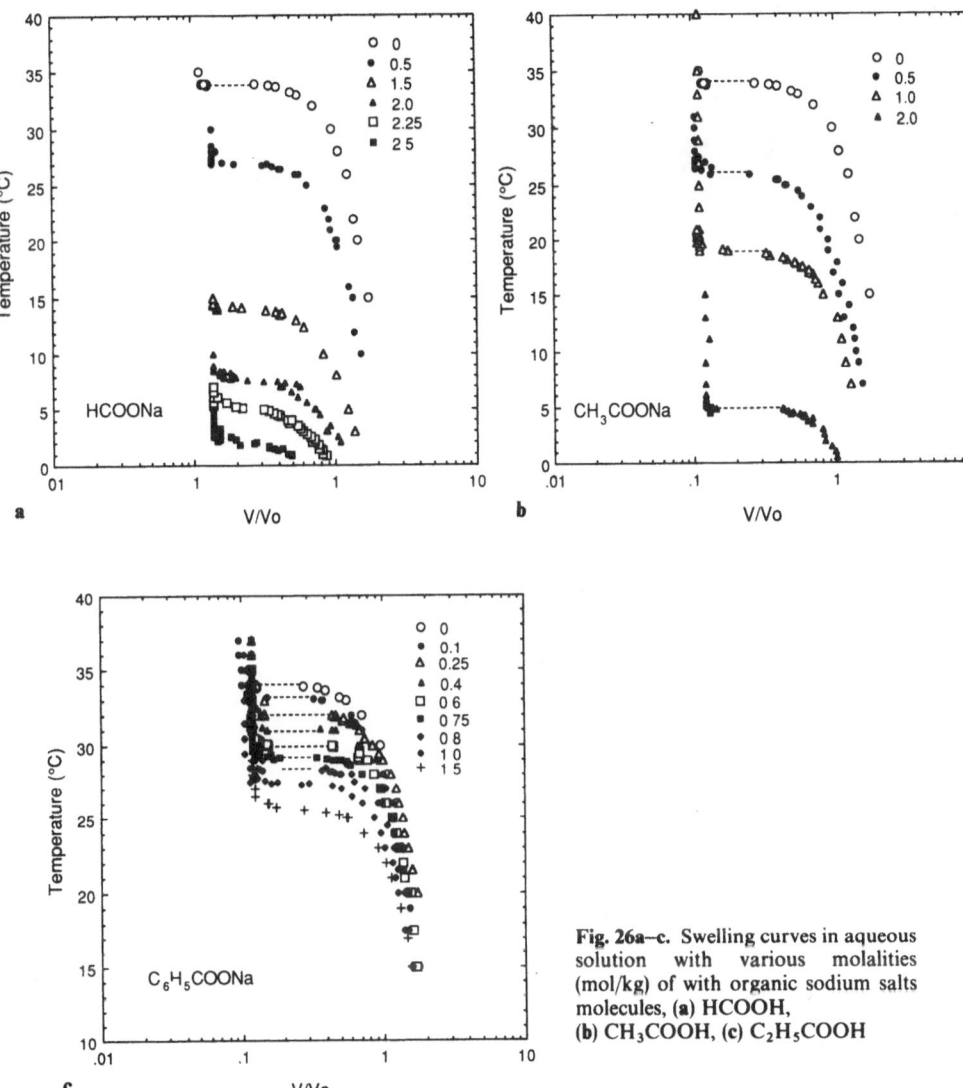

Fig. 26a–c. Swelling curves in aqueous solution with various molalities (mol/kg) of with organic sodium salts molecules, (a) HCOOH, (b) CH₃COOH, (c) C₂H₅COOH

temperature side. The diameter jump at the transition point increases in the case of NaCl and NaBr, and decreases in NaF and NaI where the initial slight discontinuous transition rapidly changes to a continuous one at higher salt molality.

Typical examples of the molality dependence of the swelling curves are shown in Figs. 26 and 27 where the swelling curves of NIPA gels immersed in four aqueous organic salt solutions are demonstrated for various molalities. With increasing salt molality, the transition temperature is shifted to lower temperatures except that of o-$C_6H_4(OH)COONa$. The diameter

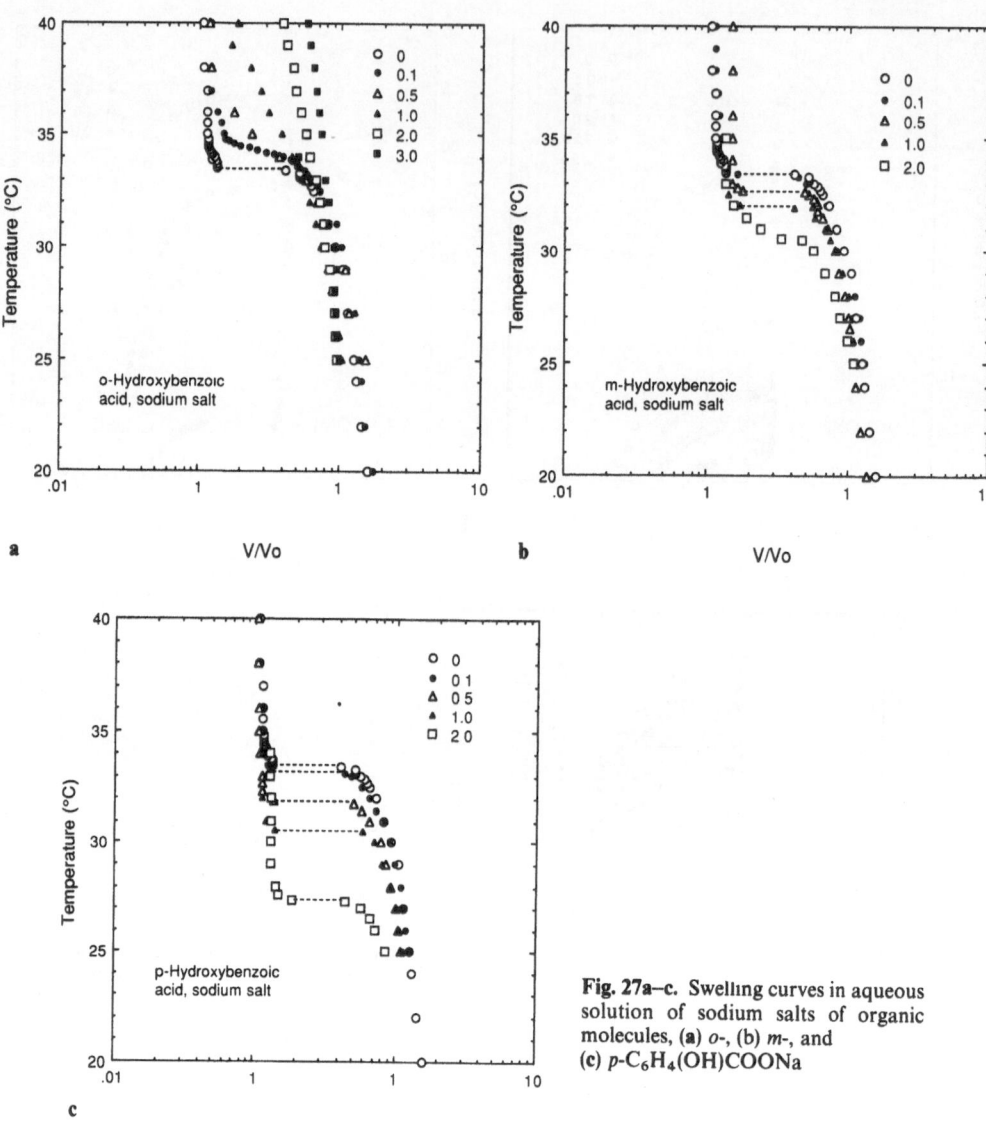

Fig. 27a–c. Swelling curves in aqueous solution of sodium salts of organic molecules, (a) *o*-, (b) *m*-, and (c) *p*-$C_6H_4(OH)COONa$

jump at the transition point increases in the case of CH_3COONa and p-$C_6H_4(OH)COONa$, the transition jump initially increases at low salt molality and then decreases to be zero. In every case, except for CH_3COONa, the discontinuous transition changes to a continuous one at high salt molalities.

It is now possible to classify the salt effects on the swelling curve. Till now four types of salt effects have been found and these are summarized in Fig. 28. These results suggest that two parameters, contact free energy and the S value in

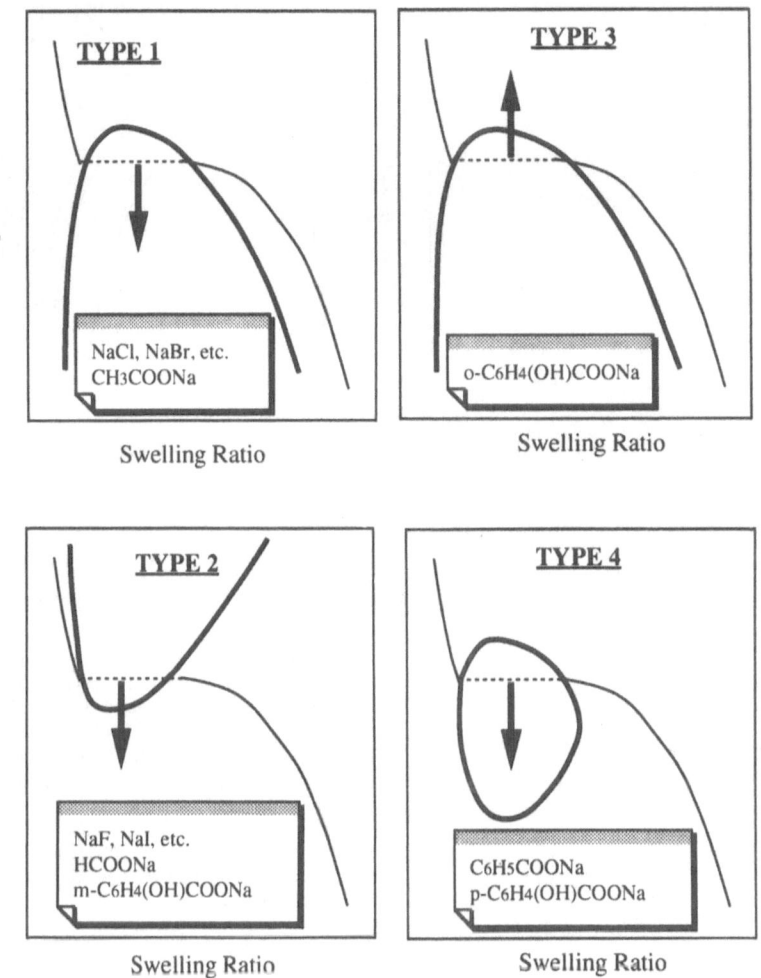

Fig. 28. Classification of salt effects on the swelling curve. Until now four types of effect have been observed, each of which has different combination of increment or decrement of the transition temperature, induction or reduction of the transition discontinuity, and their combinations

the scaling equation determine the transition temperature and discontinuity, which is strongly affected by individual anions.

4.6 Equation of State of Gels in Aqueous Salt Solution

Using the Flory–Huggins derivation, the phase transition of non-ionic gels can be described by Eq. (1). In the present case, the equation includes two variables of ϕ and ΔF. In order to know the effect of the salt molality on ϕ, it is necessary

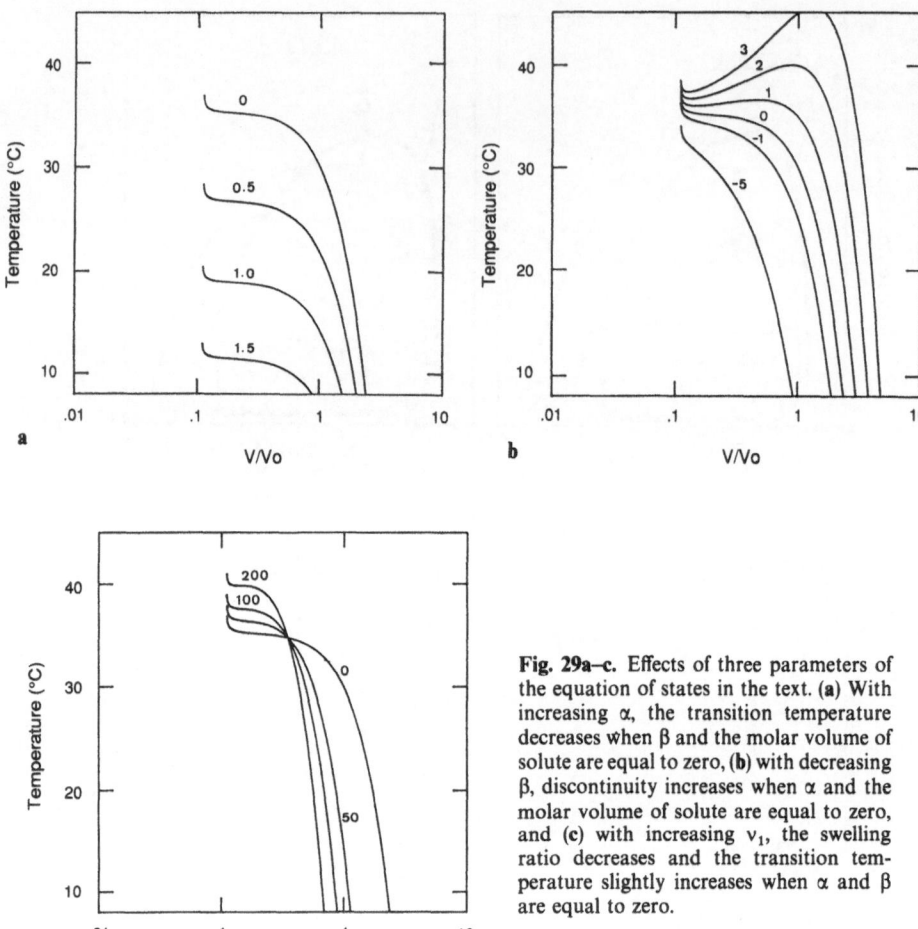

Fig. 29a–c. Effects of three parameters of the equation of states in the text. (a) With increasing α, the transition temperature decreases when β and the molar volume of solute are equal to zero, (b) with decreasing β, discontinuity increases when α and the molar volume of solute are equal to zero, and (c) with increasing v_1, the swelling ratio decreases and the transition temperature slightly increases when α and β are equal to zero.

to determine clear expression of ΔF. Now, ΔF is assumed to be described as follows;

$$\Delta F = \Delta F_0 + \Delta F_1 + \Delta F_2 \qquad (14)$$

where ΔF_0 is the change of contact free energy of chains in pure water, and ΔF_1 is the change of contact free energy (entropy term) due to the disturbance of structured water molecules, and ΔF_2 is that (enthalpy term) due to the disturbing or inducing the contacts of polymers by hydrated ions. Upon addition of a salt, ΔF_1 changes as a function of salt concentration, C, since ions destroy the water structure around polymers where the number of ions around polymers is important [49]. At the same time, ΔF_2 changes as functions of C and ϕ, since

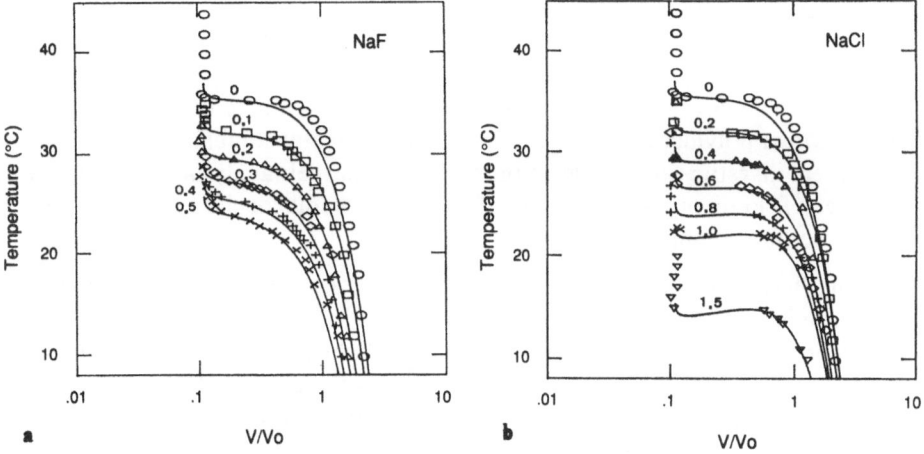

Fig. 30a, b. Theoretical fittings of temperature-swelling curves, diameter vs temperature of NaF (**a**) and NaCl (**b**) in various salt molalities (mol/kg)

ions interact with polymer chains where the number of ions per chain is essential. In this model, ΔF_1 compensate for ΔF_2 such that the total change in free energy should be minimized which are characterized by two parameters. In the first approximation, they can be assumed as follows;

$$\Delta F_1 = \alpha C, \quad \Delta F_2 = \beta \frac{C}{\phi} \tag{15}$$

where α and β are materials constants, and α is proportional to the dehydration capability of ions and β is proportional to solubility. Moreover, it should be necessary to take into account the molar volume of solvent, v_1. The relation between T and ϕ can be calculated using Eqs. (14) and (15), and the change in v_1. As illustrated in Fig. 29, the combinations of α, β, and v_1 can bring the gel state into or out of the unstable region in the T–ϕ space, inducing or restricting discontinuous phase transition as well as the transition width. The present model predicts that α, β, and v_1 determine the transition concentration, transition discontinuity, and the swelling ratio, respectively. The theoretical swelling curves were calculated; in order to determine the materials constants in Eq. (1), the calculations on the results of the basic swelling curve in pure water (C = 0) were carried out for various sets of the parameters, then, using the determined constants C was increased. Typical examples of the theoretical curves are shown in Fig. 30. The results agree excellently with the experimental observations. In order to make clear the physical as well as chemical pictures of the parameters, several approaches for providing extra physical insight should be developed.

5 Summary and Future Problems

In order to demonstrate a general scheme for understanding the volume phase transition of polymer gels in response to environmental stimuli, some results of the recent investigations by the author's group and their analysis on the basis of the mean field theory have been presented here. Several conclusions and general principles are summarized as follows:

(1) Phase transition in gels, which consist of a thermoresponsive polymer and a light absorbable chromophore, can be induced by local heating of the polymer network under illumination of visible light due to the absorption and subsequent thermal dissipation of light energy by the chromophore. The results were qualitatively described by the mean field equation of state of gels.

(2) Phase transition in gels can be affected by applying uniaxial stress. With increasing stress in the region below $1 \times 10^4 \, N \cdot m^{-2}$ at gelation, the effects of uniaxial stress was qualitatively described by the mean field theory. The present results clearly indicate the possibility of a uniaxial stress-induced phase transition of gels.

(3) Phase transition in non-ionic gels, the transition mechanism of which is mainly due to the hydrophobic interactions, can be controlled by adding salts. Transition temperatures and discontinuity are strongly affected by individual ions, especially by anions. Till now, four types of salt effects have been observed and these have been classified by two parameters of the contact free energy and the dimesion of segments. An example of clear expression of ΔF in the equation of states of gels was proposed, which describes successfully the observed swelling curve.

For future implications of the present investigations, particular emphasis can be placed on pursuing the research in the following areas.

(1) The technique of local heating of polymer networks by visible light makes it possible to carryout molecular designing of several types of photo-responsive gels, e.g., gels can be synthesized that swell discontinuously upon illumination by light. This method could be of technological importance in developing various optical devices, such as optical shutters, switches, and display units. Moreover, the increase of temperature in the immediate vicinity of polymer chains is an ideal trigger for induction of a kinetic process that should improve the kinetic experiments.

(2) High uniaxial stress experiments can contribute to a better understanding of polymer structure as well as fracture of gels which has not been extensively studied. It is also important to measure the effects of uniaxial stress on ionic gels, since stress can be expected to affect the degree of ionization which may induce the change of pH in gels.

(3) Salt effects have been well classified by two parameters, up till now, into four types. In addition to the simple types obtained by the combinations of

the parameters (where the transition temperature and discontinuity change monotonically as increasing salt molality), the mixed types (e.g., transition temperature decreases after slightly increasing at low salt molality, and discontinuity decreases after increasing) were detected in the present experiments. In order to elucidate the physical and chemical pictures of the role played by these parameters, several approaches to the investigation of the hydration of polymers and ions should be applied.

It should be noted that the present results could not be easily obtained, if gels of submillimeter size were not used. This simple technological improvement to reduce the size of the gels has not only opened up a wide variety of new investigations, such as those presented in Sect. 2, but has also demonstrated the necessity of reexamination of experiments previously conducted where one might expect to find new phenomena. The recent discovery of multiple-phase transitions, by measuring the volume as a function of pH may serve as an example. Such research will not only establish fundamental concepts of polymer gels, but may also assist in the development and application of new technologies.

Acknowledgments: The present author would like to thank Drs. T. Tanaka, E. Kokufuta, M. Tokita and S. Gorti for valuable discussions. This work was supported in part by a Grant for International Joint Research Project from the NEDO, Japan.

6 References

1. Tanaka T (1978) Phys Rev Letters 40: 820
2. Tanaka T et al. (1980) Phys Rev Letters 45: 1636
3. Hrouz J et al. (1981) J Eur Polym J 17: 361
4. Tanaka T (1981) Sci Am 244: 124
5. Ilavsky M (1982) Macromolecules 15: 782
6. Tanaka T et al. (1982) Science 218: 467
7. Ohmine I, Tanaka T (1982) J Chem Phys 77: 5725
8. Hirokawa Y, Tanaka T (1984) J Chem Phys 81: 6379
9. Katayama S et al. (1984) 17: 2641
10. Hirotsu S et al. (1987) J Chem Phys 87: 1392
11. Mamada A et al. (1990) Macromolecules 23: 1517
12. Suzuki A, Tanaka T (1990) Nature 346: 345
13. Tanaka T et al. (1973) J Chem Phys 59: 5151
14. Tanaka T et al. (1977) Phys Rev Letters 38: 771
15. Dusek K, Patterson (1968) J Polym Sci 6: 1209
16. Hirokawa Y et al. (1984) In: Life sciences research rep. 31: 177, Springer, Berlin Heidelberg New York
17. Tanaka T (1987) In: Structural and dynamics of biopolymers, NATO ASI Series E: 237, Martinus Nijhoff, Dordrecht
18. Li Y, Tanaka T (1992) Annun Rev Mater Sci 22: 243
19. DeRossi D et al. (eds) (1991) Polymer gels. Plenum, New York
20. Osada Y (1987) Adv Polym Sci 82: 1
21. Ilmain F et al. (1991) Nature 349: 400
22. Okano T et al. (1990) J Contrl Rel 11: 255
23. Otake K et al. (1990) Macromolecules 23: 283

24. Annaka M, Tanaka T (1992) Nature 355: 430
25. Tanaka T (1990) a private communication
26. Tanaka T, Fillmore DJ (1979) J Chem Phys 70: 1214
27. Kokufuta E, Tanaka T (1991) Macromolecules 24: 1605
28. Kokufuta E et al. (1991) Nature 351: 302
29. Suzuki A et al. (to be published)
30. Suzuki· A et al. (to be published)
31. Irie M et al. (1986) Macromolecules 19: 2476
32. Flory PJ (1953) Principles of polymer chemistry. Cornell University Press, Ithaca, New York
33. Huggins ML (1941) J Chem Phys 9: 440
34. Flory PJ (1941) J Chem Phys 10: 51
35. Li Y, Tanaka T (1989) J Chem Phys 90: 5161
36. Tanaka T et al. (1985) Phys Rev Letters 55: 2455
37. Matsuo ES et al. (1988) J Chem Phys 89: 1695
38. Li Y, Tanaka T (1990) J Chem Phys 92: 1365
39. Horkay F, Zrinyi M (1989) Makromol Chem, Macromol Symp 30: 133
40. Hirotsu S, Onuki A (1989) J Phys Soc Jpn 58: 1508
41. Onuki A (1988) J Phys Soc Jpn 57: 699 and 57: 1868
42. Erman B, Flory PJ (1986) Macromolecules 19: 2342
43. Long FA, McDevid WF (1952) Chem Rev 51: 119
44. Kollins KD, Washabaugh MW (1985) Quart Rev Biophys 18: 323
45. Inomata H et al. (1992) Langmuir 8: 687
46. Freitas RFS, Cussler EL (1987) Chem Eng Sci 42: 97
47. Huang X et al. (1988) J Chem Eng Jpn 21: 10
48. Bailey Jr FE, Callard RW (1959) J Appl Polym Sci 1: 56
49. Luck WAP (1976) Topics Current Chem 64: 113
50. Debye P, McAulay J (1925) J Physik Z 26: 22

Received October 22, 1992

An Approach to Artificial Muscle Using Polymer Gels Formed by Micro-Phase Separation

M. Suzuki and O. Hirasa
Biomechanics Division, Mechanical Engineering Laboratory, AIST,
1-2 Namiki, Tsukuba City 305, Japan
Head of Department of Material Design and Engineering, Research Institute for
Polymers and Textiles, 1-1-4 Higashi, Tsukuba City 305, Japan

Polymer gels react to diverse stimulations such as chemical, thermal, electrical and photonic excitation by deformation, volume change, hardening/softening and permeability changes. This paper presents advances made in chemomechanical energy conversion by polymer gels aiming to produce artificial muscle. Porous gel films made from poly(vinyl alcohol), poly(acrylic acid) and poly(allylamine) mixed solutions by subjecting them to freezing and thawing generate a mechanical power density close to skeletal muscle by changing the solvent from water to acetone. Porous gel fibers made from poly(vinyl methyl ether) solution crosslinked by γ-ray irradiation shrink as fast as skeletal muscle when the temperature is raised above 37 °C. These polymer gels swell and shrink reversibly by chemical or thermal stimulations by means of mainly solubility phase transition of polymers. Those advances can be attained by producing porous structures by micro-phase separation.

1 Introduction

Polymer gels have a versatile potential based on chemomechanical energy conversion functions. Especially the functions of movement such as deformation, volume change, hardening/softening, force generation etc. are similar to those of biological system. Mechanical force generation is one of the most attractive functions since biological motions are far different from those of manmade machines, flexible, smooth, smart and dexterous. The origin of those motions is of course due to the biological motors like muscles or flagella [1].

A polymer gel is a three-dimensional network of crosslinked polymers containing considerable amounts of solvent. The basic idea to use polymer gels as a muscle-like actuator originates from Kuhn and Katchalsky's work [2] on polyelectrolyte gels in 1950. The thermodynamic theory of gels was extensively studied in the 1950s especially by Flory [3]. Up to this point, various kinds of polymer gels have been studied as chemomechanical energy conversion materials [4–6]. However in the application of gels, researchers are facing difficulties such as slow response, weak force, poor durability and unfamiliar energy sources.

Many recent studies on polymer gels are related to volume phase transition phenomena of poly(acrylamide) PAAm gel [7] and poly(N-isopropyl acrylamide) PNIPAAm gel [8, 9]. The volume phase transition in gels was extensively studied by Tanaka and his coworkers [10, 11].

As an actuator material, poly(acrylonitrile) PAN gel fibers were examined by Umemoto et al. [12] and showed high tensile strength and large shrinkage. On the other hand, in 1982, Nanbu [13] found that poly(vinyl alcohol) PVA aqueous solution more than 5 wt % becomes a strong hydrogel by repetitive freezing and thawing. One of the authors tried to enhance chemomechanical properties of PVA hydrogel by intermixing a PVA solution with polyelectrolyte solutions before repetitive freezings [14–17]. Thermosensitive gels of poly(vinyl methyl ether) PVME which has a lower critical solution temperature LCST in water has been developed as a chemomechanical actuator material by Hirasa [18, 19].

Photochemomechanical systems have also been studied using gels and photochemical reactions such as photochromism of spiropyrans [20], *trans-cis* transition of azobenzene [21] and photo-dissociation of triphenylmethane leuco cyanide [22]. It is an attractive approach to utilize a molecular level conformational change.

Electrochemomechanical systems might be a straightforward approach to realize an electrically driven muscle-like actuator. An electric field induces a bending motion or shrinking of polyelectrolyte gels [23–29]. A review on electrochemomechanical systems is available [6].

In this paper we introduce some advancements attained in studies of gel-muscle where the phase change of a polymer solution was effectively utilized.

2 Mechanisms of Chemomechanical Reactions

There are several mechanisms which are essentially different in chemomechanical reactions of polymer molecular level structure changes, polymer conformational change, change of polymer–polymer interactions and the change of crosslinked structure in polymer gels.

As for the change at the molecular level, azobenzene is known to show an isomerization change from *trans* to *cis* by photostimulation. This reaction causes a shortening of the molecule and a change of interaction with the neighboring molecules [21].

As a type of conformational change of polymer chains there is well known helix-coil transition of polyelectrolyte and polypeptide. This type of transition is due to the change of hydrogen bonds depending on the solvent pH, ionic strength and temperature and does not show a sharp-cut change but a smooth change in a fairly narrow region. Poly(L-glutamic acid) PLGA in water changes its state from helix at pH < 6 to coil at pH > 7. This kind of transition could be also applied [30].

Polyelectrolytes change their conformation with the degree of dissociation which is varied with pH, temperature, polarity of solvent and ionic strength.

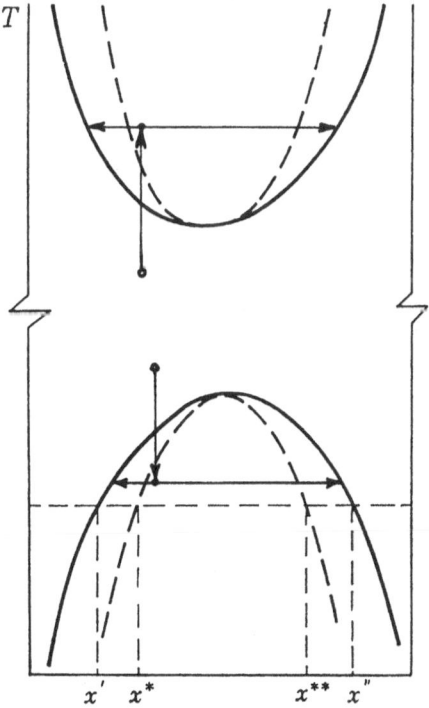

Fig. 1. Phase diagram of a polymer solution. The upper curve shows LCST and the lower curve UCST behavior. *Solid lines*: binodal, *dotted lines*: spinodal T: temperature, x: polymer concentration

Poly(acrylic acid) is a typical polyelectrolyte and shows a dissociation reaction. When pH is high, the polymer chain is charged negatively and extends itself due to coulombic repulsion and thermal motion of counter ions which pull the charged side chains.

Flory [3] formalized the equation of state for equilibrium swelling of gels. It consists of four terms: the term of rubber-like elasticity, the term of mixing entropy, the term of polymer solvent interaction and the term of osmotic pressure due to free counter ions. Therefore, the gel volume is strongly influenced by temperature, the kind of solvent, free ion concentrations and the degree of dissociation of groups on polymer chains.

The term of polymer–solvent interaction plays an important role. The stability of polymer–solvent systems abruptly changes at the binodal and spinodal (dotted) lines in the phase diagram in Fig. 1. The lower curve shows a phase diagram having an upper critical solution temperature UCST as we see in the case of a PVA aqueous solution and the other having a lower critical solution temperature LCST as in the case of a poly(vinyl methyl ether) (PVME) aqueous solution. The mechanism of solubility change of polymers is predominant in the mechanism of swelling and shrinking of the gels which will be introduced here.

Furthermore the processes of physical gelation may be due to such a phase change across spinodal lines since micro-phase separation may induce the formation of a porous structure which could have high chemo-mechanical response.

3 Amphoteric Poly(vinyl alcohol) Hydrogel

3.1 Preparation of Amphoteric PVA Hydrogel

The process of repetitive freezing and thawing of a poly(vinyl alcohol) aqueous solution produced cooperative hydrogen bonding between hydroxyl groups and gave a strong three dimensional network. The strength of PVA gel attained more than 2 MPa in the swollen state [13]. This method was considered very effective to obtain artificial muscle by one of the authors [14].

PVA: $-(-CH_2-CH-)_x-$ PAA: $-(-CH_2-CH-)_y-$ PAlAm: $-(-CH_2-CH-)_z-$
$\qquad\qquad\quad$ OH $\qquad\qquad\qquad\qquad\quad$ COOH $\qquad\qquad\qquad\qquad\; CH_2$
$\qquad\qquad\qquad\qquad\qquad\qquad\qquad\qquad\qquad\qquad\qquad\qquad\qquad\quad NH_2$

First by mixing poly(vinyl alcohol) PVA (Kuraray 117H; Mw: 74 800; saponification: 99.6%), poly(acrylic acid) PAA (SP2; Mw: 170 000), and poly(allylamine) hydrochloride PAlAm.HCl (Nittobo; Mw: 60 000) a homogeneous aqueous solution of PVA : PAA : PAlAm was obtained at concentrations of

1.74 M, 0.245 M and 0.26 M in base moles, respectively. With different spacers, the solution was sandwiched between two plates of optically ground heat resistant glass 10 mm thick. Amphoteric PVA hydrogel films were then obtained by subjecting the solution to repetitive freezing and thawing. The temperature range was between $-50\,°C$ and room temperature around $23\,°C$, with temperature falling and rising at a rate of $3.0\,°C\,min^{-1}$ just before the freezing and thawing points. The gel films were then washed with 1 N KOH, water, 1 N HCl, and water in sequence three times. Some gel films were stretched uniaxially during the process of freezing. These gel films had different Young's moduli, 0.477 MPa in the stretched direction, called "parallel type", and 0.246 MPa in the perpendicular direction to the stretching, called "perpendicular type" [16, 17].

The equilibrium swelling characteristics of these gels in a mixed solvent of acetone and water are shown in Fig. 2a and in HCl/KOH aqueous solution without extra salts in Fig. 2b. Inside the amphoteric PVA gel, ionization of polyelectrolytes occurs since poly(acrylic acid) dissociates in the basic region and poly(allylamine) dissociates in acidic region. Therefore, the gel swells both in acidic and basic solutions and shrinks in neutral pH regions because of polyanion–polycation association which effectively strengthens the gel to give it a tensile strength of 0.5 MPa in water. In acetone, these polyelectrolytes are

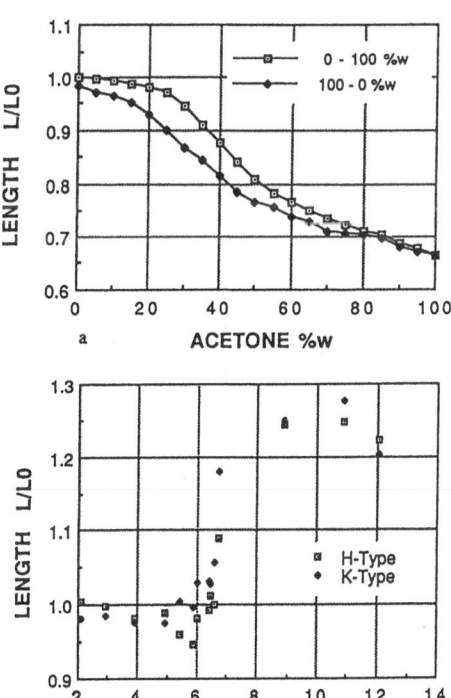

Fig. 2a, b. Equilibrium swelling of amphoteric PVA hydrogel in a. water/acetone mixture, b. HCl/KOH aqueous solution. The H-type was pretreated by 1N HCl and K-type by 1N KOH

soluble without dissociation and PVA is not. With increasing concentration of acetone the ionization of polyelectrolytes decreases and the pressure due to thermal motion of free ions decreases and the polymer–solvent interaction parameter increases. This causes gel contraction.

To examine the effect of the freezing rate on the internal structure and the mechanical properties of hydrogels we made three types of gel film A, B and C by changing the freezing rate from 3.78, 0.0766 to 0.0329 K s^{-1}, respectively, at the freezing point.

3.2 Porous Structure of Amphoteric PVA Hydrogel

The hydrogel films were first quickly dipped in liquid nitrogen precooled below $-208\,°C$ to fix its natural structure and then fractured. For etching of the fracture surfaces the frozen samples were removed on a copper block precooled to $-15\,°C$, then evacuated for a day at a pressure below 133 Pa in order to sublime the water directly from ice. By this treatment the structural deformation of the hydrogel network due to surface tension was avoided. The samples were then coated with gold and examined with a usual scanning electron microscopy technique [31]. Figure 3 shows the cross sectional views of the hydrogel films A, B and C (see Sect 3.1). With increasing freezing rate the pore size became smaller

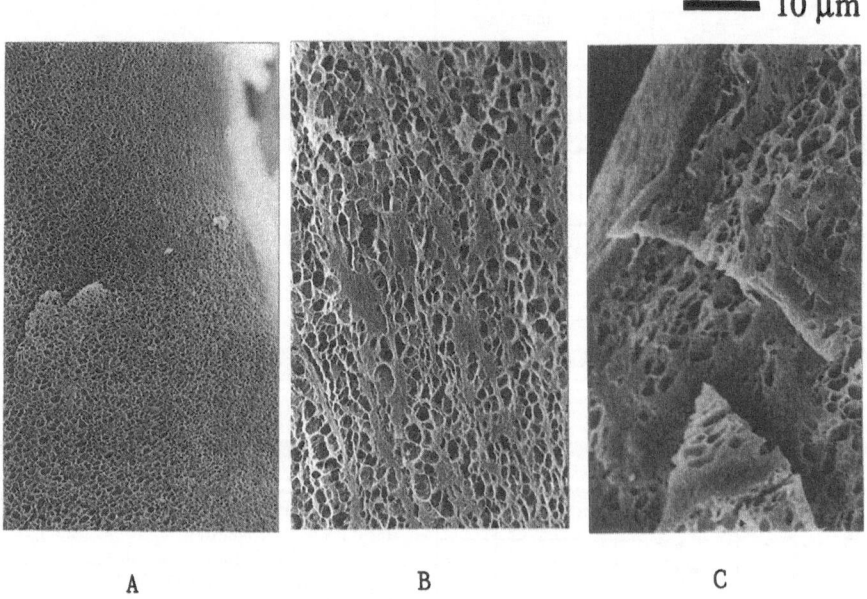

━━━ 10 μm

A   B   C

Fig. 3A–C. Cross sectional views of gels A, B and C obtained by scanning electron microscopy. Freezing rates were, A: 3.78, B: 0.0766, C: 0.0329 in $°C\,s^{-1}$

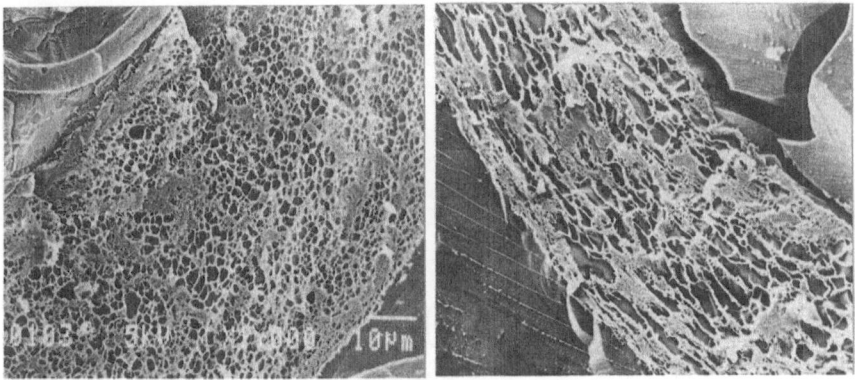

Fig. 4a, b. CryoSEM images of cross section of **a.** a non-stretched gel, **b.** a stretched gel

and smaller. In the gels frozen more slowly there were porous regions and dense regions which should lead to slow responses.

After quick freezing in liquid nitrogen and fracture, another etching method was used at $-90\,^{\circ}$C for less than 1 min in a 1.33×10^{-4} Pa chamber. The etched sample was cooled to $-130\,^{\circ}$C and coated with platinum and carbon in the same vacuum chamber and transferred to the scanning electron microscope at $-130\,^{\circ}$C. The observed structure was closer to reality than that obtained by the method previously described. This is called the cryo-SEM technique [32]. Cryo-SEM images are shown in Fig. 4 which presents the structural change by stretching [16].

3.3 Mechanical Passive Properties

3.3.1 Measurement of Modulus and Poisson Ratio

After fifteen strips of gel film were dipped in solvents of different mixing ratio of acetone to water for a sufficient time to reach their equilibrium state, the films were pulled in the longitudinal direction for the modulus measurement by using a balance-bar type apparatus shown in Fig. 5 [31]. The thickness of the films was less than 100 μm to shorten the relaxation time. At one end of the bar a weight was hung and the gel film at the other end. The Young's modulus of the gel film was measured for each acetone concentration. The modulus changed from 0.2 MPa in water to 10 MPa in acetone as shown in Fig. 6. The Young's moduli for gels A, B and C were 0.049, 0.101 and 0.245 MPa, respectively. By increasing the acetone concentration in water the Young's modulus sharply increased at 70 wt %, while a pure poly(vinyl alcohol) aqueous solution of 20 mM in base moles had a cloud point at 35 wt % of acetone. One might say that this is a result of a phase change of the polymer gel solution although there

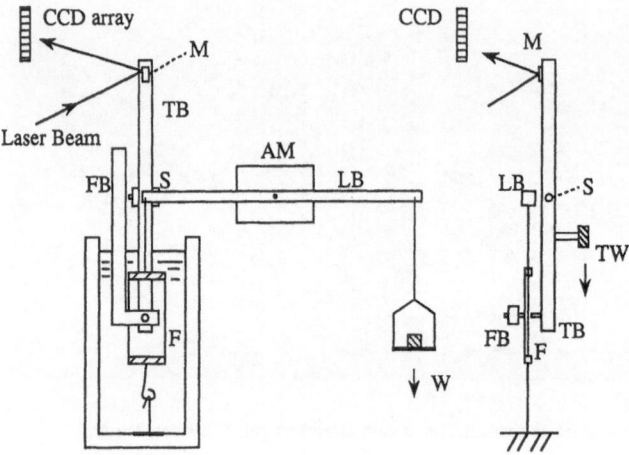

Fig. 5. Apparatus used to measure Young's modulus and Poisson ratio of gel films. *AM*: anglemeter, *LB*: bar to apply a longitudinal force, *W*: weight, *TB*: bar to detect thickness change, *M*: mirror, *S*: fulcrum, *FB*: fixed bar, *F*: gel film, *TW*: weight to apply preload on gel film through pins

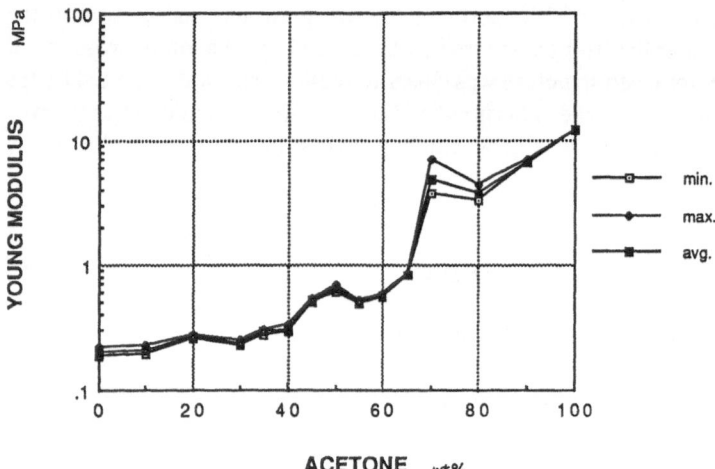

Fig. 6. Young's modulus of amphoteric PVA gel in various acetone/water mixtures

was no sharp size change in Fig. 2a. The Poisson ratio was measured from the thickness change of the gel film obtained with the apparatus shown in Fig. 5. The gel film was set between two pins, one of which was fixed at the bottom end of the balance bar (TB) and the other at the end of fixed bar (FB). Simultaneously the gel film was subjected to a longitudinal tension with the LB bar. The thickness change of the gel film causes change in the angle of the balance bar

(TB) which can be detected with a CCD array and a laser beam reflection technique. This system was able to resolve easily a 100 nm change in thickness and was also able to measure the dynamic change of thickness. The preload applied to the pin on TB of 1 mm diameter was varied from 0.08 g to 1.0 g with a weight (TW). From this, zero-preload deformation was obtained by extrapolation. The Poisson ratio was about 0.08 in water at the longitudinal load of 0.12 MPa, which was close to or lower than that obtained from the width change.

3.3.2 Elasticity Increase Due to Repetitive Freezing

By increasing the number of freezing repetitions the elasticity (Young's modulus) increased from 0.02 MPa up to 0.56 MPa as shown in Fig. 7. It is attributed to increased physical crosslinking due to cooperative hydrogen bonding between poly(vinyl alcohol) chains. There is a discontinuity in the dependence of elasticity at about six freezing cycles above which this dependence flattens out significantly.

3.4. Mechanical Power Generation

3.4.1 Dynamic Response

The effect of the thickness of gel film on the contraction response time is shown in Fig. 8. When the thickness was 10 μm the contraction response time was about 0.2 s which is almost the same speed as the contraction of skeletal muscle in animals. The typical curve of contraction is also shown. Figure 9 shows the result of contraction ratio vs load for gels made under different conditions. The higher the elasticity the higher is the contraction ratio under loading. "Perpendicular" type gels showed a comparatively high contraction ratio, and "Parallel"

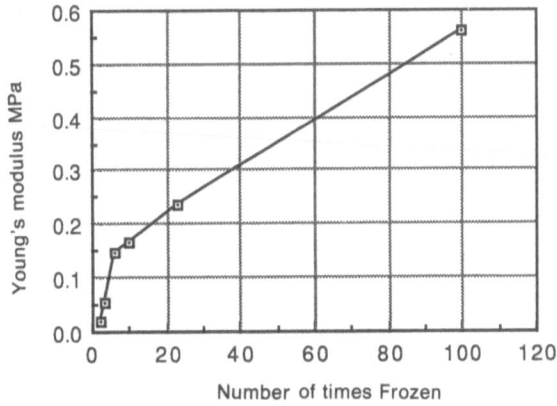

Fig. 7. Relation between Young's modulus and number of times frozen

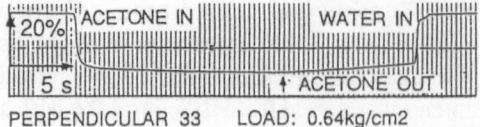

PERPENDICULAR 33   LOAD: 0.64kg/cm2

The typical curve of contraction

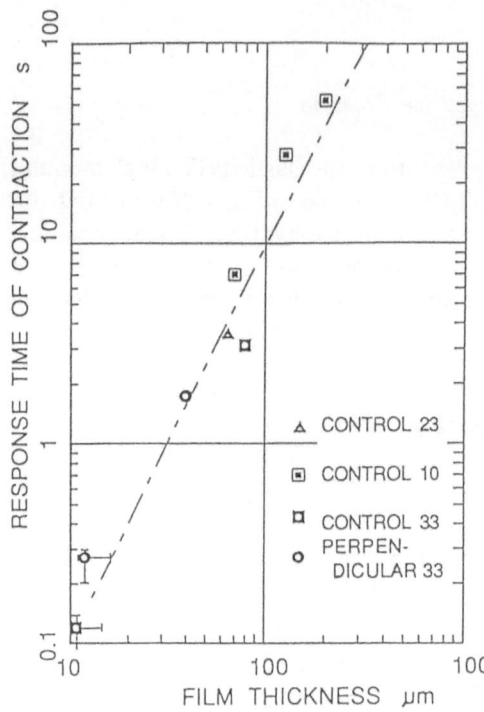

Fig. 8. Relation between contraction response time vs film thickness. A typical curve of contraction has been added

type gels had high durability for high load. However the response was not clearly improved. Figure 10 shows the typical contraction curves for A, B and C. Under loading condition the largest initial contraction speed was found in gel C but the longest settling time was also from gel C. This result tells us that the larger the pore size and the thinner the dense region the faster is the response. Figure 11 shows a model for an artificial arm with a polymer gel-muscle which was about 10 cm long and able to raise a weight of 2 g on the palm by pouring acetone onto the gel-muscle causing it to contract and it relaxed when water was poured onto it.

3.4.2 Output Power

The mechanical power generated by hydrogel films was evaluated from contraction force multiplied by contraction speed. The speed v was obtained from the

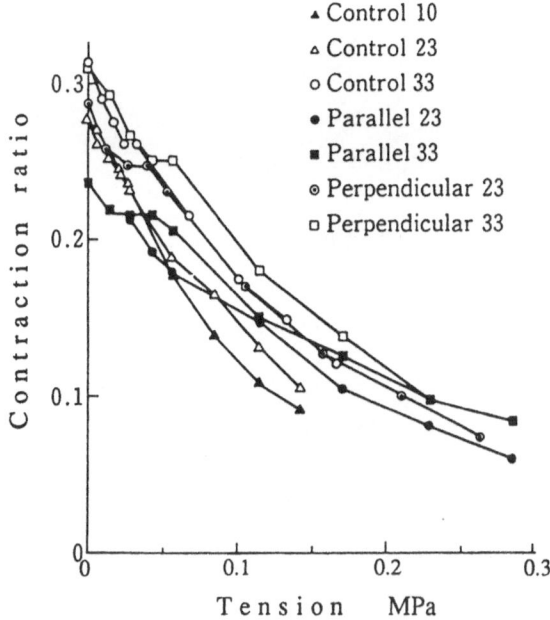

▲ Control 10
△ Control 23
○ Control 33
● Parallel 23
■ Parallel 33
⊙ Perpendicular 23
▫ Perpendicular 33

Fig. 9. Mechanical active properties of amphoteric PVA gels. Contraction ratio vs load

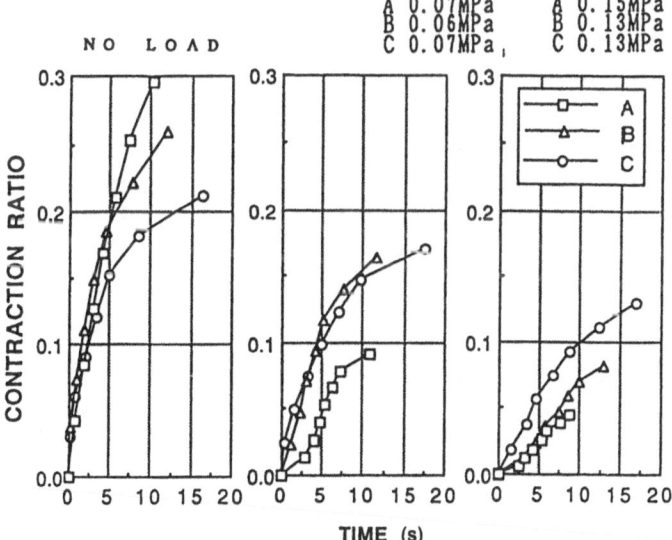

A 0.07MPa   A 0.15MPa
B 0.06MPa   B 0.13MPa
C 0.07MPa ,   C 0.13MPa

NO LOAD

— □ — A
— △ — B
— ○ — C

Fig. 10. Contraction curves for A, B and C (see Sect. 3.1) under several loading conditions. The data are limited to 7/8 of the final contraction ratio

time τ at 50% of total length change $(L' - L)$ as $v = (L' - L)/2\tau$. Figure 12 shows the generated power density for various amphoteric PVA gel films. The film which was 10 μm thick could generate power as high as 0.1 watt/g which is more than frog muscle. The freezing rate of this gel was 0.05 °C s^{-1}. In Fig. 12

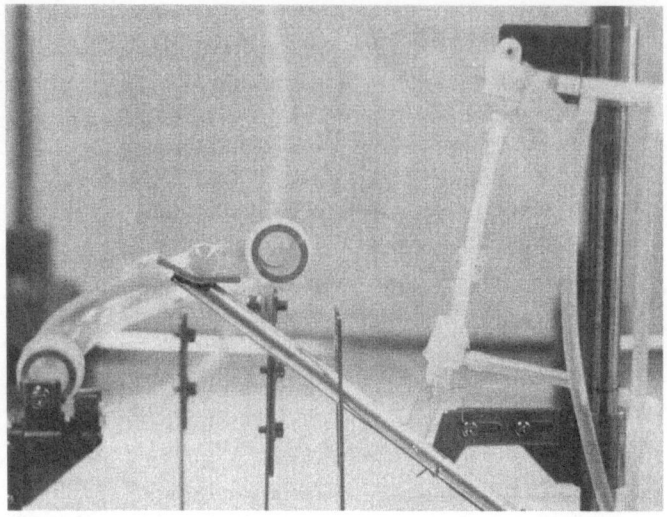

Fig. 11. Artificial muscle model using amphoteric PVA gel film of 50 μm thick. The model length was 12 cm. This model was able to raise a ball to the upper position within 10 s [15]

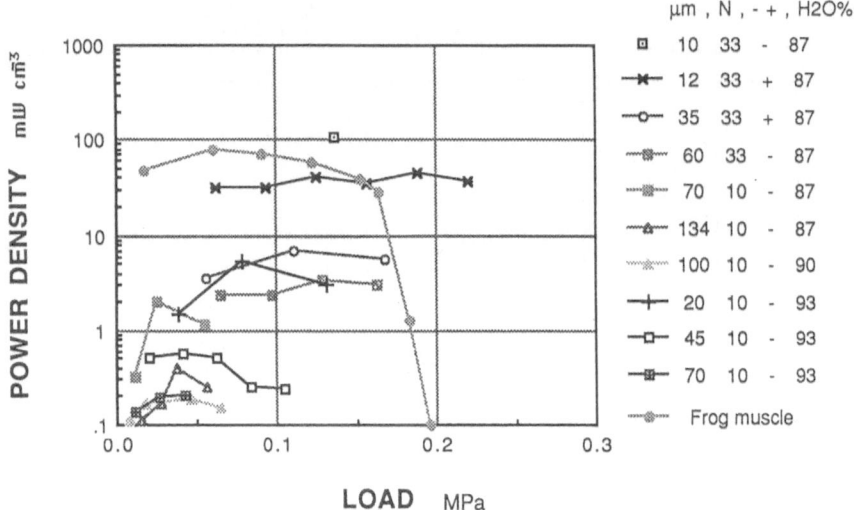

Fig. 12. The generated power density vs load for various gel films made under different conditions. The power density of frog muscle is indicated for comparison

the numbers listed under μm denote thickness, those under N the number of times the gel was frozen, the plus/minus signs show anisotropy ($-$/no, $+$/yes), and under H_2O% show the water content of the hydrogel in the initial solution state before gelation.

4 Poly(vinyl methyl ether) Hydrogel

Poly(vinyl methyl ether), PVME, is a thermo-sensitive polymer. The aqueous solution has a Lower Critical Solution Temperature (LCST) of 37 °C. Therefore, PVME is soluble in water below its LCST, but insoluble above its LCST. When an aqueous solution of PVME is irradiated with γ-rays the solution becomes PVME hydrogel [18, 19]. The gel shows thermo-sensitivity similar to the solution, and swells below 37 °C and shrinks above this temperature. It is important to form a fine porous gel structure to obtain quick response gels. There are two methods for the purpose. One is a method using micro-phase separation by heating. The other is a method using micro-phase separation by blending of polymer solutions.

4.1 Porous PVME Gel Produced by Micro-Phase Separation Brought about by Heating

4.1.1 The Heating Method of Micro-Phase Separation

PVME molecules dissolve in water with hydrogen bonding formation between methoxyl groups and water molecules at low temperatures. At high temperatures, since the hydrogen bonds are decomposed by the thermal motion of

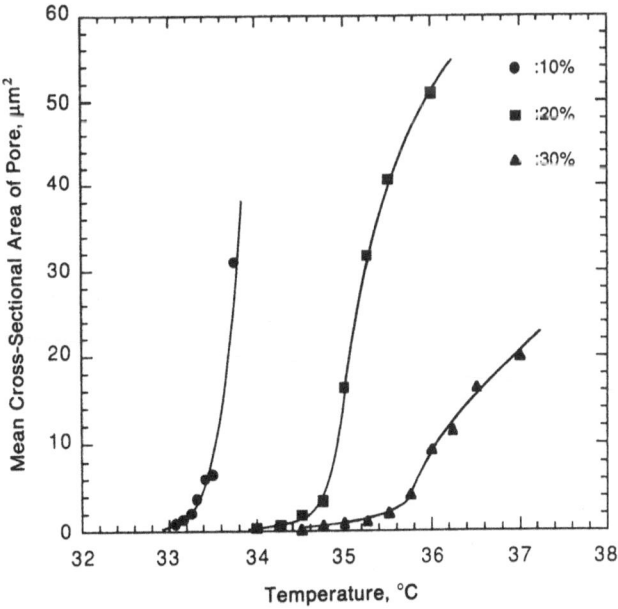

Fig. 13. Micro-phase separation of PVME aqueous solution by heating

water molecules, PVME molecules are dehydrated and aggregate together. Then micro-phase separation of PVME occurs. Effects of temperature on micro-phase separation of PVME aqueous solutions are shown in Fig. 13. The heating rate was $0.5\,°C\,min^{-1}$. Weight average and number average molecular weights of PVME used are 95 600 and 48 400, respectively. The pore size increased as the temperature was raised. The phenomenon was remarkably dependent on the concentration of PVME. An increasing pore size was measured with a temperature jump from $20\,°C$ to $37\,°C$. The results obtained are shown in Fig. 14. The growth rates also depended on the concentration of PVME. The rate at low concentration was higher than that at high concentration. Under the same conditions it was expected that the pore size of the gel should depend on the concentration.

PVME gel was made from the aqueous solution by γ-ray irradiation with Co^{60}. The gelation processes are considered to be as follows:

1) OH radicals are generated from water molecules by γ-ray irradiation.
2) The OH radicals attack PVME molecules. The radicals remove hydrogen atoms from their position on the main chains of the PVME molecules and PVME radicals are generated.
3) The polymer radicals combine and the PVME molecule is crosslinked to form the gel.

Viscosity changes of PVME aqueous solutions irradiated by γ-rays are shown in Fig. 15. The viscosity was measured with a rotational viscometer. The

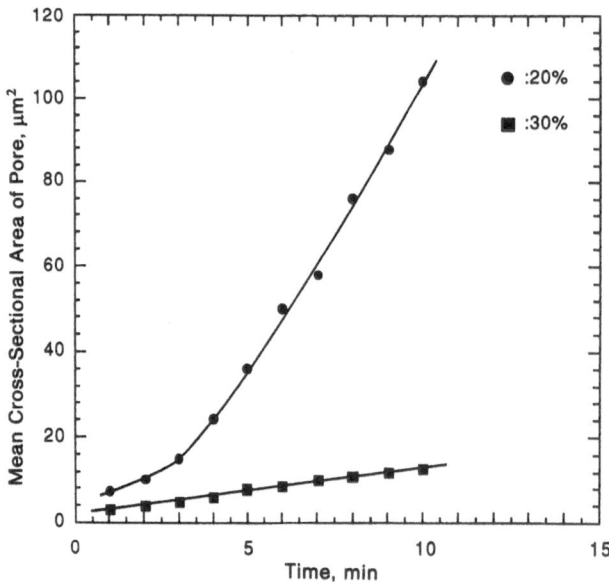

Fig. 14. Rate of phase separation of a PVME aqueous solution with a temperature jump from 23 to $37\,°C$

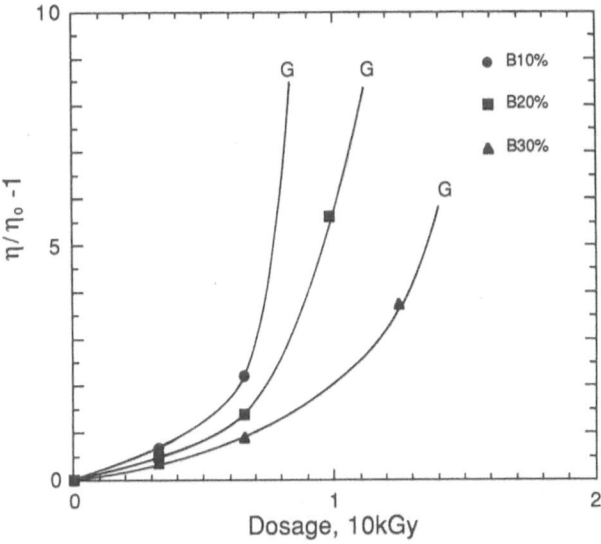

Fig. 15. Viscosity change of PVME aqueous solutions with changing dosage of γ-ray irradiation. η, η₀: viscosity of solution irradiated and nonirradiated, respectively. *G*: gel point

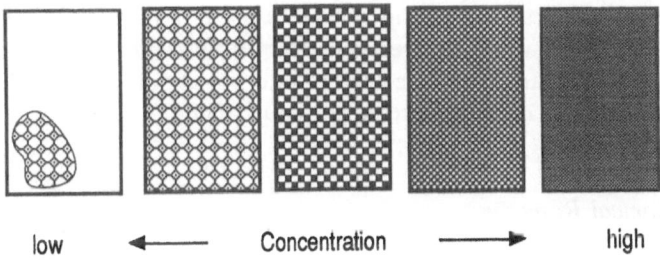

low   ◀—— Concentration ——▶   high

Fig. 16. Schematic illustration of morphology of porous PVME gels formed from solution by heating

viscosity values increased in proportion to the dosages of γ-rays exponentially until gelation. By γ-ray irradiation at the temperature below LCST, a transparent homogeneous gel was formed. However, by irradiation at temperatures just above LCST, gels with various pore sizes were formed by controlling the heating rate, the concentration of PVME and the dose rates.

The morphology of porous PVME gels formed by controlling the concentration is illustrated in Fig. 16. Under low polymer concentration conditions, the pore size of the gel was of the order of millimeter, and the gel could not occupy the space of the vessel used for irradiation. The pore size became smaller with increasing concentrations of PVME. The pore size occasionally reached sizes of the order of μm at higher concentration. The typical examples of SEM micro-

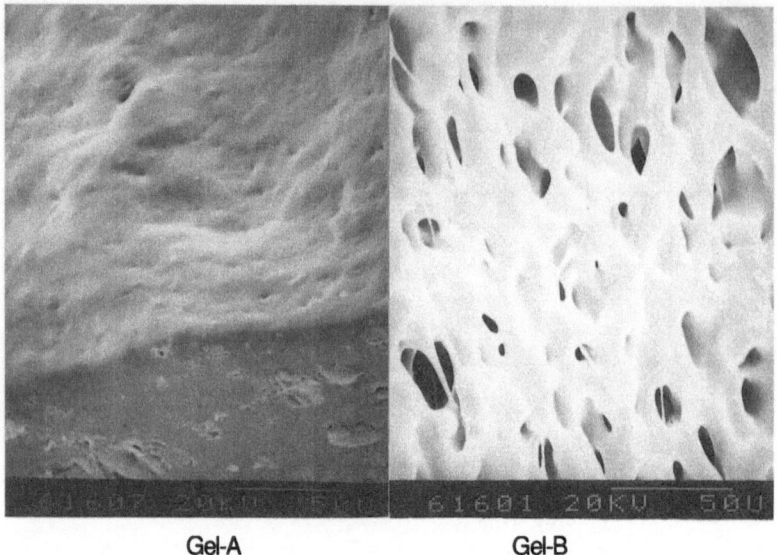

Fig. 17. SEM images of PVME hydrogels formed by γ-ray irradiation. *Gel-A*: at 23 °C, *Gel-B*: with heating

photographs of these gels are shown in Fig. 17. The concentration of PVME was 30 wt %, and the dosage of γ-ray irradiation was 100 kGy in both cases. "Gel-A" was formed at 23 °C and was homogeneous. "Gel-B" was formed under heating conditions from 23 °C to 50 °C. The heating rate was about 5 °C min^{-1}. "Gel-B" was spongy. The size of micropores was about 10 μm.

4.1.2 Thermo-Mechanical Response

The PVME hydrogel showed phase transition in the same manner as the aqueous solution by heating. The gel swelled and shrank in response to temperature change reversibly. The thermo-mechanical responses of these gels are shown in Fig. 18. "Gel-B" responded to temperature changes quickly. The response time of the swelling and shrinking processes was about 30 and 100 seconds, respectively. But "Gel-A" did not respond on the same time scale. Thus this preparation method of thermo-responsive hydrogel is useful for improving the responsiveness.

4.2 Porous PVME Gels by Blending Polymer Solutions

4.2.1 Method of Micro-Phase Separation by Blending Polymer Solutions

The phase separation of PVME aqueous solutions occurs not only by heating, but also by blending with other materials such as salts, urea, organic solvents

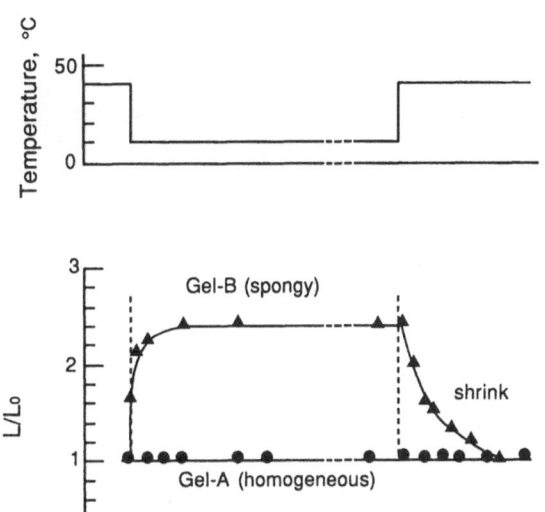

Fig. 18. Thermo-mechanical response of PVME hydrogels
Lo: gel length at 40 °C,
L: gel length at time t, the size of the swollen gel is a 1-cm cube

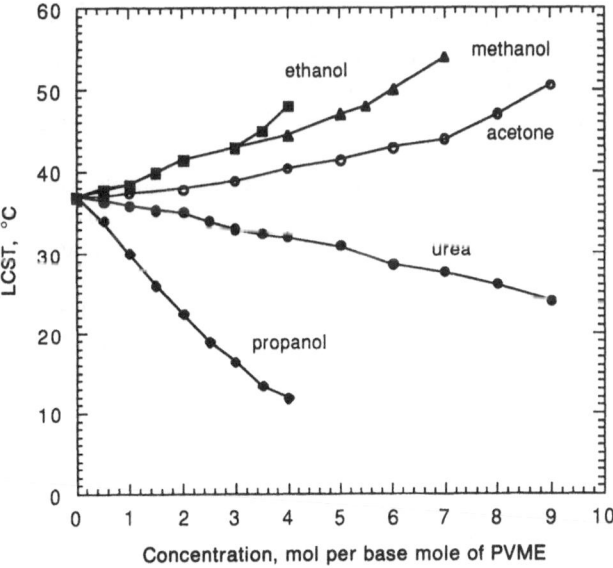

Fig. 19. The effect of additives on the LCST of PVME solutions

and polymer solutions, etc. The examples are shown in Fig. 19. The LCST of the blended polymer solution changed with the species and the concentrations of additives. In the method using phase separation by heating, the pore size depends on the rate of heating and irradiation. Therefore, it is difficult to control

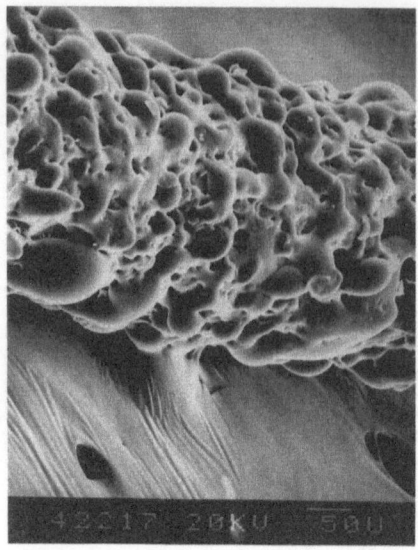

Fig. 20. SEM image of fibrous PVME hydrogel formed by γ-ray irradiation after wet spinning

the separation state. For the applications of the thermo-responsive gel, it is important to know how to form the gel into a desired shape such as a film, beads or fibers, etc. For this purpose, blending polymer solutions enables us to fix the phase separation structure of thermo-sensitive polymers. Since an aqueous solution of PVME is not miscible with an aqueous solution of sodium alginate the blend solution of these polymers shows micro-phase separation. The blend solution is extruded through a nozzle into a calcium chloride solution. Alginate anion in the solution combines with Ca^{++}, and forms fibers of calcium alginate gel. The phase separation state of PVME is fixed by the gelation of alginate with Ca^{++}, then a fibrous pre-gel is formed. In order to crosslink PVME molecules in the fibrous pre-gel after spinning, the fibrous pre-gel is irradiated with γ-rays. Since calcium alginates are decomposed by irradiation, they are easily removed from the gel by washing with water. Finally we have fibrous PVME hydrogel. A SEM microphotograph of the fibrous gel is shown in Fig. 20. Diameters of the gel at swollen and shrunk state are 400 and 200 μm, respectively. The gel has a fine porous structure.

4.2.2 Thermo-Mechanical Response

The relation between the response time and reciprocal of the heating rate of the gel is shown in Fig. 21. The response time at an infinite heating rate is estimated to be about 0.25 s and is very fast. These gels swell and shrink responding to temperature change isotropically. Mechanical properties of the gel are shown in Table 1. The mechanical strength increases remarkably by phase transition of

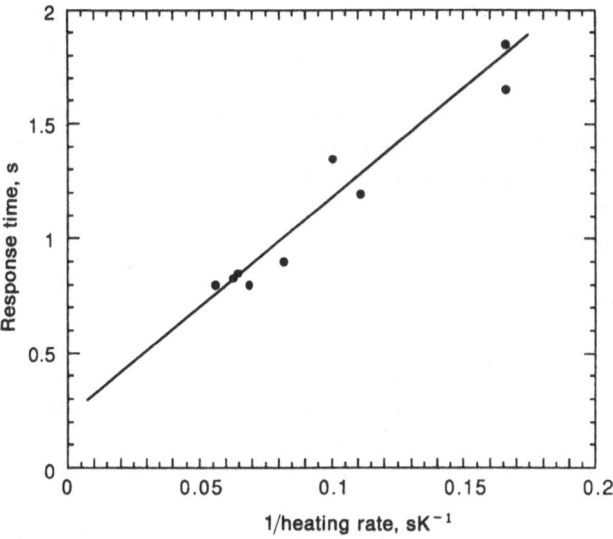

Fig. 21. Thermo-mechanical response of fibrous PVME hydrogel

Table 1. Mechanical properties of fibrous PVME hydrogel

Mechanical property	20 °C	40 °C	Dry
Maximum tensile strength (kPa)	32.0	290	920
Maximum tensile strength (mN fiber^{-1})	3.9	8.8	25.4
Maximum elongation (%)	44.0	114	450
Young's modulus (kPa, 10% elongation)	92.0	470	490

Rate of extension is 10 mm min^{-1}

gel. The generated tension due to the phase transition by heating is 3×10^{-3} N per a fibrous gel or 98 kPa. The value is about 1/3 to 1/10 as much as the stress of natural muscles. The gel forming method using micro-phase separation of blended polymer solutions can be used to form responsive gels in various shapes quickly.

5 Discussion

As seen in the above, many approaches are now appearing aiming at providing a new driving source like muscles. Responses of thermosensitive gels are mainly governed by thermal diffusivity, and the size of gel as seen from a thermal diffusion equation. In general, the thermal diffusivity of a solvent such as water or ethyl alcohol is on the order of 10^{-3} cm^2 s^{-1}, which is smaller than those of solids. On the other hand, the chemical diffusion coefficient in water is of the order of 10^{-5} cm^2 s^{-1} while those in solids are many orders of magnitude smaller than those in liquids. Therefore thermosensitive gels can respond faster to the environmental change than chemo-sensitive gels.

The response time to the chemomechanical volume change of a hydrogel is said to be governed by the motion of network segments. It can be described by a cooperative diffusion equation derived by Tanaka [33] and Tokita [34], if the chemical reaction inside the network is very fast. It should be noted here that the chemical response is not always faster than mechanical deformation or diffusion. For example amphoteric PVA hydrogel has polyanions and polycations which associate with each other to make a complex in solvents with a low acetone concentration. However, they are neutralized and then release each other in solvents with a comparatively high acetone concentration. It should produce metastable states in its shape and cause slow response depending on loading conditions.

Nevertheless, the characteristic time constant is roughly proportional to the square of the typical size and to the inverse of collective diffusion coefficient D which is given by the modulus divided by the friction. The porous structures presented here are one of the solutions to achieve a high response material.

As noted above, some gels such as PVME gel, PAN gel [12] and PVA-PAA-PA1Am gel have been able to reach the level of muscle in some ways – contraction speed, force or power density. If the frequency of use is not high, polymer gels can be readily applied.

Further improvements must still be made to the energy supplying system, synthetic methods for high density energy conversion which could be realized by regular assembly of polymer segments, a chemomechanical reaction system which can utilize high density chemical energy sources and control systems of acutation.

6 References

1. Stebbings H, Hyams JS (1979) Cell motility, Longman, London, p 1, p 68
2. Kuhn W, Hargitay B, Katchalsky A, Eisenberg H (1950) Nature 165: 514
3. Flory PJ (1953) Principles of polymer chemistry. Conrnell University Press, New York, chap 13
4. Osada Y (1987) In: Olive S, Henrici-Olive G (eds) Adv Polym Sci vol 82. Springer, Berlin Heidelberg New York, p 1
5. DeRossi D, Kajiwara K, Osada Y, Yamauchi A (eds) (1991) Polymer gels. Plenum, New York
6. DeRossi D, Suzuki M, Osada Y, Morasso P (1992) J Intell Mater Syst and Struct 3: 75
7. Tanaka T (1978) J Chem Phys 40: 820
8. Hirokawa Y, Tanaka T (1984) J Chem Phys 81: 6379
9. Fujishige S, Kubota K, Ando I (1988) J Phys Chem 93: 3311
10. Tanaka T, Fillmore D, Sun ST, Nishio I, Swislow G, Shah A (1980) Phys Rev Lett 45: 1636
11. Katayama S, Hirokawa Y, Tanaka T (1984) Macromolecules 17: 2641

12. Umemoto S, Okui N, Sakai T (1991) In: DeRossi D, Kajiwara K, Osada Y, Yamauchi A (eds) Polymer gels. Plenum, New York, p 257
13. Nanbu M (1983) Polymer Appl 32: 523
14. Suzuki M, Ushida T, Fujishige S, Takeishi T (1986) Biorheology 23: 274
15. Suzuki M (1987) In: IUPAC CHEMRAWN VI, World Conf Adv Mat 17–22 May 1987 Tokyo, p IB11
16. Suzuki M (1989) Kobunshi Ronbunshu 46: 603
17. Suzuki M (1991) In: DeRossi D, Kajiwara K, Osada Y, Yamauchi A (eds) Polymer gels. Plenum, New York, p 221
18. Hirasa O, Morishita Y, Onomura R, Ichijo H, Yamauchi A (1989) Kobunshi Ronbunshu 46: 661
19. Hirasa O, Ito S, Yamauchi A, Fujishige S, Ichijo H (1991) In: DeRossi D, Kajiwara K, Osada Y, Yamauchi A (eds) Polymer gels. Plenum, New York, p 247
20. Smets G (1975) J Polym Sci, Polym Chem Ed 13: 2223
21. Lovrien R (1967) Proc. Nat Acad Sci 57: 236
22. Irie M, Kunwatchakun D (1986) Macromolecules 19: 2476
23. Tanaka T, Nishio I, Sun ST, Nishio SU (1982) Science 218: 467
24. Osada Y, Hasebe M (1985) Chem Lett 1285
25. Kurauchi T, Shiga T, Hirose Y, Okada A (1991) In: DeRossi D, Kajiwara K, Osada Y, Yamauchi A (eds) Polymer gels. Plenum, New York, p 237
26. Shiga T, Kurauchi T (1990) J Appl Poly Sci 39: 2305
27. Suzuki M (1990) Proc. 12th Ann Int Conf IEEE EMBS, 12: 1913
28. DeRossi DE, Chiarelli P, Buzzigoli G, Domenici C, Lazzeri L (1986) Trans Am Soc Artif Intern Organs. 32: 157
29. Chiarelli P, Umezawa K, DeRossi D (1991) In: DeRossi D, Kajiwara K, Osada Y, Yamauchi A (eds) Polymer gels. Plenum, New York, p 195
30. Minoura N, Aiba S, Fujiwara Y (1986) Kobunshi Ronbunshu, 43: 803
31. Suzuki M, Matsuzawa M, Saito M, Tateishi T (1992) In: Karalis TK (ed) Mechanics of swelling, NATO ASI Series H, 64: 705. Springer, Berlin Heidelberg New York
32. Fujikawa S (1988) Electron Microsc. Rev 1: 113
33. Tanaka T, Fillmore D (1979) J Chem Phys 70: 1214
34. Tokita M, Tanaka T (1991) J Chem Phys 95: 4613

Received January 1993

Author Index Volumes 101-110

Author Index Vols. 1-100 see Vol. 100

Subject Index